Nonlinear Functional Analysis
and Its Applications

NATO ASI Series

Advanced Science Institutes Series

A series presenting the results of activities sponsored by the NATO Science Committee, which aims at the dissemination of advanced scientific and technological knowledge, with a view to strengthening links between scientific communities.

The series is published by an international board of publishers in conjunction with the NATO Scientific Affairs Division

A Life Sciences B Physics	Plenum Publishing Corporation London and New York
C Mathematical and Physical Sciences	D. Reidel Publishing Company Dordrecht, Boston, Lancaster and Tokyo
D Behavioural and Social Sciences E Engineering and Materials Sciences	Martinus Nijhoff Publishers The Hague, Boston and Lancaster
F Computer and Systems Sciences G Ecological Sciences	Springer-Verlag Berlin, Heidelberg, New York and Tokyo

Series C: Mathematical and Physical Sciences Vol. 173

Nonlinear Functional Analysis and Its Applications

edited by

S. P. Singh

Dept. of Mathematics and Statistics,
Memorial University of Newfoundland,
St. John's, Newfoundland, Canada

D. Reidel Publishing Company

Dordrecht / Boston / Lancaster / Tokyo

Published in cooperation with NATO Scientific Affairs Division

Proceedings of the NATO Advanced Study Institute on
Nonlinear Functional Analysis and Its Applications
Maratea, Italy
April 22 - May 3, 1985

Library of Congress Cataloging in Publication Data

NATO Advanced Study Institute on Nonlinear Functional Analysis and Its Applications (1985 : Maratea, Italy)
Nonlinear functional analysis and its applications.

(NATO ASI series. Series C, Mathematical and physical sciences; vol. 173)
"Proceedings of the NATO Advanced Study Institute on Nonlinear Functional Analysis and Its Applications, Maratea, Italy, April 22–May 3, 1985"—Verso t. p.
"Published in cooperation with NATO Scientific Affairs Division."
Includes index.
1. Nonlinear functional analysis—Congresses. I. Singh, S. P. (Sankatha Prasad), 1937- . II. North Atlantic Treaty Organization. Scientific Affairs Division. III. Title. IV. Series: NATO ASI series. Series C, Mathematical and physical sciences; vol. 173.
QA321.5.N38 1985 515.7 86-488
ISBN 90-277-2211-0

Published by D. Reidel Publishing Company
P.O. Box 17, 3300 AA Dordrecht, Holland

Sold and distributed in the U.S.A. and Canada
by Kluwer Academic Publishers,
190 Old Derby Street, Hingham, MA 02043, U.S.A.

In all other countries, sold and distributed
by Kluwer Academic Publishers Group,
P.O. Box 322, 3300 AH Dordrecht, Holland

D. Reidel Publishing Company is a member of the Kluwer Academic Publishers Group

Printed in The Netherlands

THIS VOLUME IS DEDICATED TO THE MEMORIES

OF

Charles C. Conley

(September 26, 1933 - November 20, 1984)

AND

James Dugundji

(August 30, 1919 - January 8, 1985)

TABLE OF CONTENTS

PREFACE

A NATO Advanced Study Institute on Nonlinear Functional Analysis and Its Applications was held in Hotel Villa del Mare, Maratea, Italy during April 22 - May 3, 1985. This volume consists of the Proceedings of the Institute.

These Proceedings include the invited lectures and contributed papers given during the Institute. The papers have been refereed. The aim of these lectures was to bring together recent and up-to-date development of the subject, and to give directions for future research. The main topics covered include: degree and generalized degree theory, results related to Hamiltonian Systems, Fixed Point theory, linear and nonlinear Differential and Partial Differential Equations, Theory of Nielsen Numbers, and applications to Dynamical Systems, Bifurcation Theory, Hamiltonian Systems, Minimax Theory, Heat Equations, Pendulum Equation, Nonlinear Boundary Value Problems, and Dirichlet and Neumann problems for elliptic equations and the periodic Dirichlet problem for semilinear beam equations.

I express my sincere thanks to Professors F. E. Browder, R. Conti, A. Dold, D. E. Edmunds and J. Mawhin members of the Advisory Committee. I also express my appreciation to A. Carbone, R. Chiappinelli, and R. Nugari for their excellent organization and smooth running of the Institute. My special thanks and appreciation to R. Guzzardi and I. Massabo for their help, cooperation and encouragement from the initial planning till the successful completion of the Institute. I extend my thanks to lecturers, whose contributions made these proceedings possible, and to members of the Department of Mathematics and Statistics at Memorial University of Newfoundland who willingly helped with the planning of the Institute.

My thanks to the NATO Scientific Affairs Division for the generous support for the Advanced Study Institute, and to Memorial University of Newfoundland and the Universita della Calabria, Italy for their financial support.

Finally my thanks go to Ms. Wanda Butler for her excellent typing of the manuscript and to the staff of D. Reidel Publishing Company, for their understanding and cooperation.

St. John's, Newfoundland, Canada S. P. Singh
October 1985.

COHOMOLOGICAL METHODS IN NON-FREE G-SPACES WITH APPLICATIONS TO GENERAL BORSUK-ULAM THEOREMS AND CRITICAL POINT THEOREMS FOR INVARIANT FUNCTIONALS*

Edward Fadell
Universitat Heidelberg
Mathematisches Institut
and
University of Wisconsin-Madison

1. INTRODUCTION

The basic source for this exposition is recent joint work with S. Husseini ([1], [2]) which in turn was motivated by our joint work with P. Rabinowitz [3]. However, we will take this opportunity to present some additional material as well as improve some of the results in [2].

We take as our starting point the following three equivalent formulations of the classical Borsuk-Ulam theorem. Recall that a map (= continuous function f is odd if $f(-x) = -f(x)$, whenever this makes sense. See §2 for additional notation.

Theorem A. Every odd map $f : S^n \to \mathbb{R}^n$ has zeros i.e., there exists an $x_0 \in S^n$ such that $f(x_0) = 0$.

Theorem B. There does not exist an odd map $f : S^n \to S^{n-1}$.

Theorem C. Every odd map $f : S^n \to S^n$ is essential, i.e. f is not homotopic to a constant.

Even though the above are all equivalent, we will examine Theorem C more closely. In the language of equivariant topology, if we let $G = \mathbb{Z}_2 = \{1, -1\}$, then S^n is a free G-space under the action $(-1)x = -x$ and an odd map is a G-map (see §2 for terminology). Thus, Theorem C says that for the free G-space S^n every G-map is essential. In §4 we will study the following general question with particular attention to G-spaces where the action is not free..

When are G-maps $f : X \to X$ essential?

§3 is denoted to the complex analogue of the classical Borsuk-Ulam theorems due to Hopf-Rueff [4]. It is probably the first example of such theorems in a non-free situation. It is not only instructive but useful when G is the circle group and the fixed point set of the action is empty. The Hopf-Rueff is also an example of showing that a G-map between different G-spaces is essential, a subject which is considered in §5.

S. P. Singh (ed.), Nonlinear Functional Analysis and Its Applications, 1–45.

Now, let us make a minor change in Theorem A as follows. Let $f : S^n \to \mathbb{R}^p$ denote an odd map with $p \leq n$, and let $Z = f^{-1}(0)$ denote the zero set which in non-empty. If $X \subset S^n$ is any set such that $x \in X \Leftrightarrow -x \in X$ (X is a $\mathbb{Z}_2$-set), set $\gamma(X) = \mathrm{cat}_Y \overline{X}$, the category of $\overline{X} = X/G$ in the orbit space $Y = S^n/\mathbb{Z}_2 = \mathbb{R}P^n$. The set functions γ is used classically to measure the "size" of the zero set Z in a manner which increases with $n - p$. Such theorems are often referred to theorems of the "Bourgin -Yang" type. For example

Theorem A'. Let f denote an odd map, $f : S^n \to \mathbb{R}^p$ $p \leq n$. Then, if $Z = f^{-1}(0)$.

$$\gamma(Z) \geq n + 1 - p \ .$$

In order to explore Bourgin-Yang theorems when the G-spaces in question are not necessarily free, we employ a cohomological analogue of category, which we refer to as cohomological index theory. This is the subject of §6 with sufficient generality to include G-pairs (X,A) as well as families of G-maps. The theory is applied in §7 to prove a very general Bourgin-Yang theorem for G-sphere bundles, $G = \mathbb{Z}_2$, S^1, or $SU(2)$.

In §8 we explore briefly geometric analogues for relative cohomological index theory (relative category) and finally in §9 we prove an abstract critical point theorem for invariant functionals based on a concept of "linking" defined in terms of either relative cohomological index theory or relative category. A direct application is the main result in [3], using computations of the relative cohomological index of appropriate G-pairs.

2. PRELIMINARIES

The basic universe of discourse for us will be the category P_G of paracompact (Hausdorff) G-pairs (X,A), where G is a compact Lie group. Recall [5] that every such G can be realized as a closed subgroup of some orthogonal group $O(n) = O(n,\mathbb{R})$ for n large. If X is a paracompact space, X is called a G-space if there is an action (map) $\mu : G \times X \to X$, with $\mu(gx)$ written gx such that $g_1(g_2x) = (g_1g_2)x$ and $e \cdot x = x$, e the identity of G. A subset $A \subset x$ is called a G-set if $gA \subset A$ for all $g \in G$. A G-pair (X,A) is a G-space X together with a closed G-set A. A G-map (morphism in P_G) $f : (X,A) \to (Y,B)$ is a map (= continuous function) such that $f(g,x) = g\, f(x)$, $x \in X$, $g \in G$. We denote the pair (X,ϕ) simply by X so that every G-space X belongs to P_G. Here is some additional basic terminology.

(a) Isotropy. If $x \in X \quad P_G$, set

$$G_x = \{g \quad G | gx = x\}$$

G_x is a closed subgroup of G, called the isotropy group at x.

(b) Free Actions. X is a free G-space (the action of G on X is free) if $G_x = e$ for all x X.

(c) Orbits. If $X \in P_G$, then if $x \in X$

$$\text{Orb } x = Gx = \{gx,\ g \in G\}.$$

It is easy to verify that the map $g \to gx$ induces a homeomorphism $G/G_x \to Gx$, where G/G_x receives the usual identification topology. Note also that orbits are necessarily closed.

(d) Orbit Space. If X P_G, we denote the set whose points are the orbits of X by X/G or $\overline{X}$. The correspondence $\eta : x \to \text{Orb } x$ is an identification map where $\overline{X}/G$ is given the identification topology. We call η the orbit map.

(e) Induced Maps. A G-map $f : X \to Y$ induces a map

$$f/G : X/G \to Y/G, \ (f/G)(Gx) = G\, f(x)$$

also denoted by $\overline{f} : \overline{X} \to \overline{Y}$.

(f) Fixed Points. If H is a subgroup of G and X is a G space, define

$$X^H = \{x | hx = x \quad \text{for all} \quad h \in H\}$$

X^H is therefore the set of points of X fixed under H. X^G is called the fixed point set of the action. Note that X^H is not a G-set. However, if H is normal in G, then

$$hgx = gh'x = gx,\ g \in G,\ h,\ h' \in H$$

thus, if NH is the normalizer of H in G, then X^H is a NH-space.

The subcategory F_G of free paracompact G-spaces X is important because it coincides with the category of principal G-bundles [14] over a paracompact base. In particular, the orbit map $\eta : X \to X/G$ is a locally trivial fiber space [5], with X/G paracompact and hence a fibre space in the sence of Hurewicz [6]. Thus, the standard tools of algebraic topology (exact sequences of homotopy groups, spectral sequences, ...) apply [5], [6], [7]. In particular, there exists universal free G-spaces in F_G

2.1 Definitions. An object E (also denoted by EG) in F_G is called universal if

(a) for each $X \in F_G$ there is a G-map $\phi : X \to E$

(b) if ϕ_1 and ϕ_2 are G-maps $X \to E$ then the induced maps $\overline{\phi}_1$ and $\overline{\phi}_2$ are homotopic from $\overline{X}$ to $\overline{E}$. ($\overline{E}$ is also denoted by BG)

A basic result [7] is that contractible free G-spaces are always universal. For example, if N is an infinite dimensional Banach space over the reals $\mathbb{R}$ or the complex numbers $\mathbb{C}$ or the quaternions $\mathbb{H}$ and S^∞ is the unit sphere in N, then S^∞ is a contractible free G-space (metric, hence paracompact) where $G = \mathbb{Z}_2$, S^1, SU(2), repsectively and the free action is given by scalar multiplication. Thus, S^∞ is a universal G-space for these three examples of compact Lie groups G. For the case of a general compact Lie group G, the classical method for obtaining such an E = EG is to first construct a free universal 0(k)-space as follows: Let [5]

$$V_{n,k} = \text{space of orthonormal k-frames in } \mathbb{R}^{n+k}$$

$$G_{n,k} = \text{space of k-planes in } \mathbb{R}^{n+k}$$

then, 0(k) acts freely on $V_{n,k}$ and the natural map $V_{n,k} \to G_{n,k}$ which assigns to a k-frame the corresponding k-plane is identified with the orbit map. The usual imbedding of $\mathbb{R}^{n+k} \subset \mathbb{R}^{n+k+1}$ gives rise to the sequence

$$\begin{array}{ccccccccc} V_{0,k} & \subset & V_{1,k} & \subset & \cdots \subset & V_{n,k} & \subset & V_{n+1,k} & \subset \cdots \\ \downarrow & & \downarrow & & & \downarrow & & \downarrow & \\ * & \subset & G_{1,k} & \subset & \cdots \subset & G_{n,k} & \subset & G_{n+1,k} & \subset \cdots \end{array}$$

and we set $V_{\infty,k} = \bigcup_n V_{n,k}$, $G_{\infty,k} = \bigcup_n G_{n,k}$.

We give this space the so-called "limit topology" e.g. $W \subset V_{\infty,k}$ is open if, and only if, $W \cap V_{n,k}$ is open for all $n \geq 0$. It then turns out that the induced map $V_{\infty,k} \to G_{\infty,k}$ can be identified with the orbit space of the free 0(k) action on $V_{\infty,k}$. Furthermore, $V_{\infty,k}$ is paracompact and contractible. Now, if G is a compact Lie group, $G \subset 0(k)$ for k large and we restrict the action of 0(k) on $V_{\infty,k}$ to G and in this manner $V_{\infty,k}$ is a free, contractible, paracompact G-space.

It follows easily that universal free G-spaces in F_G are unique in the following sense. If E and E' are universal G-spaces in F_G, there is a G-map $\phi : E \to E'$ such that $\overline{\phi} : E/G \to E'/G$ is a homotopy equivalence.

Thus, E/G, which is also called the classifying space for G and denoted by BG, is unique up homotopy equivalence. In particular, concepts defined in terms of homotopy invariants do not depend on BG.

Let us extend F_G to include all free paracompact G-pairs (X,A), A a closed G-set in X. Then, if $E \in F_G$, there is a useful functor $F : P_G \to F_G$ define as follows. Let $E = V_{\infty,k}$ denote the free contractible G-space (hence universal) previously discussed and set

$$F(X,A) = (E \times X,\ E \times A)$$

where E x X is the G-space with the coordinate-wise action g(u,x) = (gu,gx). Th since E is σ-compact (E x X, E x A) is a free, paracompact G-pair, and projection E x X → E provides a G-map into the universal G-space. The other projection E x X → X is also a G-map, giving rise to a diagram

$$\begin{array}{ccccc} X & \xleftarrow{proj_2} & E \times X & \xrightarrow{proj_1} & E \\ \downarrow & & \downarrow & & \downarrow \\ X/G & \xleftarrow[\eta_X]{} & E \times_G X & \xrightarrow[q_X]{} & BG \end{array}$$

where $E \times_G X = (E \times X)/G$, and η_X and q_X are induced on orbit spaces by projections. If $x \in X$, the mapping $E \to \text{Orb } x$ given by $u \to [u,x]$, where $[u,x] \in E \times_G X$ is the orbit containing (u,x) induces a homeomorphism $E/G_x \to \eta_X^{-1}(\text{Orb } x)$.

The cohomology employed here will be Alexander-Spanier [6]. Occasionally, the techniques of singular cohomology [6] will be used and justified when the spaces involved are such that the two cohomology theories are naturally isomorphic (e.g. in the category of locally contractible spaces). Coefficients will usually be from a field $\mathbb{K}$ and not necessarily displayed in the notation when the concept is clear. If $(X,A) \in P_G$, we define (coefficients in $\mathbb{K}$)

$$H_G^*(X,A) = H^*(E \times_G X)$$

called the Borel cohomology [5] of (X,A). It is an example of a G-cohomology theory. When, $f : (X,A) \to (X',A')$ is a G-map, then

$$1 \times f : E \times (X,A) \to E \times (X',A')$$

induces on orbit spaces

$$f_G : E \times_G (X,A) = (E \times_G X,\ A \times_G X) \to E \times_G (X',A')$$

and hence

$$f_G^* : H_G^*(X',A') \to H_G^*(X,A)$$

H_G^* is just the composition of the functors F and H^*. f also induces $\bar{f} : (X/G,A/G) \to (X'/G,A'/G)$ and hence

$$\bar{f}^* : H^*(X'/G,A'/G) \to H^*(X/G,A/G).$$

Thus, cohomology of the orbit space pair is another example of a G-cohomology theory defined in P_G. The map

$$\eta_{(X,A)} : E \times_G (X,A) \to (X/G,A/G)$$

induced by projection $E \times (X,A) \to (X,A)$ induces

$$\eta^*_{(X,A)} : H^*(X/G,A/G) \to H_G^*(X,A)$$

which serves to compare the two cohomology theories.

2.2. Theorem. Let $\mathbb{K}$ denote a field such that $H^*(BG_x,\mathbb{K}) = H^*(pt,\mathbb{K})$ for every $x \in X - A$, then

$$\eta^*_{(X,A)} : H^*(X/G,A/G,\mathbb{K}) \to H_G^*(X,A;\mathbb{K})$$

is an isomorphism.

This theorem is an extended version of the Vietoris-Begle mapping theorem [6]. An alternate way of stating the assumption is that [since $\eta^{-1}_{(X,A)}$ (Orb x) = E/G_x and $E/G_x = BG_x$, the classifying space for G_x] the preimages under $\eta_{(X,A)}$ are acyclic outside of A. For details see [5].

When the action is free on X - A, i.e., $G_x = e$ for all $x \in X - A$, then $BG_x = E$ for $x \in X - A$ and E is contractible.

2.3. Corollary If the action is free outside of A then, $\eta^*_{(X,A)}$ is an isomorphism for any coefficient field $\mathbb{K}$.

2.4. Corollary If G_x is a finite group for $x \in X - A$, then, $\eta^*_{(X,A)}$ is an isomorphism over the rational field $\mathbb{Q}$. More generally, over any field of characteristic 0.

Proof. $H^*(BG_x, \mathbb{Z})$ with integral coefficients is all torsion [8] and hence $H^q(BG_x; \mathbb{K}) = 0$, $q > 0$ for any $\mathbb{K}$ of characteristic 0.

We will find it useful to define the Lefschetz-Hopf number of a map $f : (X,A) \to (X,A)$ in both singular and Alexander-Spanier cohomology. H^* will denote Alexander-Spanier cohomology as before, while $\overline{H}^*$ will denote singular cohomology (coefficients in a field $\mathbb{K}$).

$$L_{(X,A)}(f;\mathbb{K}) = L_{(X,A)}(f) = \sum_q (-1)^q t_r[f^{*q} : H^q(X,A) \to H^q(X,A)]$$

$$\overline{L}_{(X,A)}(f;\mathbb{K}) = \overline{L}_{(X,A)}(f) = \sum_q (-1)^q tr[f^{*q} : \overline{H}^q(X,A) \to \overline{H}^q(X,A)]$$

where $H^*(X,A)$ and $\overline{H}^*(X,A)$ are finitely generated (f.g.), respectively, and tr = trace.

It is easy to verify that when defined

$$L_X(f) = L_A(f) + L_{(X,A)}(f)$$

$$\overline{L}_X(f) = \overline{L}_A(f) + \overline{L}_{(X,A)}(f)$$

where $L_X(f) = L_X(f|X)$, $L_A(f) = L_A(f|A)$, etc.

Another very useful tool is the following.

2.5. Proposition [ ] Let $F \to X \xrightarrow{p} Y$ denote a Hurewicz fibration orientable over $\mathbb{K}$ and

$$\begin{array}{ccc} X & \xrightarrow{f} & X \\ p\downarrow & \overline{f} & \downarrow p \\ Y & \to & Y \end{array}$$

a fiber preserving map. Let $F = p^{-1}(y_0)$ and assume $\overline{f}(y_0) = y_0$ (for simplicity), $y_0 \in Y$. Then, if $\overline{H}^*(Y)$ and $\overline{H}^*(F)$ are f.g. over $\mathbb{K}$ and $f_0 = f|F$.

$$\overline{L}_X(f;\mathbb{K}) = \overline{L}_Y(\overline{f};\ \mathbb{K})\,\overline{L}_F(f_0;\mathbb{K})$$

2.6. Remark When X, Y and F are lc the formula remains valid for L.

Also, if Y is compact, the above formula is also valid for L without the lc condition using the same proof which involves a Leray spectral sequence argument together with an appropriate Universal Coefficient theorem [6] for Alexander-Spanier cohomology. Finally, it is not necessary to assume that $f(F) \subset F$ because one can always follow $f|F$ by a translation along a path in Y, back to F, when Y is 0-connected.

Another useful result is the so-called Comparison Theorem [9].

2.7 Theorem Let $F \to X \xrightarrow{p} Y$ and f, $\overline{f}$, f_0 be a in theorem 2.5. Then, if any two of the induced homomorphisms f^*, $\overline{f}^*$, f_0^* in singular homology are isomorphisms so is the third.

2.8 Remark The remarks in 2.6 also indicate when the corresponding result for Alexander-Spanier cohomology is valid.

Now we give some relevant notation. $\mathbb{R}$, $\mathbb{C}$, $\mathbb{H}$ will denote the reals, complex numbers, and quaternions and $\mathbb{F}$ will denote one of the three. $S(\mathbb{F})$ is the unit sphere in $\mathbb{F}$ so that $S(\mathbb{F}) = \mathbb{Z}_2, S^1$, or $S^3 = SU(2)$. $\mathbb{F}^k$ will denote the vector space over $\mathbb{F}$ of algebraic dimension k and $S(\mathbb{F}^k)$ the corresponding unit sphere. Thus, $S(\mathbb{F}^k) = S^{k-1}, S^{2k-1}$, or S^{4k-1}, where S^n is, of course, the unit sphere in $\mathbb{R}^n$. $S(\mathbb{F})$ operates freely on $S(\mathbb{F}^{k+1})$ by scalar multiplication and the corresponding projective space is given by

$$\mathbb{F}P^k = S(\mathbb{F}^{k+1})/S(\mathbb{F})$$

thus, $\mathbb{F}P^k = \mathbb{R}P^k$, $\mathbb{C}P^k$, or $\mathbb{H}P^k$ and these are manifolds of (real) dimension k, 2k and 4k, respectively.

3. THE HOPF-RUEFF THEOREM

The setting for the Hopf-Rueff theorem [4] is the following. S^{2n+1} is the unit sphere in $\mathbb{C}^{n+1}$ and $f : S^{2n+1} \to \mathbb{C}^n$ is a map given by n component functions $(f_1, f_2, \ldots, f_n)$ where $f_i : S^{2n+1} \to \mathbb{C}$. Each f_i is assumed to be homogeneous of degree m_i, m_i an non-zero integer, i.e.

$$f_i(\zeta z) = \zeta^{m_i} f_i(z), \quad z \in \mathbb{C}^{n+1}$$

3.1. Theorem [Hopf-Rueff]. The system of equations $f_i(z) = 0$, $i = 1, \ldots, n$ has a solution $\hat{z} \in S^{2n+1}$, i.e. there exists a $\hat{z}$ such that $f(\hat{z}) = 0$.

In the language of group actions, let $G = S^1$ and let S^{2n+1} denote the G-space with the usual scalar multiplication as action and $\hat{\mathbb{C}}^n$ the G-space with the action $\zeta(z_1, \ldots, z_n) = (\zeta^{m_1}z_1, \ldots, \zeta^{m_n}z_n)$. There we may restate Theorem 3.1 as

Theorem A'. Every S^1-map $f : S^{2n+1} \to \hat{\mathbb{C}}^n$ has a zero.

This is a complex analogue of theorem A in §1.

The analogues of theorems B and C (§1) are:

Theorem B'. There does not exist an S^1-map $f : S^{2n+1} \to \hat{S}^{2n-1}$, where $\hat{S}^{2n-1}$ is the unit sphere in $\hat{\mathbb{C}}^n$ with the induced action.

Theorem C'. Every S^1-map $f : S^{2n+1} \to \hat{S}^{2n+1}$ is essential, where the action on $\hat{S}^{2n+1}$ is given by $\zeta(z_1, \ldots, z_{n+1}) = (\zeta^{m_1}z_1, \ldots, \zeta^{m_{n+1}}z_{n+1})$, m_i non-zero integers.

It is easy to see that Theorem C' implies Theorem B' implies Theorem A'. In fact, they are equivalent. It is interesting to note that, while as spaces S^{2n+1} and $\hat{S}^{2n+1}$ are the same, the actions are different. Furthermore, the action in $\hat{S}^{2n+1}$ is not free since if $\mathbb{Z}_q$ is the group of z-th roots of unity, $\mathbb{Z}_q$ appears as isotropy for $q = m_1, \ldots, m_{n+1}$. However, S^1 does not appear as an isotropy subgroup precisely because the $m_i \neq 0$. Thus, no fixed point set is allowed.

Hopf and Rueff prove Theorem 3.1 by proving a special case of Theorem C', namely when all the m_i have the same value m. They compute $\deg f = m^{n+1}$. Actually implicit in their proof is that in Theorem C', $\deg f = m_1 m_2 \ldots m_{n+1} \neq 0$.

One may consider the Hopf-Rueff result as a special case of the general question as to when a G-map $f : X \to Y$ is essential. Here Y may be X as a space but with a different action. Some initial results on this problem will be found in §5.

The extension of Theorem C' to the corresponding situation with a fixed point set is in [3]. An alternative proof of that result may also be found in [10] where as in Hopf-Rueff the degree is explicity computed.

§4. ESSENTIAL G-MAPS

The first general principal is that all G-maps $f : X \to X$ of a free G-space X are essential if G is a non-trivial compact Lie group.

4.1 Theorem (Dold [11]). Let X denote a finite dimensional, free G-space with G a non-trivial finite group. Then, if $f : X \to X$ is a G-map, f is essential.

4.2 Corollary The above theorem obtains for G any compact Lie group.

Proof. If dim G > 0, choose a maximal torus $T \subset G$ and then a $\mathbb{Z}_p \subset T$.

While Dold's proof is quite simple, based on elementary considerations in the classification theory of bundles and some local degree theory, one can give alternative proofs which give a bit more information. Local contractibility (lc) will be assumed so that we may use either singular or Alexander-Spanier cohomology.

4.3 Theorem Let X denote a 0-connected paracompact, lc space such that $H^*(X)$ is f.g. over a field $\mathbb{K}$. Then if a compact, connected, non-trivial Lie group G asts freely on X and $f : X \to X$ is a G-map, the Lefschetz number L(f) = 0 over $\mathbb{K}$; in particular f must be essential.

Proof. f induces a diagram

$$\begin{array}{ccc} X & \xrightarrow{f} & X \\ p \downarrow & & \downarrow p \\ X/G & \xrightarrow{\bar{f}} & X/G \end{array}$$

which is fibre-preserving map of principal G-bundles. We may assume without loss of generality that $\bar{f}$ has a fixed point x_0 and identify the fiber over x_0 with G. Then, if $f_0 = f/G$ and $f_0(1) = g_0$, $f(g) = g\, g_0$ and f_0 is right translation by g_0. Using a path from g_0 to 1 we see that $f_0 \simeq id : G \to G$ and hence $L(f_0) = \chi(G) = 0$, over $\mathbb{K}$. But by the Product theorem

$$L(f) = L(\bar{f})L(f_0) = 0$$

over $\mathbb{K}$.

4.4 Remark If G is a compact Lie group of dimension ≥ 1 and X is a free G-space, then X is a free G_0-space where G is the connected component of the identity and the above theorem may be applied using G_0. Also, we will consider the analogue of Theorem 4.3 for finite groups (in a more general context) later on.

When X, is a sphere, the only non-trivial compact, connected Lie groups that can act freely on X are S^1 and SU(2) (see Bredon [5]), so that Theorem 4.3 has its limitations. In particular, we wish to investigate the situation when there is non-trivial isotropy present.

The following example is instructive. Let S^2 denote the unit sphere in $\mathbb{R}^3$ and let S^1 act on $\mathbb{R}^3$ by rotation around one of the axes so that the fixed point set of the action is the axis (see diagram below).

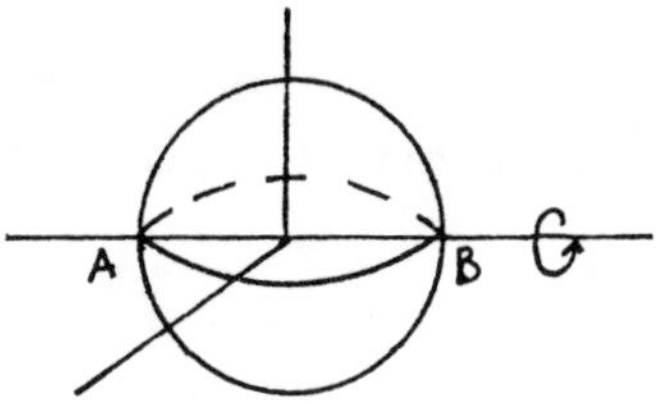

Let $X = S^2$ and $G = S^1$, then $X^G = A \cup B$. Let $f : S^2 \to S^2$ denote the constant map $f(S^2) = A$. Then, clearly

$$f : S^2 \to S^2 \tag{1}$$

is a G-map and not essential. Furthermore, f may also interpret as a G-map

$$\bar{f} : S^2 \to \mathbb{R}^1 = \text{axis of rotation} \tag{2}$$

with no zeros. However, suppose we require that in (2), $\bar{f}$ be the identity on $A \cup B$, the fixed point set. Then, clearly one must have zeros for $\bar{f}$. Furthermore, if in (2) we require the same condition, the one can also prove that f is essential. We are thus led to consider G-pairs (X,A) and G-maps $f : (X,A) \to (X,A)$ with appropriate conditions on $f|A : A \to A$.

4.5 Definition Let (X,A) denote a G-pair. (X,A) will be called a quasi-free G-pair over $\mathbb{K}$ if for each $x \in X - A$

$$H_G^q(BG_x;\mathbb{K}) = 0, \; q > 0,$$

i.e. BG_x is acyclic over $\mathbb{K}$.

Here are some examples

(a) Suppose the action of G on X is semi free, i.e., G_x is either G or e. If we set $A = X^G$, then (X,A) is a quasi-free pair over any $\mathbb{K}$.

(b) Let $G = S^1$ and X and S^1 space. Then, again set $A = X^G$ so that for $x \in X - A$, the isotropy subgroup G_x is finite. Then, (X,A) is a quasi-free G-pair over $\mathbb{Q}$

(c) More generally, let X denote any G-space (G connected) and A a G-subset so that G_x is finite for all $x \in X - A$. Then, (X,A) is a quasi-free G-pair over $\mathbb{Q}$.

(d) Let G denote a finite group and X a G-space. Let A denote a G-subset of X such that G_x has order not divisible by a prime p if $x \in X - A$. Then, (X,A) is a quasi-free G-pair over the field $\mathbb{Z}_p$.

4.6 <u>Theorem</u> Let G denote a compact, connected Lie group (non trivial) and (X,A) a quasi-free lc G-pair over $\mathbb{K}$ such that $H^*(\overline{X},\overline{A})$ is f.g. over $\mathbb{K}$. Let

$$f : (X,A) \to (X,A)$$

denote a G-map. Then, $L_{(X,A)}(f,\mathbb{K}) = 0$ and hence

$$L_X(f,\mathbb{K}) = L_A(f,\mathbb{K})$$

<u>Proof</u>. Consider the fiber map of pairs

$$G \to E \times (X,A) \to E \times_G (X,A), \quad E = EG$$

Observe that $H_G^*(X,A) \cong H^*(\overline{X},\overline{A})$ by the Victoris mapping theorem (§2) and hence $H_G^*(X,A)$ is f.g. Thus, we have a diagram of orientable G-bundles (G is connected)

$$\begin{array}{ccc} E \times (X,A) & \xrightarrow{1\times f} & E \times (X,A) \\ \downarrow & & \downarrow \\ E \times_G (X,A) & \xrightarrow{f_G} & E \times_G (X,A) \end{array}$$

and the Product theorem for Lefschetz-Hopf numbers applies so that

$$L_{(X,A)}f = L(1 \times f) = L(f_G)L(1 \times f|G) = 0$$

Since

$$L_X(f,\mathbb{K}) = L_A(f,\mathbb{K}) + L_{(X,A)}(f,\mathbb{K}) = L_A(f,\mathbb{K}) ,$$

we have the desired result.

4.7 <u>Corollary</u> Suppose in the above theorem, $L_A(f,\mathbb{K}) \neq 1$ then $L_X(f,\mathbb{K}) \neq 1$ and if X is connected, $f : X \to X$ must be essential.

4.8 <u>Corollary</u> Suppose in the above theorem X and A are spheres and

$\mathbb{K} = \mathbb{Q}$. Then, if $\deg f|A \neq 0$, f_0^* is an isomorphism over $\mathbb{Q}$ and this forces f_1^* to be an isomorphism over $\mathbb{Q}$. In fact, $\deg f|A = \pm \deg f$.

4.9 Remark The hypothesis that $H^*(\overline{X},\overline{A})$ is f.g. over $\mathbb{K}$ may seem a bit awkward to apply since the orbit pair $(\overline{X},\overline{A})$ may be rather complicated. However, it is usually easy to establish that $\dim \overline{X} \leq \dim X$, so that $H^q(\overline{X},\overline{A}) = 0$ for q large when dim X is finite. On the other hand, assuming $H^q(X,A)$ is f.g. for each q implies $H^q_G(X,A) \cong H^q(\overline{X},\overline{A})$ is f.g. for each q using the fibration of pairs.

$$G \to E \times (X,A) \to E \times_G (X,A).$$

The result corresponding to Theorem 4.6 for a finite group G is the following.

4.10 Theorem Let G denote a finite group and (X,A) a lc G-pair such that for $x \in X - A$, $(|G_x|,p) = 1$ for some prime p which divides $|G|$. Let

$$f : (X,A) \to (X,A)$$

denote a G-map such that X is finite dimensional and $H^*(X,A,\mathbb{Z}_p)$ and $H^*(X;\mathbb{Z}_p)$ are f.g. Then,

$$L_{(X,A)}(f,\mathbb{Z}_p) = 0$$

and hence

$$L_X(f,\mathbb{Z}_p) = L_A(f,\mathbb{Z}_p)$$

The proof will require the following algebraic technique. Let $G = \{1, g, \ldots, g^{p-1}, g^p = 1\}$ denote the cyclic group of order p and generator g and $\mathbb{Z}_p[G]$ the corresponding group ring over $\mathbb{Z}_p$. In $\mathbb{Z}_p[G]$, let

$$b_0 = h(g) = 1 + g + g^2 + \ldots + g^{p-1}$$

$$b_1 = h'(g)$$

$$\vdots$$

$$b_{p-1} = h^{(p-1)}(g) = (p-1)!$$

This basis for $\mathbb{Z}_p$ [G] induces a filtration

$$A_0 \supset A_1 \supset \quad \supset \dots A_{p-1} \supset 0$$

of $A_0 = \mathbb{Z}_p[G]$ by setting $A_i = \mathbb{Z}_p[b_i, \dots, b_{p-1}]$, If we let $D_i = \mathbb{Z}_p[b_i]$ we have exact sequences

$$0 \to D_0 \to A_0 \to A_1 \to 0$$

$$0 \to D_1 \to A_1 \to A_2 \to 0$$

$$\vdots$$

$$0 \to D_{p-1} \to A_{p-1} \to 0$$

where G acts <u>trivally</u> on each $D_i \subset A_i$ and of course $g(A_i) \subset A_i$. Notice that if

$$f : G \to G$$

is a G-map and $f(1) = g$, then,

$$f : \mathbb{Z}[G] \to \mathbb{Z}[G]$$

will preserve the previous filtration.

<u>Proof of Theorem 4.10</u>. Consider the diagram

$$\begin{array}{ccc} E \times (X,A) & \xrightarrow{1xf} & E \times X,A \\ q \downarrow & & \downarrow q \\ E \times_k (X,A) & \xrightarrow{f_k} & E \times_k (X,A) \end{array}$$

where $K \subset G$ is a subgroup of order p. Since X is finite dimensional and $H^*(X,A;\mathbb{Z}_p)$ is f.g. and we have $H^*(\overline{X},\overline{A};\mathbb{Z}_p) \simeq H^*_K(X,A;\mathbb{Z}_p)$, we see that $H^*_K(X,A,\mathbb{Z}_p)$ is f.g. over $\mathbb{Z}_p$.

Consider now the spectral sequence of the covering map q where the non-trivial E_2 terms are

$$E_2^{p,0} \simeq H^p(Y,\mathbb{Z}_p(K))$$

in the sense of local coefficients, where $Y = E \times_K (X,A)$, and $\mathbb{Z}_p[K]$

$= H^0$ (fiber). We also have a map (induced by f)

$$f_2^{p,0} : E_2^{p,0} \to E_2^{p,0}$$

and we wish to determine the Lefschetz number $L(f_2)$ of

$$f_2 : H^*(Y;\mathbb{Z}_p(K)) \to H^*(Y;\mathbb{Z}_p(K))$$

To simplify matters we assume $x_0 \in Y$ is a fixed point of f_K so that $1 \times f$ takes the fiber K to K and because $1 \times f$ is a K-map, $1 \times f$ takes 1 to g for some $g \in K$. We apply the algebra previously outlined and obtain a filtration

$$\mathbb{Z}_p(K) = A_0 \supset A_1 \supset \dots \quad A_{p-1} \supset 0$$

and exact sequences

$$0 \to D_i \to A_i \to A_{i+1} \to 0$$

which gives rise to exact sequences

$$\begin{array}{ccccccc} \to H^q(Y,D_i) & \to & H^q(Y,A_i) & \to & H^q(Y,A_{i+1}) & \to & \dots \\ f_K^* \downarrow & & \downarrow & & \downarrow & & \\ \to H^q(Y,D_i) & \to & H^q(Y,A_i) & \to & H^q(Y,A_{i+1}) & \to & \dots \end{array}$$

of cohomology groups with the indicated local coefficients, with D_i being coefficients. It now follows easily that $L(f_2)$ is defined and

$$L(f_2) = p\ L(f_K^*;\mathbb{Z}_p) = 0 \text{ (in } \mathbb{Z}_p)$$

thus,

$$L_{(X,A)}(f,\mathbb{Z}_p) = 0$$

and our theorem follows.

4.11 <u>Corollary</u> If $G = \mathbb{Z}_p$, and X is an lc G-space with $A = X^G$ lc, we may apply the above theorem to a G-map $f : (X,A) \to (X,A)$ to conclude that

$$L_X(f,\mathbb{Z}_p) = L_A(f,\mathbb{Z}_p)$$

and hence f is essential if $L_A(f,\mathbb{Z}_p) \neq 1$. If $A = \phi$ then

$$L_X(f;\mathbb{Z}_p) = 0$$

4.12 Corollary If X is an lc connected free G-space, G finite ($G \neq 1$), where $H^*(X,\mathbb{Z}_p)$ is f.g. and X is finite dimensional, then any G-map

$$f : X \to X$$

is essential, in fact $L_X(f;\mathbb{Z}_p) = 0$.

4.13 Remark This is a variation of Dold's theorem [11]. The hypothesis is slightly stronger on X, but the conclusion $L_X(f,\mathbb{Z}_p) = 0$ gives a bit more information,

Before leaving this technique we will comment briefly on the extension to infinite dimensional situations when the G-map $f : X \to X$ is compact. We consider only the free case and assume G is a non-trivial compact, connnected Lie group. Recall that if $f : V \to V$ is a linear transformation of a vector space over a field $\mathbb{K}$, we let f^n denote the n-th iterate of f and Ker f^n the kernel of f^n. Then

$$\text{Ker } f \subset \text{Ker } f^2 \subset \ldots \subset \text{Ker } f^n \subset \ldots$$

and we let

$$K(f) = \bigcup_n \text{Ker } f^n, \quad \tilde{V} = V/K(f).$$

f then induces an injection $f : \tilde{V} \to \tilde{V}$. f is called algebraically compact if $\tilde{V}$ is finite dimensional over $\mathbb{K}$. When f is algebraically compact the Leray trace is defined by

$$\mathrm{Tr}_\lambda(f) = \text{trace } \tilde{f} \quad .$$

We can then define the Lefschetz-Hopf number of a linear transformation of a graded vector space $V = \{V^q\}$

$$f^q : V^q \to V^q, \; q \geq 0$$

if we assume $f^q : V^q \to V^q$ is algebraically compact for all q and $\tilde{V}_q = 0$

for q sufficiently large. Set

$$L_\lambda(f) = \sum_q (-1)^q \, Tr_\lambda(f) .$$

In particular, if X is ANR (metric) and $f : X \to X$ is a compact map, then $L_\lambda(f)$ is defined.

4.14 Theorem Let X denote an ANR (metric) and suppose X is a free G-space. Then, if $f : X \to X$ is a compact G-map, we have

$$\begin{array}{ccc} X & \xrightarrow{f} & X \\ \downarrow & & \downarrow \\ \overline{X} & \xrightarrow{\overline{f}} & \overline{X} \end{array}$$

where $\overline{X} = X/G$ and $L_\lambda(f) = L_\lambda(\overline{f}) \cdot L(f_0)$, where $f_0 : G \to G$ is f restricted to a fiber. Since $L(f_0) = 0$ $L_\lambda(f) = 0$.

Some easy consequences of theorem 4.14 are

1) If X is the unit sphere in a Banach space over $\mathbb{C}$ and $G = S^1$ acting by scalar multiplication, there is no compact G-map $f : X \to X$.

2) The map f in theorem 4.14 is essential if X is connected.

It is sometimes possible to filter a G-space X

$$X_0 \subset X_1 \subset \dots \subset X_n = X$$

and show that a G-map $f : X \to X$ is essential by assuming $f|X_0 : X_0 \to X_0$ is and working up the filtration to X. We make this technique precise as follows. Let G denote our compact Lie group and take a given filtration

$$F : G = G_0 \supset G_1 \supset \dots \supset G_n = e$$

of G into closed subgroups. F induces a filtration

$$F(X) : X^G = X_0 \subset X_1 \subset \dots \subset X_n = X$$

where $X_i = X^{G_i} = \{x : gx = x, \ g \in G_i\}$, the points of X left fixed under the action of G_i. Then, a G-map $f : X \to X$ preserves this filtration F(X), i.e. $f(x_i) \subset X_i$ and if NG_i is the normalizer of G_i in G_{i-1}, then $W_i = NG_i/G_i$ acts on X_i so that under appropriate conditions

(X_i, X_{i-1}) is a quasi-free W_i-space over some field $\mathbb{K}_i$

4.15 Definition Fix a filtration F of G as above, and let $\mathbb{K} = (\mathbb{K}, \ldots, \mathbb{K}_n)$ denote a sequence of coefficient fields. Let X denote a G-space. We call X an F-quasi-free G-space over $\mathbb{K}$ provided (X_i, X_{i-1}) is a quasi-free W_i-space over $\mathbb{K}_i$, i.e. BW_{ix} is a cyclic over $\mathbb{K}_i$ for every $x \in X_i - X_{i-1}$, $1 \leq i \leq n$.

4.16 Example Let G = SU(2) and let S^1 denote a maximal torus. Let F denote the filtration

$$SU(2) \supset S^1 \supset e$$

Suppose X is a G-space such that the normalizer of S^1, NS^1 does not appear as an isotropy subgroup. Let $\mathbb{K} = (\mathbb{Z}_2, \mathbb{Q})$. Then, the induced filtration

$$X_0 \subset X_1 \subset X_2 = X$$

has the property that (X_1, X_0) is a quasi-free $\mathbb{Z}_2$-space over $\mathbb{Z}_2$, where $\mathbb{Z}_2 = NS^1/S^1$ and (X_2, X_1) is a quasi-free S^1-space over $\mathbb{Q}$. Let $f : X \to X$ denote an SU(2) map, and $f_i = f/X_i : X_i \to X_i$.

Also, assume X is finite diminsional, the X_i and lc, and $H^*(X_1, X_0)$ and $H^*(X_2, X_1)$ are finitely generated over $\mathbb{Z}_2$ and $\mathbb{Q}$, respectively. Supppose $L_{X_0}(f_0, \mathbb{Z}_2) = 0$, then using Theorem 4.10, $L_{X_1}(f_1, \mathbb{Z}_2) = 0$. But

$$L_{X_1}(f_1; \mathbb{Q}) \bmod 2 = L_{X_1}(f, \mathbb{Z}_2)$$

This forces $L_{X_1}(f_1, \mathbb{Q})$ to be even. Now apply Theorem 4.6 to conclude

$$L_X(f, \mathbb{Q}) \text{ is even.}$$

In particular, when X is connected, f must be essential. If X is a sphere, the degree of f is odd.

This example generalizes to obtain the following theorem.

4.17 Theorem Fix a filtration F of G as above and let $\mathbb{K} = (\mathbb{K}_1, \ldots, \mathbb{K}_n)$ denote a sequence of coefficient fields. Let X denote a F-quasi-free G-space over $\mathbb{K}$ with induced filtration

$$F(X) : X_0 \subset X_1 \subset X_2 \subset \dots \subset X_n = X$$

such that each X_i is lc and

(a) X is finite dimensional and $H^*(X_i)$ is f.g. over $\mathbb{K}_i$ and $\mathbb{K}_{i+1}$.

(b) If $f_i : X_i \to X_i$ is the restriction of f to X_i then we assume

$$L_{X_i}(f_i, \mathbb{K}_i) \neq 1 \Rightarrow L_{X_i}(f_i, \mathbb{K}_{i+1}) \neq 1$$

(c) $L_{X_0}(f_0, K_1) \neq 1$

Then, we may conclude that

$$L_X(f, \mathbb{K}_n) \neq 1$$

and in particular if X is connected, f must be essential.

§5 ESSENTIAL MAPS-SECOND METHOD

Our objective here is to consider G-maps $f : X \to Y$ between possibly different G-spaces and determine when f is essential. In particular X could be the same as Y but the actions different which is the Hopf-Rueff situation. Let G denote a compact Lie group acting on $\mathbb{R}^{n+1}$ as a group of unitary transformations and let $\mathbb{R}^{k+1}$ denote a G-subspace so that the corresponding sphere pair (S^n, S^k) is quasi-free over a field $\mathbb{K}$, in particular $H_G^*(S^n, S^k)$ has finite rank as a $\mathbb{K}$-module. We assume also that S^k contains elements fixed under the action of G, i.e. $(S^k)^G \neq \emptyset$. Then the map $S^n \to x_0$, x_0 a point, admits a G-section and hence

$$S^n \xrightarrow{i} E \times_G S^n \to E \times_G b_0 = BG$$

$$S^k \xrightarrow{i} E \times_G S^k \to BG$$

admit sections. This situation allows one to conclude [2], that as $\Lambda = H^*(BG, \mathbb{K})$-modules

$$H_G^*(S^n) \cong \Lambda \otimes H^*(S^n) \qquad \otimes = \otimes_{\mathbb{K}}$$

$$H_G^*(S^n) \cong \Lambda \otimes H^*(S^n)$$

and hence both are free Λ-modules. The sequence

$$S^k \xrightarrow{i} S^n \xrightarrow{i} (S^n, S^k)$$

induces an exact sequence

$$\to H_G^q(S^n) \xrightarrow{i^*} H_G^q(S^k) \xrightarrow{\delta^*} H_G^{q+1}(S^n, S^k) \xrightarrow{j^*}$$

where the homomorphisms are Λ-module homomorphisms. This forces i^* to be injective for each q as follows. If $i^*(u) = 0$, then $u = j^*(v)$. But $\lambda v = 0$ for some $\lambda \in \Lambda$ and hence, $j^*(\lambda v) = \lambda u = 0$. This forces $\lambda = 0$ because $H_G^*(S^n)$ is a free Λ-module. Thus, we have short exact sequences

$$0 \to H_G^q(S^n) \to H_G^q(S^k) \xrightarrow{\delta} H_G^{q-1}(S^n, S^k) \to 0$$

Now let us assume Λ is monogenic, eg. $G = \mathbb{Z}_2$ and $\mathbb{K} = \mathbb{Z}_2$ or $G = S^1$, SU(2) and $\mathbb{K} = \mathbb{Q}$. Then, $H_G^*(S^n, S^k)$ as a $\mathbb{K}$-module is generated by images of

$$1 \otimes u,\ \alpha \otimes u,\ \alpha^2 \otimes u,\ \ldots,\ \alpha^m \otimes u$$

α the generator of Λ, u the generator of $H^k(S^k)$ and

$$(m + 1)|\alpha| + |u| = n$$

Notice that if (Σ^n, Σ^k) is another quasi-free G-pair where Σ^n and Σ^k are cohomology spheres over $\mathbb{K}$, the same computation obtains and $H_G^*(\Sigma^n, \Sigma^k)$ has the same rank (namely m + 1). Suppose now that

$$f : (S^n, S^k) \to (\Sigma^n, \Sigma^k)$$

is a G-map such that over $\mathbb{K}$

$$\overline{f}^* : H^*(\Sigma^k) \to H^*(S^k)$$

is an isomorphism, where $\overline{f} = f/K$ then, applying the Comparison Theorem for spectral sequences to the situation

$$\begin{array}{ccc} S^k & \xrightarrow{\bar{f}} & \Sigma^k \\ \downarrow & & \downarrow \\ E \times_G S^k & \xrightarrow{\bar{f}_G} & E \times_G \Sigma^k \\ & \searrow \quad \swarrow & \\ & BG & \end{array}$$

we see that in the diagram

$$\begin{array}{ccccccccc} 0 & \to & H_G^q(\Sigma^n) & \to & H_G^q(\Sigma^k) & \to & H_G^{q+1}(\Sigma^n,\Sigma^k) & \to & 0 \\ & & \hat{f}_G^* \downarrow & & \bar{f}_G^* \downarrow & & \downarrow f_G^* & & \\ 0 & \to & H_G^q(S^n) & \to & H_G^q(S^k) & \to & H_G^{q+1}(S^n,S^k) & \to & 0 \end{array}$$

$\bar{f}_G^*$ is an isomorphism. But this forces f_G^* to be surjective. But since $H_G^*(\Sigma^n,\Sigma^k)$ and $H_G^*(S^n,S^k)$ have the same finite rank, it follows that f_G^* is also an isomorphism. This, in turn forces $\hat{f}_G^*$ to be an isomorphism and hence $\hat{f}^* : H^*(\Sigma^n) \to H^*(S^n)$ is also by the aforementioned Comparison Theorem. In particular $f : S^n \to \Sigma^n$ induces isomorphisms over $\mathbb{Q}$ when $G = S^1$ and hence f has non-zero degree, which is a generalization of the Hopf-Rueff result without knowing specifics about the actions involved. We collect the above in the following theorem.

5.1 Theorem Let G denote S^1 or SU(2) and (S^n,S^k), (Σ^n,Σ^k) denote G-pairs such that isotropy groups G_x are finite for $x \in S^n - S^k$ and $x \in \Sigma^n - \Sigma^k$. Let

$$f : (S^n,S^k) \to (\Sigma^n,\Sigma^k)$$

denote a G-map such that (over $\mathbb{Q}$).

$$\bar{f}^* : H^*(\Sigma^k) \to H^*(S^k)$$

is an isomorphism. Then, f : $S^n \to \Sigma^n$ has non-zero degree, or equivalently, f is essential.

5.2 <u>Remark</u> The corresponding theorem for $G = \mathbb{Z}_2$ has less content and is left to the reader.

One can obtain slightly more general results and the following is representative.

5.3 <u>Theorem</u> Let $G = SU(2)$ and S^n, Σ^n cohomology spheres on which G acts. Consider the filtration

$$SU(2) \supset S^1 \supset e$$

where S^1 is a maximal torus with induced filtrations

$$S^k \subset S^\ell \subset S^n$$

$$\Sigma^k \subset \Sigma^\ell \subset \Sigma^n$$

where all of the above are cohomology spheres of the indicated dimensions. Let $f : S^n \to \Sigma^n$ denote a G-map such that

$$\bar{f} : S^k \to \Sigma^k$$

has odd degree. Then, if neither action has NS^1 as an isotropy subgroup,

$$f : S^n \to \Sigma^n$$

has non-zero degree and is, therefore, essential

5.4 <u>Remark</u> Theorem 5.3 is also valid for $G = SO(3)$.

The properties possessed by the G-pairs (S^n,S^k), (Σ^n,Σ^k) in the previous discussion lead to the following concept.

5.5 <u>Definition</u> A G-pair (X_1,X_0) will be called <u>nilpotent</u> over $\mathbb{K}$ if it has the following properties relative to the field $\mathbb{K}$ of coefficients

N.1 $H_G^*(X_1,X_0)$ is a torsion module over $\Lambda = H^*(BG)$

N.2 $H_G^*(X_1)$ is a torsion free Λ-module.

N.3 $X_i \to E \times_G X_i \to BG$ are orientable G-bundles over $\mathbb{K}$ (automatic if G is connected).

This property is studied in [2] and we will content ourselves with one more example which will be useful later on. Let $\mathbb{F}$ denote the field $\mathbb{R}$, $\mathbb{C}$, or $\mathbb{H}$ (reals, complex numbers or quaternions) and $p : V \to B$ an orientable vector bundle over $\mathbb{F}$. Our coefficients $\mathbb{K}$ are $\mathbb{Z}_2$ if $\mathbb{F} = \mathbb{R}$ and $\mathbb{Q}$ otherwise. Let $G = \mathbb{Z}_2$, S^1 or $SU(2)$, in each respective case. Let $p' : V' \to B'$ denote another such vector bundle and let $V_0 = V^G$ and $V_0' = (V')^G$, the points fixed under G, forming subbundles. Suppose the fiber dimensions of V, V', V_0 are n, k, and k_0, respectively. We also assume that p and p' are G-maps with B given the trivial action so that we are in the class of G-bundles over B. We wish to consider in the next section a Bourgin-Yang type theroem applied to the situation

$$\begin{array}{ccc} S(V) & \xrightarrow{f} & V' \\ p \downarrow & & \downarrow p' \\ B & \xrightarrow{\bar{f}} & B' \end{array} \qquad (1)$$

where S(V) is the corresponding sphere bundle in V and f is a G-map, and investigate the "size" of the zero set $Z(f) = \{x | f(x) = 0\}$. By replacing (V', p', B') by the induced bundle (by $\bar{f}$) over B we may assume the simpler situation,

$$\begin{array}{ccc} S(V) & \xrightarrow{f} & V' \\ & p \searrow \quad \swarrow p' & \\ & B & \end{array} \qquad (1^*)$$

5.6 Lemma Suppose p admits a section $\sigma : B \to S(V_0)$ and B is finite dimensional. If $(S(V),S(V_0))$ is a quasi-free pair over $\mathbb{K}$, then $(S(V),S(V_0))$ is a nilpotent pair over $\mathbb{K}$.

Proof Consider the sphere bundles (orientable)

$$S^{n'} \to E \times_G S(V) \to E \times_G B = BG \times B$$

$$S^{k_0'} \to E \times_G S(V_0) \to E \times_G B = BG \times B$$

which admit sections. Then, over $\mathbb{K}$

$$H_G^*(S(V)) \cong H^*(\Lambda) \otimes H^*(B) \otimes H^*(S^{n'})$$

$$H_G^*(S(V_0)) \cong \Lambda \otimes H^*(B) \otimes H^*(S^{k_0'})$$

as modules over $\Lambda \otimes H^*(B)$, where n' and k_0' are the real dimensions of the fibers in $S(V)$ and $S(V_0)$, respectively. Thus, both $H_G^*(S(V))$ and $H_G^*(S(V_0))$ are free Λ-modules.

If we consider, the projection

$$(S(V),S(V_0) \leftarrow E \times (S(V),S(V_0))$$

which induces on the orbit pairs

$$(\overline{S}(V),\overline{S}(V_0) \overset{\alpha}{\leftarrow} E \times_G (S(V),S(V_0))$$

then γ induces (Vietoris Mapping theorem) isomorphisms

$$\gamma^* : H^*(\overline{S}(V),\overline{S}(V_0)) \to H_G^*(S(V),S(V_0)$$

and hence, since B is finite dimensional, $H_G^*(S(V),S(V_0))$ vanishes in sufficiently high dimensions and $H_G^*(S(V),S(V_0))$ is a torsion Λ-module. The orientability requirement N.3 is a simple exercise.

Returning to situation (1^*), to show that the zero set is non-zero under appropriate conditions we consider the following theorem.

5.7 Proposition Suppose V_0' is a subbundle of V' of dimension k_0 and f in (1^*) induces isomorphisms

$$f_0^* : H^*(V_0' - 0) \to H^*(S(V_0))$$

where $(S(V),S(V_0)$ and $(S(V'),S(V_0')$ are quasi-free over $\mathbb{K}$. Then, the zero-set $Z(f) \subset S(V) - S(V_0)$ is non-empty in every fiber.

Proof Let F, F', F_0, F_0' denote fibers in V, V', V_0, V_0' and $S(F)$, $S(F')$, etc., the corresponding unit spheres. Consider the restriction

$$\overline{f} : S(F) \to F'$$

with

$$\overline{f}_0^* : H^*(F_0' - 0) \to H^*(S(F_0))$$

$\overline{f}_0^*$ is an isomorphism (by the Comparison theorem). If $\overline{f}$ has no zeroes, we obtain a G-map $\overline{f} : S(F) \to S(F')$ where $\dim S(F) > \dim S(F')$, contradicting Theorem 5.1.

§6. RELATIVE G-COHOMOLOGICAL INDEX THEORIES OVER B

Cohomological replacements for the notion of Ljusternik-Schnirelmann category in the case of free $\mathbb{Z}_2$-actions began with Yang [12] and more recently extensions to non-free actions of a compact Lie group actions have played an interesting role (eg. [1], [3], [13], [14]). We refer the reader to [15] for a brief historical discussion as well as variations of LS-category in the category of G-spaces. We will restrict discussion here to the cohomological variety with an emphasis on the relative theory as well as the "theory over B" or parameterized families of G-spaces.

First we review the relative theory and give only a simplified version (sufficient for our purposes) of the material in [1] and [2].

Fix a compact Lie group G and let P denote the category of paracompact G-pairs (X,A), A closed in X. $p_G : EG \to BG$ denotes the universal principal G-bundle and EG is denoted by E. Then $E \times (X,A)$ is a free G-pair and the projection $E \times X \to E$ induces

$$q_X : E \times_G X \to BG$$

and letting $\Lambda = H^*(BG,\mathbb{K})$, $\mathbb{K}$ a fixed field of coefficients, we have

$$q_X^* : \Lambda \to H_G^*(X) \qquad \text{(coef. in } \mathbb{K})$$

And, as we have already seen the cup product

$$H_G^*(X) \otimes H_G^*(X,A) \to H_G^*(X,A)$$

makes $H_G^*(X,A)$ into a left Λ-module, namely

$$\lambda u = q_X^*(\lambda) \cdot u, \quad \lambda \in \Lambda, \quad u \in H_G^*(X,A)$$

Let $\mu = \mu(X,A)$ denote the annihilator of $H_G^*(X,A)$. Note that when $A = \phi$, μ is just the kernel of

$$q_X^* : \Lambda \to H_G^*(X)$$

6.1 Definition

$$\mathrm{Index}_G(X,A) = \dim_{\mathbb{K}} \Lambda/\mu$$

where Λ/μ is a vector space (module) over $\mathbb{K}$.

Remark Index_G (X,A) depends on the coefficient field $\mathbb{K}$ but $\mathbb{K}$ is not displayed in the notation.

We are already in a position to make a basic computation. Let $G = S^1$ and $\mathbb{K} = \mathbb{Q}$ and suppose (S^n,S^k) is a G-pair such that G_x is a finite group for $x \in S^n - S^k$ and S^k contains at least one point in $(S^n)^G$. Then, as we have seen

$$H_G^*(S^n) \simeq \Lambda \otimes H^*(S^n) \text{ as } \Lambda\text{-modules}$$

and hence $\mathrm{Index}_G\ (S^n) = +\infty$, by virtue of the presence of points in $(S^n)^G$. We have also seen the exact sequence

$$0 \to H_G^q(S^n) \to H_G^q(S^k) \to H_G^{q+1}(S^n,S^k) \to 0$$

in section 5 with the map

$$\Lambda \to H_G^{q+1}(S^n,S^k)$$

given by $\lambda \to \lambda\ \delta\ (1 \quad u)$

being surjective. Thus,

$$\mathrm{Index}_G\ (S^n,S^k) = \frac{n-k}{2}$$

We can formulate this computation more generally as follows.

6.2 Proposition Let $G = \mathbb{Z}_2$, S^1, SU(2) and $\mathbb{K} = \mathbb{Z}_2$ if $G = \mathbb{Z}_2$, otherwise $\mathbb{K} = \mathbb{Q}$. Suppose (S^n,S^k) is a quasi-free G-pair of spheres over $\mathbb{K}$, with $S^k \cap (S^n)^G \neq 0$. Then

1) $\mathrm{Index}_G\ (S^n,S^k) = n - k$ if $G = \mathbb{Z}_2$

2) $\mathrm{Index}_G\ (S^n,S^k) = \frac{n-k}{2}$ if $G = S^1$

3) $\mathrm{Index}_G\ (S^n,S^k) = \frac{n-k}{4}$ if $G = SU(2)$

Definition 6.1 extends to G-pairs over B as follows. Let (X,A) denote a paracompact G-pair and

$$p : X \to B$$

a G-map, where G acts trivially on B. p induces

$$p_G : E \times_G X \to E \times_G B = BG \times B$$

and hence for any field of coefficients $\mathbb{K}$

$$p_G^* : \Lambda[B] \to H_G^*(X)$$

where $\Lambda[B] = H_G^*(B) = \Lambda \otimes H^*(B)$. Thus, $H_G^*(X)$ and $H_G^*(X,A)$ are left $\Lambda[B]$-modules.

6.3 <u>Definition</u> Let $\mu \subset \Lambda[B]$ denote the annihilator of $H_G^*(X,A)$. Then, set

$$\text{Index}_B\ (X,A) = \dim_{\mathbb{K}}\ \Lambda[B]/\mu\ .$$

6.4 <u>Remarks</u> Note that when B is a point Index_B (X,A) coincides with Index_G (X,A). Also, the definition makes sense when the action of G on B is not trivial. However, we will not consider this generality at this time. Notice also that p is not necessarily a bundle map or fiber map. The notation $p : (X,A) \to B$ will be employed for a G-pair over B where $p|A : A \to B$.

We will not cover all the properties we know about Index_B (X,A) but discuss those we will need in our example of a Bourgin-Yang type theorem as well as those needed in our discussion of the Ljusternik-Schnirelmann method in critical point theory. First two results about its computation.

6.5 <u>Proposition</u> Let $p : (X,A) \to B$ denote a G-pair over B such that for $x \in X - A$, BG_x is acyclic over $\mathbb{K}$. Then if $(\overline{X},\overline{A})$ are the orbit spaces (X/G), A/G) and $H^*(\overline{X},\overline{A})$ vanishes in sufficiently high dimension,

$$\text{Index}_B(X,A) < \infty \qquad (\text{over } \mathbb{K}\,).$$

<u>Proof</u> $H^*(\overline{X},\overline{A}) \cong H_G^*(X,A)$ by the usual Vietoris Mapping theorem argument and hence the annihilator $\mu \subset \Lambda$ of $H_G^*(X,A)$ contains all of Λ except for a submodule which is finitely generated.

6.6 Theorem Let $p : (X,A) \to B$ denote a G-pair over B where G is connected. Let T denote a maximal torus in G. Then if $W = NT/T$ is the Weyl group of G and $\mathbb{K}$ has characteristic 0

$$\mathrm{Index}_G^T(X,A) = |W| \ \mathrm{Index}_B^G(X,A)$$

where $|W|$ is the order of G.

Proof See [2].

The monotone property for $\mathrm{Index}_B\,(X,A)$ unfortunately requires a stringent assumption. If $p : (X,A) \to B$ and $p' : (X',A') \to B$ are G-pairs over B, a G-map of G-pairs over B is a G-map $f : (X,A) \to (X',A')$ such that $p'f = p$, or in a commutative diagram

$$\begin{array}{ccc} (X,A) & \xrightarrow{f} & (X',A') \\ & {\scriptstyle p}\searrow \quad \swarrow{\scriptstyle p'} & \\ & B & \end{array} \qquad (*)$$

6.7 Proposition (Monotone Property) Let f be a G-map of G-pairs as in (*) such that

$$f_G^* : H_G^*(X',A') \to H_G^*(X,A)$$

is surjective. Then,

$$\mathrm{Index}_B\,(X,A) \le \mathrm{Index}_B\,(X',A')$$

If $A' = A = \phi$,

$$\mathrm{Index}_B\, X \le \mathrm{Index}_B\, X'$$

without the surjective condition on f_G^*.

Proof Since f_G^* is a $\Lambda[B]$-homomorphisms, if λ annihilates $H_G^*(X',A')$, it annihilates the image of f_G^*. When $A = A' = \phi$, we have a commutative diagram

$$\begin{array}{ccc} E \times_G X & \xrightarrow{p_G} & E \times_G B = BG \times B \\ & {\scriptstyle f_G}\searrow \quad \swarrow & \quad {\scriptstyle p'_G} \\ & E \times_G X' & \end{array}$$

and $\ker(p'_G)^* \subset \ker p_G^*$.

When $A = \emptyset$ the surjective condition above can be replaced by one on the restriction $f|A : A \to A'$ (which is more natural if one thinks of A and A' as fixed points of the action and one known or stipulates the behaviour of f on A) by altering the definition of Index_B somewhat.

When $A \neq \emptyset$, let $M^q(X,A) = \text{image } \delta : H_G^{q-1}(A) \to H_G^q(X,A)$, $q \geq 1$. $M^*(X,A)$ is a $\Lambda[B]$-submodule of $H_G^*(X,A)$. Let $\mu^\delta \subset \Lambda[B]$ denote the annihilator of $M^*(X,A)$.

6.8 Definition When $A \neq \emptyset$,

$$\delta\text{-Index}_B(X,A) = \dim_{\mathbb{K}} \Lambda[B]/\mu^\delta$$

and when B is a point

$$\delta\text{-Index}_G\,(X,A) = \dim_{\mathbb{K}} \Lambda/\mu^\delta.$$

For completeness, when $A = \phi$, set $\delta\text{-Index}_B\, X = \text{Index}_B\, X$.

6.9 Proposition (Monotone Property for δ-Index) Let f be a G-map of G-pairs over B as in (*) such that

$$(f|A)_G^* : H_G^*(A') \to H_G^*(A)$$

is surjective. Then,

$$\delta\text{-Index}_B\,(X,A) \leq \delta\text{-Index}_B\,(X',A').$$

Proof A look at the diagram

$$\begin{array}{ccc} H_G^*(X',A') & \xrightarrow{f_G^*} & H_G^*(X,A) \\ \delta \uparrow & & \delta \uparrow \\ H_G^*(A') & \xrightarrow{(f|A)_G^*} & H_G^*(A) \end{array}$$

implies $M^*(X,A) \subset f_G^*(M^*(X',A')$ and hence $\text{Annih } M^*(X',A') \subset \text{Annih } M(X,A)$.

6.10 Remark In general, δ-Index$_B$ (X,A) $\leq$ Index$_B$ (X,A). However, there are interesting situations when δ is surjective and they are equal.

The "continuity property" for Index$_B$ we will need here is only for single G-spaces. The situation for G-pairs over B can be found in [2].

6.11 Proposition (Continuity) Let $Y \to B$ denote a G-space over B with $H^*(B)$ f.g. and X a closed G-set in Y. Then, given a G-neighborhood U of X, there is a closed G-neighborhood N of X such that

$$\text{Index}_B\ N = \text{Index}_B\ X$$

and the same result holds for δ-Index$_B$.

There are also several forms of the "additivity property'. The first we give is for the special case B = point and $\Lambda = H^*(BG)$ monogenic.

6.12 Proposition (Additivity) Let $X = X_1 \cup X_2$, where X_1, X_2 are closed G-sets in the G-space X whose interiors cover X, $A \subset X_1$ a closed G-set and $\Lambda = H^*(BG)$ monogenic. Then,

$$\text{Index}_G\ (X,A) \leq \text{Index}_G\ (X_1,A) + \text{Index}_G\ X_2$$

and the same result holds for δ-Index$_G$.

This proposition is also valid in special circumstances over B which we proceed to explain. Let $\mathbb{F} = \mathbb{R}$, $\mathbb{C}$, or $\mathbb{H}$ and $p : V \to B$ a vector bundle over B with fiber $\mathbb{F}^n$, endowed with a Riemannian metric, and suppose $G = S(\mathbb{F}) = \mathbb{Z}_2$, S^1, or $S^3 = SU(2)$ acts unitarily on V. Suppose also that p is a G-map when G acts trivially on B and $V_0 = V^G$ the fixed point set of the action is non-trivial, i.e. the fiber $\mathbb{F}^\ell$ in V_0 is non-zero. Let S(V), $S(V_0)$ denote the corresponding sphere bundles over B. We wish first to compute (δ-) Index$_B$ $(S(V),S(V_0))$ under the additional assumption that $p|S(V_0) : S(V_0) \to B$ admits a section. As coefficients $\mathbb{K}$ we use $\mathbb{Z}_2$ for $\mathbb{F} = \mathbb{Z}_2$ and $\mathbb{Q}$ otherwise. Note that $\Lambda = H^*(BG)$, is monogenic with generator in dimension d = 1, 2, or 4.

6.12 Lemma $H_G^*(S(V)) = \Lambda[B] \otimes H^*(S^{n-1})$ as a module over $\Lambda[B] = \Lambda \otimes H^*(B)$.

The proof of this lemma consists of applying the Leray-Hircsh theorem [6] to the fibration

$$S^{n-1} \to E \times_G S(V) \to E \times_G B = BG \times B$$

6.13 Lemma Suppose that $(S(V),S(V_0))$ is a quasi-free G-pair, i.e. BG_x is acyclic for $x \in V - V_0$. Then, if $H^q(B) = 0$ for q sufficiently large, $H^q_G(S(V),S(V_0)) = 0$ for q sufficiently large. In particular,

$$\text{Index}_B\ (S(V),S(V_0)) < \infty$$

The proof of this lemma uses the extended Vietoris-Begle Theorem,

$$H^*_G(S(V),S(V_0)) = H^*(\overline{S(V)},\overline{S(V_0)})\ .$$

If we consider the inclusion induced map, with $(S(V),S(V_0))$ quasi-free,

$$H^*_G(S(V),S(V_0)) \xrightarrow{i^*_G} H^*_G(S(V)$$

we see that the Λ module on the left is all Λ-torsion, while the one on the right is free over Λ and hence $i^*_G = 0$. Thus we have short exact sequences (compare §4) of $\Lambda[B]$-modules

$$0 \to H^q_G(S(V)) \to H^q_G(S(V_0)) \xrightarrow{\delta} H^{q+1}_G(S(V),S(V_0)) \to 0$$

or

$$0 \to \Lambda[B] \times H^*(S^{n+1}) \to \Lambda[B] \times H^*(S^{1-1}) \xrightarrow{\delta} H^*_G(S(V),S(V_0)) \to 0$$

6.14 Proposition $\delta\text{-Index}_B\ (S(V),S(V_0)) = \text{Index}_B\ (S(V),S(V_0))$

6.15 Lemma Define a map

$$\psi : \Lambda[B] \to H^*_G(S(V),S(V_0))$$

by $\psi(w) = \delta(w \otimes \alpha)$, α generates $H^{\ell-1}(S^{\ell-1})$. Then,

$$\ker \psi = \text{Annih}\ H^*_G(S(V),S(V_0)).$$

Proof A simple exercise

6.16 Proposition When $(S(V),S(V_0))$ is quasi-free and $\dim H^*(B) = |H^*(B)| < \infty$, then

$$\mathrm{Index}_B\ (S(V),S(V_0)) = |H^*(B)|\ \mathrm{Index}_G\ (S^{n-1},S^{\ell-1})$$

$$= |H^*(B)|\ (n - \ell)/d$$

$d = 1, 2, 4$. Moreover, the Annih $H_G^*(S(V),S(V_0))$ has the form $\tilde{\Lambda}^s \quad H^*(B)$, where $\tilde{\Lambda}^s$ is the Λ-module generated by $\lambda^s \otimes H^*(B)$, $s = (n - 1)/d$, λ generator of Λ.

Proof An application of the Leray-Hirsch theorem [6] gives

$$|H_G^*(S(V),S(V_0))| = |H_G^*(S^{n-1},S^{\ell-1})| \cdot |H^*(B)|$$

and two applications of Lemma 6.15 gives

$$\mathrm{Index}_B(S(V),S(V_0)) = |H_G^*(S(V),S(V_0))|$$

$$\mathrm{Index}_G(S^{n-1},S^{\ell-1}) = |H_G^*(S^{n-1},S^{\ell-1})|.$$

Hence, using Proposition 6.2, we have the first part of the proposition. To see the second part of the proposition we consider the diagram.

$$\begin{array}{ccccccccc} 0 \to & H_G^q(S^{n-1}) & \to & H_G^q(S^{\ell-1}) & \xrightarrow{\delta} & H_G^{q+1}(S^{n-1},S^{\ell-1}) & \to 0 \\ & i_G^* \uparrow & & \uparrow i_G^* & & \uparrow k_G^* & \\ 0 \to & H_G^q(S(V)) & \to & H_G^q(S(V_0)) & \xrightarrow{\delta'} & H_G^{q+1}(S(V),S(V_0)) & \to 0 \end{array}$$

where i_G^*, j_G^*, and k_G^* are all surjective and induced by inclusions. We first observe that under the identification $H_G^*(S^{\ell-1}) = \Lambda \times H^*(S^{\ell-1})$, $H_G^*(S^{n-1}) = \Lambda \otimes H^*(S^{n-1})$, the kernel of δ has the form Λ-multiples of

$$a(\lambda^s \otimes \alpha) + b(\lambda^q \otimes 1)$$

where $ds + (\ell - 1) = n - 1$ and $dq = n - 1$. If we restrict δ to $\Lambda \otimes H^{\ell-1}(S^{\ell-1})$, the kernel is just $\tilde{\Lambda}^s \times H^{\ell-1}(S^{\ell-1})$, or $\tilde{\Lambda}^s$ is the annihilator of $H_G^*(S^{n-1}, S^{\ell-1})$. Let θ_0 and θ_1 denote sections for j_G^* and k_G^*, respectively so that $\sigma_1\delta = \delta'\sigma_0$. Let $p_0 = p|S(V_0)$, where, $p : S(V) \to B$. Then, applying the Leray-Hirsch theorem

$$\delta'(p_0^*(u)\ \theta_0(v)) = p_0^*(u)\ \delta'\theta_0(v) = p_0^*(u)\ \theta_1\ \delta(v)$$

where, $u \in H^*(B)$, $v \in H_G^*(S^{\ell-1})$. Thus,

$$H^*(B) \otimes \ker \delta \subset \ker \delta'$$

On the other hand $\delta'(p_0^*(u)\ \theta_0(v)) = 0$ implies $u \otimes \delta v = 0$. But since $H_G^*(S(V), S(V_0))$ is free over $H^*(B)$, $\delta v = 0$. Thus,

$$H^*(B) \otimes \ker \delta \supset \ker \delta'$$

and

$$\tilde{\Lambda}^s \otimes H^*(B) = \text{Annih } H_G^*(S(V) \otimes S(V_0))$$

6.17 Remark The second part of the above proof did not require that $|H^*(B)| < \infty$.

In Proposition 6.16, the structure of Annih $H_G^*(S(V), S(V_0))$ motivates the next additivity result which will be useful in the next section. First we need a general lemma.

6.18 Lemma Let $X = X_1 \cup X_2$, $A \subset X_1$ with $p : (X,A) \to B$, $p|X_1 : (X_1, A) \to B$, G-pairs over B and $p|X_2 : X_2 \to B$ a G-space over B. Let μ = Annih $H_G^*(X,A)$, μ_1 = Annih $H_G^*(X_1, A)$, μ_2 = Annih $H_G^*(X_2)$, over $\Lambda[B]$. Then, $\mu_1\ \mu_2 \subset \mu$

Proof See [1].

6.18 Proposition (Additivity over B) Let $X = X_1 \cup X_2$, A and $\Lambda = H^*(BG)$ be as in Proposition 6.12. Suppose that $p : X \to B$ makes X a G-space over B, $|H^*(B)| < \infty$, and

$$\mu = \text{Annih } H_G^*(X,A) = \tilde{\Lambda}^s \otimes H^*(B)$$

$$\mu_1 = \text{Annih } H^*_G(X_1, A) \supset \tilde{\Lambda}^t \times H^*(B), \qquad t \leq s$$

then,

$$\text{Index}_B\ (X,A) \leq \text{Index}_B\ (X_1,A) + \text{Index}_B\ X_2$$

<u>Proof</u> It suffices to show that $\text{Index}_B\ X_2 \geq (s - t)\,|H^*(B)|$. If $\text{Index}_B\ X_2 < (s - t)\,|H^*(B)|$, there must be an element u of the form

$$u = \lambda^{i_1} \otimes b_1 + \ldots + \lambda^{i_k} \otimes b_k \in \mu_2 = \text{Annih } H^*(X_2)$$

where $i_1 \leq i_2 \leq \ldots \leq i_k \leq s - t - 1$, $b_j \in H^*(B)$. Multiply u by $\lambda^t \otimes 1$ in μ_1 and since $\mu_1\mu_2 \subset \mu$, the element, $(\lambda^t \otimes 1)u \in \mu$, contradicting $\mu = \tilde{\Lambda}^s \times H^*(B)$.

6.19 <u>Proposition</u> When $A \neq \emptyset$, Proposition 6.18 obtains for δ-Index_B.

§7. A BOURGIN-YANG THEOREM

Our setting will be the following:

1. $p : V \to B$, $p' : V' \to B$ will be vector bundles over $\mathbb{F} = \mathbb{R}$, $\mathbb{C}$, or $\mathbb{H}$ with fibers $\mathbb{F}^n$, $\mathbb{F}^k$, $k < n$, respectively, (as in §6).
2. Both V and V' are G-spaces over B, $G = \mathbb{Z}_2$, S^1, or SU(2).
3. $V^G = V_0$ and $(V')^G = V'_0$ are non-trivial of the same dimension ℓ and $p : S(V_0) \to B$ admits a section.
4. $|H^*(B)| < \infty$ and both $(S(V), S(V_0))$ and $(S(V'), S(V'_0))$ are quasi-free.
5. Coefficients $\mathbb{K} = \mathbb{Z}_2$ or $\mathbb{Q}$ (as in §6).

7.1 <u>Theorem</u> Let $f : S(V) \to V'$ denote a G-map over B and $Z = \{v \in S(V),\ f(v) \in \overline{0} = 0\text{-section}\}$. Suppose $f(S(V_0)) \subset V'_0 - \overline{0}$, and

$$f^* : H^*(V'_0 - \overline{0}) \to H^*_G(S(V_0))$$

is an isomorphism. Then,

$$\text{Index}_B\ Z \geq (n - k)|H^*(B)|$$

Proof Let $\overline{U}$ denote a G-neighborhood of Z such that $\text{Index}_B\ \overline{U} = \text{Index}_B\ Z$ and W a G-neighborhood of Z such that $\overline{W} \subset U$. Then,

$$f : (S(V) - W, S(V_0)) \subset (V' - \overline{0}, V'_0 - \overline{0})$$

Observe that

$$\delta\text{-Index}_B(V' - \overset{)}{0}, V'_0 - \overset{)}{0}) = \text{Index}_B\ (S(V'), S(V'_0))$$

$$\delta\text{-Index}_B(S(V), S(V_0)) = \text{Index}_B\ (S(V), S(V_0))$$

and since $f_G^* : H_G^*(V'_0 - \overline{0}) \to H_G^*(S(V_0))$ is an isomorphism by the Comparison theorem, we have by Proposition 6.9.

$$\delta\text{-Index}_B\ (S(V) - W, S(V_0)) \leq \text{Index}_B\ (S(V'), S(V'_0)).$$

Then, by the addivitiy property in Proposition 6.19

$$\text{Index}_B\ (S(V), S(V_0)) \leq \text{Index}_B\ \overline{U} + \text{Index}_B\ (S(V'), S(V'_0))$$

or

$$\text{Index}_B\ Z \geq (n - k)|H^*(B)|\ .$$

7.2 Remark If V^G and $(V')^G$ are empty, then one can take the direct sum of V and V' with a trivial bundle with the trivial action and extend f using the identity on the trivial factor. This gives the corresponding result for the case when BG_x is acyclic for $x \in V - \overline{0}$, and $x \in V' - \overline{0}$.

7.3 Remark Theorem 7.1 in the special case when $\mathbb{F} = \mathbb{R}$, $G = \mathbb{Z}_2$, the $\mathbb{Z}_2$-action is free on S(V), and B is an ENR is essentially a result of Jaworowski [16] and Nakaoka [17], which does not use index theory. In their case Z is a free $\mathbb{Z}_2$-space and if $\gamma : \overline{Z} \to \mathbb{R}P^\infty$ is the classifying map, they show that for $i \leq n - k - 1$ the mappings

$$\psi^i : H^q(B, \mathbb{Z}_2) \to H^{q+i}(B, \mathbb{Z}_2)\ ,$$

given by $\psi^i(u) = \gamma^*(\lambda^i)\overline{p}^*(u)$, $\overline{p} : \overline{Z} \to B$

are isomorphisms for all q. This can be seen to imply that the kernel of

$$H_G^*(Z) \leftarrow \Lambda \otimes H^*(B)$$

does not contain $\lambda^i \otimes u$ for all $i \leq n - k$ 1 and all u which implies

$$\text{Index}_B\ Z \geq (n - k)|H^*(B)|$$

On the other hand, we have the following simple corollary of Theorem 7.1 (and its proof) which includes the Jaworowski-Nakaoka result as a very special case.

7.4 <u>Theorem</u> Under the hypothesis of Theorem 7.1, the homomorphism

$$\phi^i : H^p(B) \to H_G^*(Z), \qquad i \leq (n - k - 1)/d$$

given by $\phi^i(u) = (1 \otimes p)_G^* (\lambda^i \otimes u)$, $1 \times p : E \times Z \to E \times B$

is an injection.

<u>Proof</u> Let $\mu = \text{Annih } H_G^*(S(V), S(V_0))$,

$\mu_1 = \text{Annih (image } \delta : H_G^*(S(V) - W, S(V_0))$, $\mu_2 = \text{kernel } (1 \times p)_G^*$
$= \text{Annih } H_G^*(Z)$, where $\mu_1\mu_2 \subset \mu$, $\mu = \tilde{\Lambda}^s \otimes H^*(B)$
$s = (n - 1)/d$ and $\mu_1 \supset \tilde{\Lambda}^t \otimes H^*(B)$, $t = (k - 1)/d$. If $\phi^i(u) = 0$, $u \neq 0$, then $\lambda^i \times u \in \mu_1$ and $(\lambda^t \otimes 1)(\lambda^i \otimes u) = \lambda^{t+i} \otimes u \in \mu$. This forces $t + i \geq s$ or $i \geq (n - k)/d$.

8. RELATIVE CATEGORY

If X is a G-space, then one possible geometric analogue of $\text{Index}_G X$ is the (Ljusternik-Schnirelmann) category of the orbit space X/G (see [15]). We will briefly explore here two possible geometric analogues of the <u>relative</u> $\text{Index}_G (X,A)$ which satisfy the basic properties usually employed in the Ljusternik-Schnirelmann Min-Max method in critical point theory and are conceptually simple. Computations, of course, are usually difficult, especially in the non-free case.

First recall that a set $U \subset Y$ is <u>categorical</u> (in Y) if U is contractible in Y to a point. Then if $X \subset Y$, $\mathrm{cat}_Y X$ is the minimum number of categorical open sets needed to cover X.

Now let (Y,A) denote a fixed topological pair with $A \neq \emptyset$ and closed in Y. Then, if $A \subset U \subset Y$, U is called <u>categorical</u> relative to A if there is a homotopy commutative diagram

$$\begin{array}{ccc} (Y,A) & \overset{i}{\leftarrow} & (U,A) \\ {}_{j}\nwarrow & & \swarrow_{\rho} \\ & (A,A) & \end{array}$$

for some map ρ and inclusions i and j, i.e. U to deformable (in Y) into A, relative to A.

8.1 <u>Definition</u> $\mathrm{cat}_Y(X,A) = n$ if there exists n + 1 open sets which cover X such that each is categorical relative to A and n is minimal with this property.

The following basic properties are easily verified for a normal space Y.

1. (Monotone)

$$\mathrm{cat}_Y(X,A) \leq \mathrm{cat}_Y(X',A) \text{ if } X \subseteq X'$$

2. (Invariance) If $\ : (Y,A) \to (Y',A')$ is a homeomorphism of pairs,

$$\mathrm{cat}_Y(X,A) = \mathrm{cat}_{Y'}(\ (X),A')$$

3. (Subadditive)

(a)

$$\mathrm{cat}_Y(X_1 \cup X_2,A) \leq \mathrm{cat}_Y(X_1,A) + \mathrm{cat}_Y(X_2,A) + 1 .$$

(b) If $\mathrm{cat}_Y(A,A) = 0$ and X_2 and A are disjoint closed sets and $X_1 \supset A$, then if Y is 0-connected

$$\mathrm{cat}_Y(X_1 \cup X_2,A) \leq \mathrm{cat}_Y(X_1,A) + \mathrm{cat}_Y X_2 .$$

4. (Continuity) If X is closed in Y, there is a closed neighborhood N of X such that $A \subset X \subset N$ and

$$\mathrm{cat}_Y(X,A) = \mathrm{cat}_Y(N,A)$$

5. (Compactness) If Y is locally contractible and 0-connected and $\mathrm{cat}_Y(A,A) = 0$, then if X is compact,

$$\text{cat}_Y(X,A) < \infty .$$

As in the case of $\text{cat}_Y X$, cohomology products can aid in estimating $\text{cat}_Y(X,A)$. For example, let $H^*(X,A)$ denote singular or Alexander-Spanier cohomology with coefficients in a field $\mathbb{K}$. We define relative cup length as follows.

8.2 Definition For $0 \neq A \subset X \subset Y$, set

$$C|_Y(X,A) = C|_Y(X,A;\mathbb{K}) = n$$

if there exists $u_i \in H^{n_i}(Y,A)$, $1 \leq i \leq n$, $n_i \geq 0$, such that

$$j^*_{(X,A)}(u_1 u_2 \ldots u_n) \neq 0$$

where $j_{(X,A)} : (X,A) \to (Y,A)$ is inclusion. If no such maximum exists and $j^*_{(X,A)}$ is non-trivial, set $C|_Y(X,A) = \infty$. If $j^*_{(X,A)}$ is trivial, $C|_Y(X,A) = 0$.

The following theorem is an easy exercise.

8.3 Theorem $\text{cat}_Y(X,A) \geq C|_Y(X,A)$, $A \neq 0$.

8.4 Remark In some cases this result doesn't help too much. For example if $Y = \mathbb{C}P^\infty$, $X = \mathbb{C}P^5$, $A = \mathbb{C}P^3$, $H^*(\mathbb{C}P^5,\mathbb{C}P^3)$ has a trivial ring structure and hence $C|_Y(\mathbb{C}P^5,\mathbb{C}P^3) = 0$. However, it does show that $C|_Y(Y,\mathbb{C}P^k) = \infty = \text{cat}_Y(Y,\mathbb{C}P^k)$ for any $k < \infty$.

An alternative definition of relative category which makes certain computations simpler is the following. As before we consider a fixed topological pair (Y,A) with A a non-empty closed set and Y is normal. We also assume that Y is locally contractible and 0-connected and that there is an open set $V \supset A$ such that V is categorical relative to A. We consider subsets X such that $Y \supset X \supset A$.

8.5 Definition An open cover Ω of X is called admissible if Ω has the form $\Omega = \{W,U_i\}$ where $W \supset A$ is categorical relative to A and each U_i is categorical.

8.6 Definition $\text{cat}^*_Y(X,A) = n$ if X admits an admissible cover

$\Omega = \{W, U_1, \ldots, U_n\}$ and n is minimal with this property. Set $\mathrm{cat}^*_Y(X,A) = \infty$ if no such finite Ω exists.

8.7 Remark Note that

$$\mathrm{cat}_Y(X,A) \leq \mathrm{cat}^*_Y(X,A) .$$

Also, when $A = \phi$, we agree to take $W = \phi$ and $\mathrm{cat}^*_Y(X,\phi) = \mathrm{cat}_Y X$.

The analogue of monotonicity, invariance, continuity and compactness are valid exactly as stated for $\mathrm{cat}_Y(X,A)$. However, subadditivity takes on a slightly different form.

3*. (Subadditivity) If $X_1 \supset A$ and X_2 is any subset of Y, then

$$\mathrm{cat}^*_Y(X_1 \cup X_2, A) \leq \mathrm{cat}^*_Y(X_1,A) + \mathrm{cat}^*_Y X_2$$

8.8 Remark Comparing this with 3b (Subadditivity) for $\mathrm{cat}_Y(X,A)$, we note that $A \cap X_2 = \phi$ is not required and X_2 is not necessarily closed.

We now give an analogue of Theorem 8.3 which is useful in estimating $\mathrm{cat}^*_Y(X,A)$. Let $\Lambda = H^*(Y)$. Then $H^*(X,A)$ is a module over Λ as follows. Let $j : X \to Y$ denote inclusion and set

$$\lambda w = j^*(\lambda)w, \quad \lambda \in \Lambda, \quad w \in H^*(X,A)$$

using the (cup) product $H^*(X) \otimes H^*(X,A) \to H^*(X,A)$. Annih $H^*(X,A)$ will denote the annihilator of $H^*(X,A)$ as a Λ-module.

8.9 Definition

(a) Set $\mathrm{Index}_Y(X,A) = 0$ if $H^*(X,A) = 0$

(b) Set $\mathrm{Index}_Y(X,A) = 1$ if $H^*(X,A) \neq 0$ and Annih $H^*(X,A)$ is $\tilde{\Lambda} = \sum_{q>0} H^*(Y)$.

(c) If $n \geq 2$, set $\mathrm{Index}_Y(X,A) = n$ if there exist $n - 1$ elements $u_1, \ldots, u_{n-1}$ in $\tilde{\Lambda}$ such that the product $u_1 u_2 \ldots u_{n-1} \notin$ Annih $H^*(X,A)$ and n is maximal with this property. As usual, if this is true for every $n \geq 2$, set $\mathrm{Index}_Y(X,A) = \infty$.

The following theorem has a straight-forward proof.

8.10 Theorem $\text{cat}^*_Y(X,A) \geq \text{Index}_Y(X,A)$.

8.11 Example $\text{cat}^*_{\mathbb{C}P^\infty}(\mathbb{C}P^n,\mathbb{C}^k) = n - k$.

9. AN ABSTRACT CRITICAL POINT THEOREM

Consider now the following setting. Let M denote a smooth, connected Hilbert or Banach manifold on which a compact Lie group G acts and let $A \neq \phi$ denote a fixed closed G-set in M. Let P_M denote the set of pairs (X,A) such that $X \supset A$ and X is a G-set in M, together with all G-sets $Y \subset M - A$. Let $\gamma : P_M \to \mathbb{Z}^+ \cup \infty$ denote a function with the following properties ($\mathbb{Z}^+$ = non-negative integers). In order to shorten the exposition, we allow A to be ϕ in what follows and keep in mind that $(X,A) \in P_M$, $A = \phi$ implies $X \subset M - A$.

γ1. (Monotone) $\gamma(X,A) \leq \gamma(X',A)$ if $X \subset X'$

γ2. (Invariance) If $\phi : (M,A) \to (M,A)$ is a G-homeomorphism, with $\phi|A$ = identity, then

$$\gamma(X,A) = \gamma(\phi(X),A)$$

γ3. (Subadditive) If X_2 and A are disjoint, X_2 closed, then

$$\gamma(X_1 \cup X_2,A) \leq \gamma(X_1,A) + \gamma(X_2)$$

γ4. (Continuity) If $X \in P_M$ is a closed G-set, $X \subset M - A$, then there is a closed neighborhood N of X, $N \subset M - A$ such that

$$\gamma(N) = \gamma(X)$$

γ5. If $X \in P_M$, $X \subset M - A$, is compact, $\gamma(X) < \infty$.

γ6. (Count) If $X \in P_M$, $X \subset M - A$, then $\gamma(X) \geq 1$ implies $X \neq \phi$ while $\gamma(X) \geq 2$ implies X/G is infinite.

9.1 Examples (Recall that $\overline{M} = M/G$, $\overline{X} = X/G$, etc)

(a) Let $\gamma(X,A) = \text{cat}_{\overline{M}}(\overline{X},\overline{A})$, $\gamma(X) = \text{cat}_{\overline{M}}\overline{X}$

(b) Let $\gamma(X,A) = \text{cat}^*_{\overline{M}}(\overline{X},\overline{A})$, $\gamma(\overline{X}) = \text{cat}_{\overline{M}}\overline{X}$

(c) Let $\mathbb{K}$ denote a coefficient field and set

$$\gamma(X,A) = \delta\text{-Index}_G(X,A),\quad \gamma(X) = \text{Index}_G X$$

provided we impose certain conditions. $\gamma.3$ will require $\Lambda = H^*(BG)$ to be monogenic. (One can however base δ-Index on a monogenic submodule of $H^*(BG)$). $\gamma.5$ will require that if $x \in M - A$, then $H^*(BG_x)$ is acyclic (over $\mathbb{K}$).

Now suppose $f : M \to \mathbb{R}$ is a G-functional, i.e., $f(gx) = f(x)$, $x \in M$, $g \in G$, which is assumed to be C^1. K will denote the critical set, i.e., $K = \{x : f'(x) = 0\}$, $K_c = f^{-1}(c) \cap K$, and $M_b = \{x : f(x) \leq b\}$. We will replace a Palais-Smale condition (PS) on f by its consequence, namely an appropriate "deformation theorem".

9.2 Definition f is said to generalized Palais-Smale (gPS) if $f|K$ is proper and for every $c \in \mathbb{R}$, $\bar{\epsilon} > 0$ and U a neighborhood of K_c, there is an $\epsilon > 0$, $\epsilon < \bar{\epsilon}$ and a G-homeomorphism $\phi : M \to M$ such that

(1) $\phi(x) = x$ if $|f(x) - c| \geq \bar{\epsilon}$

(2) $\phi(M_{c+\epsilon} - U) \subset M_{c-\epsilon}$, with the convention that $K_c = \phi$ implies $U = \phi$.

9.3 Remark There are numerous references for the fact a (PS)-condition implies (gPS), e.g., [3], [18], [19], [20], [21]. The result for the G-case is obtained from the case of no G-action present by averaging the group G (e.g. see [21]).

9.4 Definition Suppose $Y \subset M - A$ is a G-set. Then, we say that A and Y γ-link if

$$\gamma(M - Y, A) < \gamma(M, A)$$

If, in addition, $\gamma(M,A) = \infty$, then we say that A and Y strongly γ-link.

9.5 Theorem (See [1], [3]) Let M, $f : M \to \mathbb{R}$, and γ be as above. Assume that f is (gPS) and for some G-set $Y \subset M - A$

(i) A and Y strongly γ-link

(ii) $\inf f|Y < \sup f|A$

(iii) for each $j \in \mathbb{Z}^+$, there is a pair $(X,A) \in P_M$ such that X is closed,

$$\sup_X f|X < \infty,$$

and $\gamma(X,A) \geq j$

Then, f posseses an unbounded sequence of critical values.

Proof Let

$$\Sigma_j = \{X : (X,A) \in P_M,\ X \text{ closed},\ \gamma(X,A) \geq j\}$$

and

$$c_j = \inf_{X\, \Sigma_j} \sup_X f(x)$$

Properties (ii) and (iii) above assure us that the c_j's are well-defined. Let $m_0 = \gamma(M - Y,A)$. If $X \in \Sigma_j$ with $j > m_0$, then $X \cap Y \neq \phi$. This forces

$$c_{m_0+1} > \max f|A.$$

Thus, in the sequence c_j we may assume without loss that $c_{m_0} < c_{m_0+1}$. Suppose

$$c_{j+1} = \dots = c_{j+\ell} = c, \; m_0 \leq j, \; 1 \leq \ell .$$

We prove $\gamma(K_c) \geq \ell$. This will show $K_c \neq \phi$ and if $\gamma(K_c) \geq 2$, K_c consists of infinitely many orbits. Since $c > \max f|A$, A and K_c are disjoint closed sets and hence there is a closed G-neighborhood U of K_c such that $U \subset M - A$ and $\gamma(U) = \gamma(K_c)$. Let W denote an open G-neighborhood of K_c such that $\overline{W} \subset \operatorname{int} U$. Let $\epsilon = \frac{1}{2}[c_{j+1} - \max f|A]$. Choose $\epsilon = (\epsilon,W)$ and a G-homeomorphism $\phi = \phi(\epsilon,W)$ satisfying the conditions in (gPS). Let X denote a closed G-set such that $\gamma(X,A) \geq j + 1$ and

$$\sup_X f(x) \leq c + \epsilon$$

Now,

$$\gamma(X,A) \leq \gamma(X - W,A) + \gamma(U \cap X)$$

and hence

$$\gamma(X - W,A) \geq \gamma(X,A) - \gamma(K_c)$$

If $\gamma(K_c) < \ell$, $\gamma(X - W,A) \geq j + 1$. But $\phi(X - W) \subset M_{c-\epsilon}$ and $\gamma(\phi(X - W),A) = \gamma(X - W,A) \geq j + 1$ and hence $c_{j+1} < c$ which is a contradiction. Thus at this point we know that K contains infinitely many orbits. Now, if the c_j's are bounded, there is a critical point $\overline{c}$ such that $c_j \leq \overline{c}$ and for every $\epsilon > 0$, $c_m > \overline{c} - \epsilon$ for m large. Let $\tilde{K} = f^{-1}[c_{m_0+1},\overline{c}] \cap K$. $\tilde{K}$ is compact, lies in $M - A$ and $\gamma(\tilde{K}) < \infty$. Let

$\bar{\epsilon} = \frac{1}{2}[\bar{c} - \max f|A]$ and U a closed G-neighborhood of $\tilde{K}$, disjoint from $\tilde{A}$, such that $\gamma(U) = \gamma(\tilde{K})$. Also let W denote an open G-neighborhood of $\tilde{K}$ such that $\bar{W} \subset \text{int } U$. Choose $\epsilon < \bar{\epsilon}$ and ϕ as in the (gPS)-condition and let $k = \gamma(\tilde{K})$. Now, take $m > m_0 + 1$ such that $c_m > \bar{c} - \epsilon$ and $X \in \Sigma_{m+k}$ such that

$$\sup_X f(x) < c_{m+k} + \epsilon \leq \bar{c} + \epsilon$$

Then,

$$m + k \leq \gamma(X,A) \leq \gamma(X - W,A) + \gamma(U)$$

Then, $\gamma(\phi(X - W),A) \geq m$, $\phi(X - W) \subset M_{\bar{c}-\epsilon}$, and we have $c_m \leq \bar{c} - \epsilon$, which is a contradiction. Thus, the c_j's are unbounded.

An application of Theorem 9.5 is the main result in [3], where the proof there employed a Borsuk-Ulam theorem for non-free S^1-actions. Let V denote a separable infinite dimensional Hilbert space where $G = S^1$ acts as unitary transformations. Let $v_1, \ldots, v_j, \ldots$ denote an orthogonal basis for B chosen so that $V_0 = \text{space}\{v_0, \ldots, v_\ell\}$ is the fixed set of the action. Then, for $x \in V - V_0$, the isotropy group G_x is finite. Let $V_m = \text{space}\{v_1, \ldots, v_\ell, \ldots, v_{\ell+2m}\}$.

9.6 Corollary Let V be as above and $f : V \to \mathbb{R}$ an S^1-functional which is C^1 and satisfies (PS). Suppose also that

(1) for each $m \geq 0$, there is an $R_m > 0$ such that $f(x) \leq 0$ for all $x \in V_m$, with $\|x\| \geq R_m$.

(2) There is an m_0 and a $\rho > 0$ such that

$$\inf f|\partial B_\rho \cap E_{m_0} > \max(0, \sup_{V_0} f)$$

where $B_\rho = \{x : \|x\| \leq p\}$.

Then, f posseses an unbounded sequence of critical values.

The proof of this corollary uses $\gamma(X,A) = \delta\text{-Index}_G(X,A)$ and $\text{Index}_G X$. The sets A and Y are chosen appropriately and the necessary computations can be made without difficulty using the techniques contained herein. The details may be found in [1], which also contains an analogue for SU(2)-actions.

References

1. Fadell, E. and Husseini, S., *Relative Cohomological Index Theories*, Preprint. To appear in Advances in Mathematics.

2. Fadell, E. and Husseini, S., *Index Theory for G-bundle Pairs With Applications to Borsuk-Ulam Theorems for G-sphere Bundles*, Preprint. To appear.

3. Fadell, E.; Husseini, S. and Rabinowitz, P., *Borsuk-Ulam Theorems for S^1-actions and Applications*, Trans. Amer. Math. Soc., 274(1982), 345-359.

4. Hopf, H. and Rueff, M., *Uber Faserung treue Abildungen der Spharen*, Comm. Helvetici 11(1938), 49-61.

5. Bredon, G.E., *Introduction to Compact Transformation Groups*, Academic Press, N.Y., 1972.

6. Spanier, E., *Algebraic Topology*, McGraw-Hill, N.Y., 1966.

7. Whitehead, G.W., *Elements of Homotopy Theory*, Springer-Verlag, N.Y., 1978.

8. Weiss, *Cohomology of Groups*, Academic Press, N.Y., 1969.

9. Zeeman, A proof of the comparison theorem for spectral sequences. Proc. Cambridge Phil. Soc. 53(1957) 57-64.

10. Nirenberg, L., *Comments on Nonlinear Problems*, Matematiche (Catania) 36 (1981), 109-119.

11. Dold, A., *Simple Proofs of Some Borsuk-Ulam Results*, Contemporary Mathematics 19 (1983), 65-69.

12. Yang, C.T., *Continuous Functions from Spheres to Euclidean Spaces*, Ann. Math 62 (1955), 284-292.

13. Fadell, E. and Rabinowitz, P., *Bifurcation for Odd Potential Operators and an Alternative Topological Index*, J. Functional Analysis 26 (1977), 48-67.

14. Fadell, E. and Rabinowitz, P., *Generalized Cohomological Index Theories for Lie Group Actions with an Application to Bifurcation questions for Hamiltonian Systems*, Invent. Math. 45 (1978), 139-174.

15. Fadell, E., *The Equivariant Ljusternik-Schnirelmann Method for Invariant Functionals and Relative Cohomological Index Theories*, Proceedings of the Montreal Conference, Summer 1983, Montreal University Press (1985), edited by A. Granas, 41-71.

16. Jaworowski, J., *Fiber Preserving Maps of Sphere Bundles into Vector Space Bundles*, Proc. Fixed Point Theory Conf., Sherbrooke 1981, Lect. Notes in Math #886, Springer-Verlag, 154-162.

17. Nakaoka, M., *Equivariant Point Theorems for Fiber-preserving Maps*, Preprint.

18. Palais, R.S., *Critical Point Theory and the Minimax Principle*, AMS Proc. of Symposia in Pure Math., 1970.

19. Browder, Felix E., *Nonlinear Eigenvalue Problems and Group Invariances*, Functional Anal. and Related Fields, Springer-Verlag, (1970), 1-58.

20. Schwartz, J.T., *Generalizing the Ljusternik-Schnirelmann Theory of Critical Points*, Comm. Pure Appl. Math, 17 (1964), 307-315.

21. Bartolo, P.; Benci, V. and Fortunato, D., *Abstract Critical Point Theorems and Applications to some Nonlinear Problems with Strong Resonance at Infinity*, Atti del' Instituto di Math. Appl., Univ. Bari, 1981.

Supported in part by the National Science Foundation under Grant No. DMS-8320099.

ON A THEOREM OF ANOSOV ON NIELSEN NUMBERS FOR NILMANIFOLDS

Edward Fadell and Sufian Husseini
University of Wisconsin - Madison

1. INTRODUCTION

If $f : M \to M$ is a self-map of a compact manifold, it rarely happens that the Nielsen number $n(f)$ is equal to the numerical value of the Lefschetz-Hopf number $L(f)$, i.e., $n(f) = |L(f)|$. For example, if M is an n-sphere, $n \geq 2$, $n(f) = 1$, while $L(f) = 1 + (-1)^n \deg f$. On the positive side, $n(f) = |L(f)|$ is valid for all tori but tori are the only compact Lie groups for which the result is valid [1]. In the early summer of 1984, D. V. Anosov (Steklov Mathematical Institute) wrote us inquiring whether it would be of interest to prove this result for compact nil-manifolds. Since nilmanifolds form a much larger class than Tori, we thought it worthwhile and wrote him this in a return letter. During the summer we worked out a proof of Anosov's conjecture based on a product theorem for Nielsen numbers in fiber spaces [2] and sent it to Anosov. Later, we received a reply from Anosov, that he had worked out his proof in a purely geometric manner, deforming f to have exactly $n(f)$ fixed points all having the same index 1 or -1.

Theorem [Anosov]. If $f : M \to M$ is a self-map of a compact nilmanifold M, then $n(f) = |L(f)|$.

In this note, we give our alternative proof of this theorem.

2. A CLASS OF MANIFOLDS

Let N denote a class of compact connected manifolds satisfying the following conditions:

N.1 N contains all tori (products of circles)

N.2 Given any map $g : M \to M$, where $M \in N$ is not a torus, there is a diagram

S. P. Singh (ed.), Nonlinear Functional Analysis and Its Applications, 47–53.

$$\begin{array}{ccc} T & \xrightarrow{f_0} & T \\ \downarrow & & \downarrow \\ M & \xrightarrow{f} & M \\ \downarrow p & & \downarrow p \\ B & \xrightarrow{\bar{f}} & B \end{array}$$

where p is a principal T-fibration, T a torus, $B \in N$ and $f \sim g$.

2.1 Definitions. Call a class of manifolds N satisfying N.1 and N.2 a nilpotent class (We shall see later that the compact nilpotent manifolds form a nilpotent class).

2.2 Proposition. Let N denote a nilpotent class. Then, every $M \in N$ is a $K[\pi,1]$, orientable manifold, and the Euler characteristic $\chi(M) = 0$.

Proof. The theorem is true for all tori and we proceed by induction on dim M. If M is not a torus, we make use of the principal fibration $T \to M \xrightarrow{p} B$ given by N.2. By induction, $\pi_i(B) = 0$ for $i \geq 2$ and the homotopy exact sequence of a fibration [3] implies that $\pi_i(M) = 0$ for $i \geq 2$. Thus, M is a $K[\pi,1]$. Since our fibration is orientable $H_m(M) = H_b(B) \otimes H_t(T)$ where m = dim M, b = dim B, t = dim T by a simple spectral sequence argument. Also, [4] $\chi(M) = \chi(B)\chi(T) = 0$.

Our next proposition requires some preliminaries. If $f : M \to M$ is a map, L(f) will denote the Lefschetz-Hopf number of f and n(f) the Nielson number of f [5]. $f_\# : \pi_1(M) \to \pi_1(M)$ will denote the induced homomorphism on the fundamental group. (We assume, without loss of generality that f preserves a base point.) Set

$$\text{Fix } f_\# = \{u \in \pi_1(M) : f_\#(u) = u\}$$

2.3 Proposition. Let N denote a nilpotent class. If $g : M \to M$ is a map, where $M \in N$ and $L(g) \neq 0$, then Fix $g_\# = 1$, the identity element.

Proof. a) If M is a torus and $L(g) \neq 0$, then all the Nielsen classes of g have the same local index (non-zero) and $n(g) = |\text{coker } (1 - g_\#)|$ (see [5]). If $u \in$ Fix $g_\#$ is non-trivial, then $1 - g_\#$ has a non-trivial kernel and coker $1 - g_\#$ is a group of infinite order. But n(g) is finite and hence u = 1.

b) To prove the general case, we proceed by induction on dim M and use N.2 with $f \sim g$.

Consider the diagram

$$\begin{array}{ccccccccc} 1 & \to & \pi_1(T) & \xrightarrow{i_\#} & \pi_1(M) & \xrightarrow{p_\#} & \pi_1(B) & \to & 1 \\ & & f_{0\#} \downarrow & & f_\# \downarrow & & \overline{f}_\# \downarrow & & \\ 1 & \to & \pi_1(T) & \to & \pi_1(M) & \to & \pi_1(B) & \to & 1 \end{array}$$

Since $L(f) \neq 0$, $L(f) = L(\overline{f})L(f_0)$ implies $L(\overline{f}) \neq 0$ and $L(f_0) \neq 0$. Thus, $\text{Fix } f_{0\#} = 1$ and, by induction, $\text{Fix } \overline{f}_\# = 1$. Suppose $f_\#(u) = u$. Then,

$$\overline{f}_\# \; p_\#(u) = p_\# \; f_\#(u) = p_\#(u)$$

and $p_\#(u) = 1$. Thus, u pulls back to v by $i_\#$ and

$$i_\# \; f_{0\#}(v) = f_\# \; i_\#(v) = u = i_\#(v)$$

and $f_{0\#}(v) = v$. Thus, $v = 1$ and hence $u = 1$. Thus, $\text{Fix } f_\# = 1$ and hence $\text{Fix } g_\# = 1$.

We now review some basic facts about tori. If $g : T \to T$ is a self-map of a torus and $L(g) = 0$ it is always possible to deform g to be fixed point free, i.e., there is an $f \sim g$ such that $\text{Fix } f = \phi$. If $\dim T = 1$, degree $g = 1$ and the result is obvious. If $\dim T = 2$, g is induced up to homotopy by a linear map given by an integer matrix

$$A = \begin{bmatrix} a & c \\ b & d \end{bmatrix}$$

where $L(g) = \det [I - A]$. A small irrational perturbation of this linear map is fixed point free. If $\dim T \geq 3$, one may invoke the theorem of Wecken [5] after observing that $L(f) = 0$ implies $n(f) = 0$. We also can note here that the 2-torus is the only compact 2-manifold in a nilpotent class N.

2.4 Theorem. Every manifold in a nilpotent class N has the property that for every self map $g : M \to M$, $n(f) = |L(f)|$.

Proof. a) The result for all tori is contained in [1].

b) to complete the proof we proceed by induction and choose $f \sim g$ together with the diagram

$$\begin{array}{ccc} T & \xrightarrow{f_0} & T \\ \downarrow & & \downarrow \\ M & \xrightarrow{f} & M \\ \downarrow & & \downarrow \\ B & \xrightarrow{\bar{f}} & B \end{array}$$

where $1 \leq \dim B < \dim M$. If $L(\bar{f}) = 0$, then $n(\bar{f}) = 0$. If $\dim B \leq 2$, B is a torus and $\bar{f}$ may be deformed to a fixed point free map. If $\dim B \geq 3$, we apply the theorem of Wecken [5] and in this case also $\bar{f}$ may be deformed to be fixed point free. If we apply the Covering Homotopy theorem [3], f is homotopic to a fixed point free map and hence $n(f) = n(g) = 0$.

We now assume $L(\bar{f}) \neq 0$. Notice first that since T is 0-connected, the loop space $\Omega(b)$ acts on T by maps homotopic to the identity. Furthermore, $\pi_1(T) \to \pi_1(M)$ injects, and Fix $\bar{f}_{\#} = 1$. This are precisely the hypotheses needed to invoke a product theorem for Nielsen numbers ([2], theorem 6.1) namely:

$$n(f) = n(f_0)\, n(\bar{f})$$

Since, $L(f) = L(f_0)\, L(\bar{f})$ we have (by induction)

$$|L(f)| = |L(f_0)|\, |L(\bar{f})| = n(f_0)\, n(\bar{f}) = n(f).$$

3. COMPACT NILMANIFOLDS

Our objective in this section is to show the class of compact nilmanifolds is indeed a nilpotent class. The basic reference is the paper of A. I. Mal'cev [6]. To review a bit, every nilpotent 1-connected Lie group G is homeomorphic (as a space) to some Euclidean space and every compact nilmanifold is of the form $M = G/\Gamma$ where Γ is a discrete subgroup. Since G is nilpotent the descending central series

$$G = G_1 \supset G_2 \supset \dots \supset G_k \supset G_{k+1} = 1$$

is finite, terminating with 1, where

$$G_2 = [G,G_1], G_3 = [G,G_2], \dots, G_{i+1} = [G,G_i]$$

All of the G_i are closed, connected subgroups (becuase G is 1-connected

[6]). Note that $[G,G_k] = 1$ implies that G_k is in the center of G. If we let $\Gamma^1 = \Gamma \cap G_k$, we have the diagram

$$\begin{array}{ccccc} \Gamma' & \to & G_k & \to & G_k/\Gamma' \\ \downarrow & & \downarrow & & \downarrow \\ \Gamma & \to & G & \to & G/\Gamma \\ \downarrow & & \downarrow & & \downarrow \\ \Gamma/\Gamma' & \to & G/G_k & \to & G/G_k/\Gamma/\Gamma' = G/\Gamma/G_k/\Gamma' \end{array}$$

Note the following:

1. G_k is a closed normal subgroup of G and G/G_k is a nilpotent Lie group.
2. Γ/Γ' is a discrete subgroup and $G/G_k/\Gamma/\Gamma'$ is a compact nilmanifold.
3. G_k is an abelian nilpotent group and hence G_k/Γ' is a compact abelian Lie group, hence a torus.
4. G_k/Γ' acts freely on G/Γ so that

$$G_k/\Gamma' \to G/\Gamma \to G/\Gamma/G_k/\Gamma'$$

is a principal torus fibration.

3.1 Proposition. If we employ the above notation and set

$$M = G/\Gamma,\ T = G_k/\Gamma',\ B = G/G_k/\Gamma/\Gamma' = G/\Gamma/G_k/\Gamma'$$

then

$$T \to M \to B$$

is a principal fiber bundle with group T.

For our next step we will need the following result of Mal'cev [6].

Theorem. Let G denote a nilpotent Lie group and Γ a discrete subgroup. Then, there exists a closed subgroup G' of G such that G'/Γ is compact and

$$G/\Gamma' \simeq G'/\Gamma \times \mathbb{R}^m,\ m \geq 0$$

where $\simeq$ indicates toplogical equivalence.

Now, returning to our previous notation let $M = G/\Gamma$ denote our compact nilmanifold where G is a 1-connected nilpotent Lie group and Γ is a discrete subgroup. We identify $\pi_1(M)$ with Γ, with Γ identified with the discrete fiber in the covering map $G \to M$. We take a map $g : M \to M$ which we may assume preserves the base point. g induces $g_\# : \pi_1(M) \to \pi_1(M)$ which may be identified with a homomorphism $\phi : \Gamma \tau \Gamma$.

3.2 Proposition. There exists a homomorphism $\Phi : G \to G$ which extends ϕ such that the induced map $\overline{\Phi} = f : G/\Gamma \to G/\Gamma$ is homotopic to g.

Proof. Let $\overline{\Gamma} = \{(x,y),\ y = \phi(x),\ x \in \Gamma\}$ = graph of Φ. We apply Mal'cev's theorem above to $\overline{\Gamma} \subset G \times G$ and conclude that there is a closed subgroup $A \subset G \times G$ containing $\overline{\Gamma}$ such that $A/\overline{\Gamma}$ is compact and $A/\overline{\Gamma} \times \mathbb{R}^m \simeq G \times G/\Gamma$. A is also 1-connected. Let p_i, $i = 1,2$, denote the projections $G \times G \to G$ and consider the diagram

$$\begin{array}{ccccc} G \times G & \supset & A & \supset & \overline{\Gamma} \\ \downarrow p_1 & & \downarrow p_1' & & \downarrow p_1'' \\ G & \supset & G & \supset & G \end{array}$$

Since p_1'' is a isomorphism, p_1' induces isomorphisms $\pi_1(A/\overline{\Gamma}) \to \pi_1(G/\Gamma)$ and applying theorem 5 in Mal'cev [6], $(p_1'')^{-1}$ admits an extension to an isomorphism

$$\gamma : G \to A$$

and $\Phi = p_2\gamma : G \to G$ extends ϕ. Since M is a $K[\Gamma,1]$, $\overline{\Phi} = f : G/\Gamma = M \to M$ is homotopic to g because $f_\# = g_\#$.

Thus, we see that Φ in Proposition 3.2 induces, using the descending central series at the beginning of this section,

$$\begin{array}{ccc} T = G/G_k & \xrightarrow{f_0} & G/G_k = T \\ \downarrow & & \downarrow \\ M = G/\Gamma & \xrightarrow{f} & G/\Gamma = M \\ \downarrow & & \downarrow \\ B = G/\Gamma/G/G_k & \xrightarrow{\overline{f}} & G/\Gamma/G/G_k = B \end{array}$$

and we may state the following.

3.3 Theorem. The class of compact nilmanifolds is a nilpotent class.

3.4 Corollary [Anosov]. If M is a compact nilmanifold and $f : M \to M$ is a self-map, then $n(f) = |L(f)|$.

REFERENCES

1. Brooks, R.B.S., Brown, R.F., Pak, J., Taylor, D.H., *Nielsen Numbers of Maps of Tori*, Proc. Amer. Math. Soc. 52 (1975), 398-400.

2. Fadell, E., *Natural Fiber Splittings and Nielsen Numbers*, Houston Jour. Math., 2 (1976), 71-84.

3. Spanier, E., *Algebraic Topology*, McGraw-Hill (1966).

4. Serre, J. P., Ann. Math., 54 (1951), 425-505.

5. Brown, R.F., *The Lefschetz Fixed Point Theorem*, Scott Foresman (1971).

6. Mal'cev, A.I., *On a Class of Homogeneous Spaces*, Amer. Math. Soc. Translation, No. 39 (1951), 276-307.

Both authors supported in part by the National Science Foundation under Grant No. DMS-8320099

GENERALIZED TOPOLOGICAL DEGREE AND BIFURCATION

K. Geba, Uniwersytet Gdanski
I. Massabó, Universita della Calabria
A. Vignoli, II Universita di Roma (Tor Vergata)

0. INTRODUCTION

The main task of this paper is to present a generalized degree theory for continuous maps $f : \overline{U} \to \mathbb{R}^n$, where $U \subset \mathbb{R}^m$, $m \geq n$, is a bounded open subset such that $f(x) \neq 0$ for all $x \in \partial U$ - the boundary of U - (as usual $\overline{U}$ stands for the closure of U).

In order to give at least a very rough idea of our approach to this problem let us consider here the case when $m = n$ (i.e. the situation in which the Brouwer topological degree can be settled). So, let $f : \overline{U} \to \mathbb{R}^n$ be nonvanishing on ∂U and let $\hat{\mathbb{R}}^n$ be the Alexandroff one-point compactification of $\mathbb{R}^n$. Let $\hat{f} : \hat{\mathbb{R}}^n \to \hat{\mathbb{R}}^n$ be an extension of f such that $\hat{f}(x) = f(x)$ for $x \in \overline{U}$ and $\hat{f}(x) \neq 0$ for $x \in \hat{\mathbb{R}}^n \backslash U$ (for details see Section 2 of this paper). Now, since $\hat{\mathbb{R}}^n$ is homeomorphic to the unit sphere S^n of $\mathbb{R}^{n+1}$, set $\sigma_n : \hat{\mathbb{R}}^n \to S^n$. Using this homeomorphism we obtain a map $\tilde{f} : S^n \to S^n$ defined by $\tilde{f} = \sigma_n \circ \hat{f} \circ \sigma_n^{-1}$. Therefore, to any map f as above we may associate the homotopy class $[\hat{f}] \in \Pi_n(S^n)$ - the nth homotopy group of S^n. It can be shown that this homotopy class does not depend on the extension $\hat{f}$ of f and we may call $[\tilde{f}]$ the <u>degree</u> of f with respect to U and denote it deg(f,U). Now, since $\Pi_n(S^n) \simeq \mathbb{Z}$ and deg(f,U) satisfies the axioms of a degree theory then, by the uniqueness of the Brouwer topological degree (see [A.W.]), it follows that our topological degree coincides with that of Brouwer.

The above construction of degree is non-standard in the sense that, instead of using homology groups of spheres (as it is usually done in most text-books of Algebraic Topology), we appeal to homotopy groups of spheres. Notice, that exactly the same construction (again, for details see Section 2 below) can be carried over to maps

S. P. Singh (ed.), Nonlinear Functional Analysis and Its Applications, 55–73.

acting from $\mathbb{R}^m$ into $\mathbb{R}^n$, $m \geq n$. Indeed, in this case, our degree will be an element of $\Pi_m(S^n)$ - the mth homotopy group of S^n. In this situation, of course, no valuable reference to $H_m(S^n)$ - the mth homology group of S^n, could be made unless $m = n$, since $H_n(S^n) \simeq \mathbb{Z}$ and $H_m(S^n)$ is trivial for $m \neq n$.

The idea of defining a (generalized) topological degree for maps acting between spaces of different dimensions is not new. The need for such a degree stems mainly from treating problems involving (non-linear) compact perturbations of linear Fredholm operators of positive index. To the best of our knowledge the first attempt to apply a generalized degree was made in [N.], where stable homotopy groups and their infinite dimensional analogues (constructed) in [G.], see also [G.G.]) were used. The results contained in [N.] have been recently extended in [D.].

Another interesting contribution to this topic is contained in [I.], where the structure of cohomotopy groups is exploited.

In our opinion the method of constructing the generalized degree presented in this paper is quite general and simple.

The structure of the present work is the following. After this introduction, Section 1 is devoted to the definition and main properties of essential maps. Section 2 is the principal part of this paper. It contains the construction of our generalized degree and all of its main properties (one of these is that if a given map has non trivial generalized degree then the map is essential).

In Section 3 we give a description of the J-homomorphism in terms which are slightly different from those already given in [A.], [A.F.], [A.Y.] and [I.]. In the final Section 4 we present an application of our degree to bifurcation problems in the spirit of [F.M.P.], [I.M.P.V.] and [R.]. We close this paper with a result related to [A.], [A.F.], [A.Y.] and [I.].

We close this introduction by adding that an extension to the context of infinite dimensional Banach spaces of the results of this work will be the subject of a forthcoming paper.

1. ESSENTIAL MAPS AND SOME OF THEIR PROPERTIES

As promised in the introduction we recall here the definition of essential maps and gather some properties which will be used in subsequent sections. Let us start with some standard definitions.

A <u>pair</u> (X,A) of topological spaces consists of a space X and a subspace A. If X and A are compact, then (X,A) is called a <u>compact pair</u>. If $(X,A,)$ and (Y,B) are pairs of spaces then a <u>map of pairs</u> $f : (X,A) \to (Y,B)$ is, by definition, a continuous map form X into Y such that $f(A) \subset B$.

Two maps of pairs $f,g : (X,A) \to (Y,B)$ are said to be <u>homotopic</u> (usually written $f \sim g$) if there exists a continuous map $k : X \times [0,1] \to Y$ such that the one-parameter family of maps of pairs

$k_t : (X,A) \to (Y,B)$, $k_t(x) = k(x,t)$ has the property $k_0 = f$ and $k_1 = g$. If $f \sim g$ and the homotopy $k : X \times [0,1] \to Y$ has the further property $k_t(x) = f(x)$ for $x \in A$ and $t \in [0,1]$, then f is said to be homotopic to g relative to A and we write $f \sim g$ rel. A.

Note, that homotopy is an equivalence relation. In what follows we let $[X,A;Y,B]$ denote the set of all homotopy classes of continuous maps from (X,A) into (Y,B). If $f : (X,A) \to (Y,B)$ is a continuous map then we let $[f] \in [X,A;Y,B]$ denote the homotopy class of f.

Definition 1.1. Let (X,A) be a compact pair. A map $f : (X,A) \to (\mathbb{R}^n, \mathbb{R}^n/\{0\})$ is called inessential if $f \sim g$, where $g(X) \subset \mathbb{R}^n/\{0\}$ (i.e. f is homotopic to a map g which is nonvanishing everywhere on X). A map which is not inessential will be called essential.

The following result shows that in Definition 1.1 we may consider relative homotopies. Namely

Lemma 1.2. If $f : (X,A) \to (\mathbb{R}^n, \mathbb{R}^n/\{0\})$ is inessential, then there exists a map $g : (X,A) \to (\mathbb{R}^n, \mathbb{R}^n/\{0\})$ such that $f \sim g$ rel. A and $g(X) \subset \mathbb{R}^n/\{0\}$.

Proof. Since f is inessential, there exists a homotopy $k : X \times [0,1] \to \mathbb{R}^n$ such that $k_0 = f$, $k_t(A) \subset \mathbb{R}^n/\{0\}$ for all $t \in [0,1]$ and $k_1(X) \subset \mathbb{R}^n/\{0\}$. Since $k(A \times [0,1]) \subset \mathbb{R}^n/\{0\}$, there exists an open subset U of X such that $A \subset U$ and $k(U \times [0,1]) \subset \mathbb{R}^n/\{0\}$ (cfr. [H.W] page 82). Let $\phi : X \to [0,1]$ be a Urysohn function such that $\phi^{-1}(0) \supset A$ and $\phi^{-1}(1) \supset X/U$. Define $h : X \times [0,1] \to \mathbb{R}^n$ by $h(x,t) = k(x,t\phi(x))$. It is an easy computation to show that $f \sim h_1$ rel. A with $h_1(X) \subset \mathbb{R}^n/\{0\}$. Q.E.D.

The following two simple lemmas are used to obtain an interesting property of essential maps (see Corollary 1.5 below).

Lemma 1.3. Let $f : (X,A) \to (\mathbb{R}^n, \mathbb{R}^n/\{0\})$ be a map of pairs and let $y_1, y_2 \in \mathbb{R}^n/f(A)$ be given. Define $f_i : (X,A) \to (\mathbb{R}^n, \mathbb{R}^n/\{0\})$ by $f_i(x) = f(x) - y_i$, $i = 1, 2$. If y_1 and y_2 belong to the same connected component of $\mathbb{R}^n/f(A)$, then $f_1 \sim f_2$.

Proof. Let $\sigma : [0,1] \to \mathbb{R}^n/f(A)$ be a path joining y_1 and y_2. Clearly, the map $h : X \times [0,1] \to \mathbb{R}^n$ defined by $h(x,t) = f(x) - \sigma(t)$ furnishes the required homotopy. Q.E.D.

Lemma 1.4. *Let* $f : (X,A) \to (\mathbb{R}^n, \mathbb{R}^n/\{0\})$ *be essential. Then* 0 *belongs to a bounded component of* $\mathbb{R}^n/f(A)$ *and* $\mathbb{R}^n/f(A)$ *is not connected.*

Proof. Assume that 0 belongs to the unbounded component $\mathbb{R}^n/f(A)$. Set $r = \sup\{|f(x)| : x \in X\}$ (notice that r is finite since X is compact) and choose $y_0 \in \mathbb{R}^n$ with $|y_0| > r$. Define $g : (X,A) \to (\mathbb{R}^n, \mathbb{R}^n/\{0\})$ by $g(x) = f(x) - y_0$. Observe that $|g(x)| = |f(x) - y_0| \geq |y_0| - |f(x)| > 0$ for all $x \in X$. Thus, g is inessential. On the other hand 0 and y_0 are in the same component of $\mathbb{R}^n/f(A)$ so that, by Lemma 1.3, $g \sim f$, contradicting the assumption on f. Hence, the component containing 0 is bounded, so that $\mathbb{R}^n/f(A)$, being unbounded cannot be connected. Q.E.D.

As a direct consequence of Lemma 1.3 and Lemma 1.4, we obtain the following

Corollary 1.5. *Let* $f : (X,A) \to (\mathbb{R}^n, \mathbb{R}^n/\{0\})$ *be essential and let* W *be the connected component of* $\mathbb{R}^n/f(A)$ *containing* 0. *Then* $W \subset f(X)$.

In order to state the next result we shall need some preliminaries.

Let U be an open subset of a normal space such that $\overline{U}$ is compact. Let $f : \overline{U} \to \mathbb{R}^m$ and $g : \overline{U} \to \mathbb{R}^n$ be continuous maps. Set $X = f^{-1}(0)$, $Y = g^{-1}(0)$ and $A = X \cap \partial U$. Assume $X \cap Y \subset U$. Then, we may consider the following maps of pairs $f \times g : (\overline{U}, \partial U) \to (\mathbb{R}^{m+n}, \mathbb{R}^{m+n}/\{0\})$, $\phi : (X,A) \to (\mathbb{R}^n, \mathbb{R}^n/\{0\})$ defined by $(f \times g)(x) = (f(x), g(x))$ and $\phi(x) = g(x)$.

In this context we have the following

Proposition 1.6. *If* ϕ *is inessential, then* $f \times g$ *is also inessential.*

Proof. By Lemma 1.2 there exists a homotopy $H : X \times [0,1] \to \mathbb{R}^n$ such that $H_0 = \phi$, $H_t(x) = \phi(x)$ for $x \in A$, $t \in [0,1]$ and $H_1(X) \subset \mathbb{R}^n/\{0\}$. Set $B = \overline{U} \times \{0\} \cup X \times [0,1]$ and define a map $\hat{k} : B \to \mathbb{R}^n$ by

$$\hat{k}(x,t) = \begin{cases} g(x) & \text{for } x \in \overline{U} \text{ and } t = 0 \\ H(x,t) & \text{for } (x,t) \in X \times [0,1] \end{cases}$$

Clearly, $\hat{k}$ is continuous. Now, let $k : \bar{U} \times [0,1] \to \mathbb{R}^n$ be any continuous extension of $\hat{k}$ and define a homotopy $h : \bar{U} \times [0,1] \to \mathbb{R}^{m+n}$ by $h(x,t) = (f(x), k(x,t))$. Then, $h_0 = f \times g$, $h_t : (\bar{U}, \partial U) \to (\mathbb{R}^{m+n}, \mathbb{R}^{m+n}/\{0\})$ and $h_1(x) \neq 0$ for all $x \in \bar{U}$ (observe that if $f(x) = 0$, then $x \in X$ and so $k_t(x) = H_t(x)$). Thus, the map $f \times g$ is inessential. Q.E.D.

From Proposition 1.6 and Lemma 1.5 we readily obtain the following

Corollary 1.7. *If* $f \times g$ *is essential, then* 0 *belongs to a bounded component of* $\mathbb{R}^n/g(A)$ *and* $\mathbb{R}^n/g(A)$ *is not connected. In particular, the set* A *is not empty.*

Remark 1.8. An easy consequence of Corollary 1.7 is the following

for $1 \le i \le n$, *the i-th component* g_i *of* g *must change sign on the set* $A \cap \{x \in \bar{U} : g_j(x) = 0, j \neq i, 1 \le j \le n\}$.

This property of essential maps has been already observed and exploited in deriving bifurcation results in [F.M.P.] and [I.M.P.V.].

2. THE GENERALIZED TOPOLOGICAL DEGREE

This is the principal part of this paper. As the title indicates, this section is devoted to the construction of a generalized degree.

In the sequel we shall use the following notations.

$E^n_+ = \{x \in S^n : x_{n+1} \ge 0\}$, $E^n_- = \{x \in S^n : x_{n+1} \le 0\}$, where S^n is the unit sphere in $\mathbb{R}^{n+1}$. $\hat{\mathbb{R}}^n = \mathbb{R}^n \cup \{+\infty\}$ - the Alexandroff one-point compactification of $\mathbb{R}^n$; $\mathbb{R}^n_+ = \{x \in \mathbb{R}^n : x_n \ge 0\}$, $\hat{\mathbb{R}}^n_+ = \mathbb{R}^n_+ \cup \{+\infty\}$, $\mathbb{R}^n_- = \{x \in \mathbb{R}^n : x_n \le 0\}$, $\hat{\mathbb{R}}^n_- = \mathbb{R}^n_- \cup \{+\infty\}$ (notice that $\hat{\mathbb{R}}^n_+$ and $\hat{\mathbb{R}}^n_-$ are homeomorphic to the unit ball of $\mathbb{R}^n$).

Let $\sigma_0 : [0,+\infty) \to [-1,1)$ be the homeomorphism defined by

$$\sigma_0(t) = \begin{cases} t - 1, & \text{if } 0 \le t \le 1 \\ 1 - \frac{1}{t}, & \text{if } 1 < t < +\infty \end{cases}$$

Define now $\bar{\sigma}_n : \mathbb{R}^n \to S^n$ by

$$\bar{\sigma}_n(x) = \begin{cases} (\sigma_0(|x|), \sqrt{1 - \sigma_0(|x|)^2} \cdot \frac{x}{|x|}) \in \mathbb{R} \times \mathbb{R}^n & \text{if } x \neq 0 \\ (-1, 0, \ldots, 0) \in \mathbb{R}^{n+1}, & \text{if } x = 0 \end{cases}$$

This map extends uniquely to a homeomorphism $\sigma_n : \hat{\mathbb{R}}^n \to S^n$ (sending the point $+\infty$ into the point $(1, 0, \ldots, 0)$). Thus, we may identify $\hat{\mathbb{R}}^n$ with S^n through σ_n. Note, moreover, that the homeomorphism σ_n has the following additional properties

(i) $\sigma_n(\hat{\mathbb{R}}^n_+) = E^n_+$, $\sigma_n(\hat{\mathbb{R}}^n_-) = E^n_-$

(ii) $\sigma_{n+1}(x) = \sigma_n(x)$, for $x \in \mathbb{R}^n$ (in this case we identify) $x = (x_1, \ldots, x_n)$ with $x = (x_1, \ldots, x_n, 0)$).

In what follows $\Pi_m(S^n)$ stands, as usual, for the mth homotopy group of S^n.

We recall the definition of the suspension homomorphism $\Sigma : \Pi_m(S^n) \to \Pi_{m+1}(S^{n+1})$ between homotopy groups of spheres.

Given a continuous map $f : S^m \to S^n$, we define $\tilde{f} : S^{m+1} \to S^{n+1}$ ($\tilde{f}$ is called the <u>suspension</u> of f) as an extension of the map f such that $\tilde{f}(E^{m+1}_\pm) \subset E^{n+1}_\pm$. An extension of this type exists since $E^{n+1}_\pm$ are homeomorphic to the unit ball in $\mathbb{R}^n$. Moreover, any two such extensions are homotopic since they agree on S^m.

Now, the assignment $f \to \tilde{f}$ induces the <u>suspension homomorphism</u> $\Sigma : \Pi_m(S^n) \to \Pi_{m+1}(S^{n+1})$ defined by $\Sigma([f]) = [\tilde{f}]$.

Notice that if $g : S^m \to \mathbb{R}^{n+1}$ is of the form $g(x_1, \ldots, x_m, x_{m+1}) = (f(x_1, \ldots, x_m), x_{m+1})$, with $f : S^{m+1} \to S^{n-1}$, then the homotopy class of the suspension $\tilde{f}$ of f coincides with the homotopy class of the map $\frac{g}{|g|} : S^m \to S^n$.

Analogously, for a map $f : \hat{\mathbb{R}}^m \to \hat{\mathbb{R}}^n$ we may define a suspension map $\tilde{f} : \hat{\mathbb{R}}^{m+1} \to \hat{\mathbb{R}}^{n+1}$ for which $\tilde{f}(\hat{\mathbb{R}}^{m+1}_\pm) \subset \hat{\mathbb{R}}^{n+1}_\pm$. Then the assignement $f \to \tilde{f}$ induces a map $S : [\hat{\mathbb{R}}^m, \hat{\mathbb{R}}^n] \to [\hat{\mathbb{R}}^{m+1}, \hat{\mathbb{R}}^{n+1}]$ where $[\hat{\mathbb{R}}^h, \hat{\mathbb{R}}^\ell]$ is the set of homotopy clases of maps acting between $\hat{\mathbb{R}}^h$ and $\hat{\mathbb{R}}^\ell$,

$h \geq \ell$. Finally, define the map $T : [\hat{\mathbb{R}}^m, \hat{\mathbb{R}}^n] \to \Pi_m(S^n)$ by $T([f]) = [\sigma_n \circ f \circ \sigma_m^{-1}]$. Taking into account the above definitions and the properties (i), (ii) of σ_m and σ_n, we obtain

Proposition 2.1. The following diagram is commutative

$$\begin{array}{ccc} [\hat{\mathbb{R}}^m, \hat{\mathbb{R}}^n] & \xrightarrow{T} & \Pi_m(S^n) \\ \downarrow S & & \downarrow \Sigma \\ [\hat{\mathbb{R}}^{m+1}, \hat{\mathbb{R}}^{n+1}] & \xrightarrow{T} & \Pi_{m+1}(S^{n+1}) . \end{array}$$

We are now in a position of defining the generalized topological degree for maps defined on the closure of an open bounded subset of $\mathbb{R}^m$ taking values in $\mathbb{R}^n$, $m \geq n$.

Assume $U \subset \mathbb{R}^m$ is open and bounded and $f : (\overline{U}, \partial U) \to (\mathbb{R}^n, \mathbb{R}^n/\{0\})$ is a continuous map. Let $f_0 : \partial U \to \mathbb{R}^n/\{0\}$ denote the restriction of f to ∂U. We may consider $\mathbb{R}^n$ and $\mathbb{R}^n/\{0\}$ as subsets of $\hat{\mathbb{R}}^n$. Since $\hat{\mathbb{R}}^n/\{0\}$ is homeomorphic to $\mathbb{R}^n$, the map f_0 extends to a continuous one $f_\infty : \hat{\mathbb{R}}^m/U \to \hat{\mathbb{R}}^n/\{0\}$. Thus, we can define a map $\hat{f} : \hat{\mathbb{R}}^m \to \hat{\mathbb{R}}^n$ by

$$\hat{f}(x) = \begin{cases} f(x) & \text{for } x \in \overline{U} \\ f_\infty(x) & \text{for } x \in \hat{\mathbb{R}}^m/U \end{cases}$$

(extensions of this form will be called admissible) .

Definition 2.2. Let $f : (\overline{U}, \partial U) \to (\mathbb{R}^n, \mathbb{R}^n/\{0\})$ be continuous and let $\hat{f} : \hat{\mathbb{R}}^m \to \hat{\mathbb{R}}^n$ be an admissible extension of f. We define the generalized degree, $d(f,U)$, of f to be

$$d(f,U) = T([\hat{f}]) \in \Pi_m(S^n).$$

Proposition 2.3. The generalized degree does not depend upon the admissible extension $\hat{f}$ of f, i.e., the generalized degree is well defined.

Proof. Let $\hat{f}_i : \hat{\mathbb{R}}^m \to \hat{\mathbb{R}}^n$, i = 1, 2, be admissible extensions of the map f, that is

$$\hat{f}_i(x) = \begin{cases} f(x) & \text{for } x \in \overline{U} \\ f_i(x) & \text{for } x \in \mathbb{R}^m/U \end{cases}$$

with $f_i(\hat{\mathbb{R}}^m/U) \subset \hat{\mathbb{R}}^n/\{0\}$, i = 1, 2. Since $\hat{\mathbb{R}}^n/\{0\}$ is contractible we can find a homotopy between f_1 and f_2. Extending the homotopy inside U to be the given map f, we obtain an admissible homotopy between $\hat{f}_1$ and $\hat{f}_2$. Thus, $[\hat{f}_1] = [\hat{f}_2]$, i.e., the generalized degree is well defined. Q.E.D.

In what follows we collect the basic properties of this generalized degree. We omit their proofs which are straightforward.

Proposition 2.4. (Homotopy invariance). Let I be a nonempty compact interval of $\mathbb{R}$, let U be an open and bounded subset of $\mathbb{R}^m$. Suppose that $h : (\overline{U}, \partial U) \times I \to (\mathbb{R}^n, \mathbb{R}^n/\{0\})$ is a continuous map. Then $d(h_t, U)$, $t \in I$, is well defined and independent of $t \in I$.

Proposition 2.5. (Excision). Let $f : (\overline{U}, \partial U) \to (\mathbb{R}^n, \mathbb{R}^n/\{0\})$ be continuous. Then, for every open subset $V \subset U$ such that the map f has no zeros in U/V, we have

$$d(f,U) = d(f,V) .$$

Proposition 2.6. (Solution property). Let 0 stand for the homotopy class of constant maps from S^m into S^n and let $f : (\overline{U}, \partial U) \to (\mathbb{R}^m, \mathbb{R}^n/\{0\})$ be continuous. If $d(f,U) \neq 0$, then f is essential, in particular, the equation $f(x) = 0$ has solutions in U.

We discuss now an important property of the generalized degree.

Assume that U is an open bounded subset of $\mathbb{R}^{m+1}$ and $f : (\overline{U}, \partial U) \to (\mathbb{R}^{n+1}, \mathbb{R}^{n+1}/\{0\})$ is a continuous map such that

$$(*)\quad f(\overline{U} \cap \mathbb{R}^{m+1}_+) \subset \mathbb{R}^{n+1}_+ \text{ and } f(\overline{U} \cap \mathbb{R}^{m+1}_-) \subset \mathbb{R}^{n+1}_- .$$

Let $U_0 = U \cap \mathbb{R}^m$. Hence, if $x \in U_0$ we have that $f(x) \in \mathbb{R}^n = \mathbb{R}^n \times \{0\}$. Let $f_0 : (\overline{U}_0, \partial U_0) \to (\mathbb{R}^n, \mathbb{R}^n/\{0\})$ denote the restriction of the map f.[1] Under these assumptions we have the following additional property of the degree.

Proposition 2.7. (Suspension property).

$$d(f,U) = \Sigma d(f_0, U_0)).$$

Proof. Let $\hat{f}_0 : \hat{\mathbb{R}}^n \to \hat{\mathbb{R}}^n$ be an admissible extension of the map f_0. Let the following two maps $f_1^{\pm} : \overline{U} \cap \mathbb{R}_{\pm}^{m+1} \cup \hat{\mathbb{R}}^m \to \mathbb{R}_{\pm}^{n+1}$ defined by

$$f_1^{\pm}(x) = \begin{cases} f(x) & \text{if } x \in \overline{U} \cap \mathbb{R}_{\pm}^{m+1} \\ \hat{f}_0(x) & \text{if } x \in \hat{\mathbb{R}}^m . \end{cases}$$

Let us take now an extension $f^+ : \hat{\mathbb{R}}_+^{m+1} \to \hat{\mathbb{R}}_+^{n+1}$ of the map f_1^+ such that $f^+(\hat{\mathbb{R}}_+^{m+1}/U) \subset \hat{\mathbb{R}}_+^{n+1}/\{0\}$ and analogously an extension $f^- : \hat{\mathbb{R}}_-^{m+1} \to \hat{\mathbb{R}}_-^{n+1}$ of the map f_1^- such that $f^-(\hat{\mathbb{R}}_-^{m+1}/U) \subset \hat{\mathbb{R}}_-^{n+1}/\{0\}$. Define a map $\hat{f} : \hat{\mathbb{R}}^{m+1} \to \hat{\mathbb{R}}^{n+1}$ as follows

$$\hat{f}(x) = \begin{cases} f^+(x) & \text{if } x \in \hat{\mathbb{R}}_+^{m+1} \\ f^-(x) & \text{if } x \in \hat{\mathbb{R}}_-^{m+1} . \end{cases}$$

From the above constructions we have at once that $T([\hat{f}]) = d(f,U)$ and $S([\hat{f}_0]) = [\hat{f}]$. Being $T([\hat{f}_0]) = d(f_0, U_0)$, the proof is complete since, by Proposition 2.1, $\Sigma T([\hat{f}_0]) = TS([\hat{f}_0])$. Q.E.D.

Another important property of the generalized degree is the following.

Proposition 2.8. (Addivity). Let $U \subset \mathbb{R}^{n+k}$ be open bounded and let $f : \overline{U} \to \mathbb{R}^n$ be continuous. Let U_1, U_2 be open subsets of U such that $U_1 \cap U_2 = \emptyset$. Assume that $f(x) \neq 0$ for all $x \in U/(U_1 \cup U_2)$. If $k \leq n - 4$, then

---

[1] Note that, if $\overline{U}$ = the unit ball of $\mathbb{R}^{n+1}$ then condition (*) implies $[f] = \Sigma([f_0])$.

$$d(f,U) = d(f,U_1) + d(f,U_2) .$$

Proof. Since $k \le n - 4$ the cohomotopy groups $\pi^q(X,A)$ for a given compact pair (X,A) with $X \subset \mathbb{R}^{n+k}$ are defined provided $q \ge n - 1$ (see e.g. S. T. Hu, Homotopy Theory, Academic Press, 1959). In this case our generalized degree can be defined through a composition of homomorphisms as follows. Let $f : (\overline{U}, \partial U) \to (\mathbb{R}^n, \mathbb{R}^n/\{0\})$ be a continuous map. Define $g : \partial U \to S^{n-1}$ by $g(x) = \frac{f(x)}{|f(x)|}$. The map g can be extended to a continuous map $\hat{g} : \hat{\mathbb{R}}^{n+k} \to S^n$ such that $\hat{g}(\overline{U}) \subset E^n_-$ and $g(\mathbb{R}^{n+k}/U) \subset E^n_+$. The assignment $g \to \hat{g}$ induces a map $D : \pi^{n-1}(\partial U) \to \pi^n(\hat{\mathbb{R}}^{n+k})$, that can be viewed as the composition of the following homomorphisms

$$\pi^{n-1}(\partial U) \xrightarrow{\delta} \pi^n(\overline{U}, \partial U) \xrightarrow{(e^*)^{-1}} \pi^n(\hat{\mathbb{R}}^{n+k}, \hat{\mathbb{R}}^{n+k}/U) \xrightarrow{j^*} \pi^n(\hat{\mathbb{R}}^{n+k}),$$

where δ is the coboundary operator, $e : (\overline{U}, \partial U) \to (\hat{\mathbb{R}}^{n+k}, \hat{\mathbb{R}}^{n+k}/U)$ is the excision map and $j : \hat{\mathbb{R}}^{n+k} \to (\hat{\mathbb{R}}^{n+k}, \hat{\mathbb{R}}^{n+k}/U)$ is the inclusion.

By Proposition 2.5 we may assume, without loss of generality, that $U = U_1 \cup U_2$, $\overline{U}_1 \cap \overline{U}_2 = \emptyset$ and $f : (\overline{U}, \partial U) \to (\mathbb{R}^n, \mathbb{R}^n/\{0\})$. Let $i_\alpha : \partial U_\alpha \to \partial U$, $\alpha = 1, 2$ be the inclusion maps. Then, $f_\alpha = f \circ i_\alpha : (\overline{U}_\alpha, \partial U_\alpha) \to (\mathbb{R}^n, \mathbb{R}^n/\{0\})$ and $g_\alpha = g \circ i_\alpha : \partial U_\alpha \to S^{n-1}$, $\alpha = 1, 2$. Define, for $\alpha = 1, 2$, the map $\tilde{g}_\alpha : \partial U \to S^{n-1}$ by

$$(*) \qquad \tilde{g}_\alpha(x) = \begin{cases} g_\alpha(x), & \text{if } x \in \partial U_\alpha \\ s_0 \quad , & \text{if } x \in \partial U/\partial U_\alpha . \end{cases}$$

We have $i^*_\alpha([\tilde{g}_\alpha]) = [g_\alpha]$, $\alpha = 1, 2$ and, by the definition of the sum in $\pi^{n-1}(\partial U)$, we get $[g] = [\tilde{g}_1] + [\tilde{g}_2]$. Thus, $d(f,U) = D([g]) = D([\tilde{g}_1]) + D([\tilde{g}_2]) = D(I_1[g_1]) + D(I_2[g_2]) = d(f,U_1) + d(f,U_2)$, where $I_\alpha : \pi^{n-1}(\partial U_\alpha) \to \pi^{n-1}(\partial U)$, $\alpha = 1, 2$, are the maps induced by the assignment $g_\alpha \to \tilde{g}_\alpha$, $\alpha = 1, 2$, via the map defined in (*). Q.E.D.

Let us remark that, at present, it is an open question for us, whether the addivity property is true in full generality.

We close this section with some remarks on the generalized degree defined above.

Remark 2.9. i) If $m = n$ then $d(f,U) = \deg_B(f,U,0)$ where $\deg_B(f,U,0)$ stands for Brouwer s topological degree.

This can be easily proven via the uniqueness of the topological degree exhibited in [A.W.].

ii) A continuous map $f : (\overline{U},\partial U) \to (\mathbb{R}^n,\mathbb{R}^n/\{0\})$ may be essential and yet $d(f,U) = 0$. Indeed, let $U = (-2,2)/\{0\})$ and $f : \mathbb{R} \to \mathbb{R}$ be defined by $f(x) = x^2 - 1$. The map f changes sign on the boundary of U and hence is essential. On the other hand, $\deg_B(f,U,0) = 0$ and hence, in view of (i), $d(f,U) = 0$.

iii) If U = unit ball of $\mathbb{R}^m$ then $d(f,U) = \Sigma([f_0/|f_0|])$ where f_0 is the restriction of f to S^{m-1}.

iv) In the case when the map f is defined on an open, not necessarily bounded, subset of W of $\mathbb{R}^m$ and $f^{-1}(0)$ is a compact subset of W, we may define $d(f,W) = d(f,U)$ being U any open bounded subset of W such that $f^{-1}(0) \subset U \subset \overline{U} \subset W$. The degree for a such f is well defined in view of the excision property. Moreover, if $d(f,W) \neq 0$ then $f : (\overline{V},\partial V) \to (\mathbb{R}^n,\mathbb{R}^n/\{0\})$ is essential provided that V is an open bounded set such that $f^{-1}(0) \subset V \subset \overline{V} \subset W$. Mappings with this property has been considered in [I.M.P.V.].

3. ON THE J-HOMOMORPHISM

In this section we recall the definition of J-homomorphism (our general reference will be [H.], especially pages 199, 213) and reformulate some of its properties in the form which will be needed in the next section.

Recall first that the suspension SX of the space X is the quotient of X x [0,1] with respect to the relation $(x,0) \sim (x^1,0)$ and $(x,1) \sim (x^1,1)$ for $x, x^1 \in X$. The coset of (x,t) is denoted by $\langle x,t\rangle$.

Recall also that the join $X * Y$ of two spaces X and Y is the quotient of X x [0,1] x Y with respect to the relation $(x,0,y) \sim (x^1,0,y)$ and $(x,1,y) \sim (x,1,y^1)$, for $x, x^1 \in X$ and $y, y^1 \in Y$. The coset of (x,t,y) is denoted by $\langle x,t,y\rangle$.

Throughout the rest of this section we fix integers k,n and use the natural identification $\mathbb{R}^k \times \mathbb{R}^n = \mathbb{R}^{n+k}$.

We let

$$B^k = \{x \in \mathbb{R}^k : |x| < 1\},\ D^k = \{x \in \mathbb{R}^k : |x| \le 1\}$$

$$C = \partial(D^k \times D^n) = S^{k-1} \times D^n \cup D^k \times S^{n-1}.$$

Define $v : S^{k-1} * S^{n-1} \to S^{n+k-1}$ by $v\langle x,t,y\rangle = ((\sin \frac{\pi}{2}t)x, (\cos \frac{\pi}{2}t)y)$ and

$$\ominus : SS^{n-1} \to S^n$$

by $\ominus\langle x,t\rangle = (\sin(\pi t)x, -\cos\pi t)$.

Let

$$f : S^{k-1} \times S^{n-1} \to S^{n-1}$$

be a continuous map. The map

$$H(f) : S^{k-1} * S^{n-1} \to S^n$$

defined by $H(f)\langle x,t,y\rangle = \ominus\langle f(x,y),t\rangle$ is called the <u>Hopf construction</u> map.

The assignment $f \to H(f) \circ v^{-1}$ induces the map

$$\chi : [S^{k-1} \times S^{n-1}, S^{n-1}] \to \pi_{n+k-1}(S^n).$$

Suppose now that we are given a continuous map

$$f : S^{k-1} \times S^{n-1} \to \mathbb{R}^n/\{0\}.$$

Let $\psi : D^k \times D^n \to \mathbb{R}$ be a continuous function satisfying the following two conditions:

(a) $\psi(x,y) > 0$ if $(x,y) \in S^{k-1} \times B^n$,

(b) $\psi(x,y) < 0$ if $(x,y) \in B^k \times S^{n-1}$

(note, that $(x,y) \in S^{k-1} \times S^{n-1}$ implies $\psi(x,y) = 0$).

Extend f to a continuous map $\hat{f} : D^k \times D^n \to \mathbb{R}^n$ and define

$$F : (D^k \times D^n, C) \to (\mathbb{R}^{n+1}, \mathbb{R}^{n+1}/\{0\})$$

by $F(x,y) = (\hat{f}(x,y), \psi(x,y))$.

Clearly the homotopy class of F depends neither on the choice of the extension $\hat{f}$ nor on ψ (if only ψ satisfies (a) and (b)).

Therefore the formula $\hat{\chi}[f] = d(F, B^k \times B^n)$ defines a map

$$\hat{\chi} : [S^{k-1} \times S^{n-1}, \mathbb{R}^n/\{0\}] \to \pi_{n+k}(S^{n+1}).$$

Let

$$i_* : [S^{k-1} \times S^{n-1}, S^{n-1}] \to [S^{k-1} \times S^{n-1}, \mathbb{R}^n/\{0\}]$$

denote the map induced by the inclusion $i : S^{n-1} \to \mathbb{R}^n/\{0\}$, i.e., $i_*[f] = [i \circ f]$. Let $\Sigma : \pi_m(S^n) \to \pi_{m+1}(S^{n+1})$ be the suspension homomorphism introduced in Section 2. Then, the following result holds.

Proposition 3.1. The following diagram is commutative

$$\begin{array}{ccc} [S^{k-1} \times S^{n-1}, S^{n-1}] & \xrightarrow{\chi} & \pi_{n+k-1}(S^n) \\ \downarrow i_* & & \downarrow \Sigma \\ [S^{k-1} \times S^{n-1}, \mathbb{R}^n/\{0\}] & \xrightarrow{\hat{\chi}} & \pi_{n+k}(S^{n+1}) \end{array}$$

Proof. Let $f : S^{k-1} \times S^{n-1} \to S^{n-1}$ be a continuous map. Define $W : D^k \times D^n/B^{k+n} \to S^{n+k-1}$ by $W(x,y) = \frac{1}{|(x,y)|}(x,y)$. Let

$$\hat{f} : H(f) \circ v^{-1} \circ W : D^k \times D^n/B^{n+k} \to S^n.$$

Clearly, the map $\hat{f}$ coincides with $\chi(f)$ on $\partial B^{n+k} = S^{n+k-1}$. Extend $\hat{f}$ to a continuous map $F : D^k \times D^n \to \mathbb{R}^{n+1}$ and let $F(x,y) = (\tilde{f}(x,y), \psi(x,y))$, where $\tilde{f} : D^k \times D^n \to \mathbb{R}^n$ and $\psi : D^k \times D^n \to \mathbb{R}$ are continuous maps. For $(x,y) \in S^{k-1} \times S^{n-1}$ we have $W(x,y) = \frac{1}{\sqrt{2}}(x,y)$ and hence $\tilde{f}(x,y) = f(x,y)$. Therefore $\tilde{f} : D^k \times D^n \to \mathbb{R}^n$ is an extension of f.

From the definition of Θ it is clear that the conditions (a) and (b) for the map ψ are satisfied. Therefore $\hat{\chi}[f] = d(F, B^k \times B^n)$. Since $F(x,y) \neq 0$ for $(x,y) \in D^k \times D^n/B^{k+n}$, we have by Proposition 2.5,

$d(F, B^{k+n}) = d(F. B^k \times B^n)$. On the other hand, by Remark 2.9 iii) $d(F, B^{k+n}) = \Sigma[F_0]$, where F_0 stands for the restriction of F to S^{n+k-1}. The proof is now complete, since $F_0 = \hat{f} = \chi(f)$ on S^{n+k-1}.
Q.E.D.

We collect now some simple facts on classical groups. These are the last ingredients we need to define the classical J-homomorphism.

Let $GL(n,\mathbb{R})$ denote the topological group of all linear automorphisms of $\mathbb{R}^n$.
Let

$$GL_+(n,\mathbb{R}) = \{A \in GL(n,\mathbb{R});\ \det A > 0\},$$

$$GL_-(n,\mathbb{R}) = \{A \in GL(n,\mathbb{R});\ \det A < 0\}.$$

It is well known that $GL_+(n,\mathbb{R})$ and $GL_-(n,\mathbb{R})$ are arcwise connected. Recall also that $O(n)$ usually denotes the subgroup of $GL(n,\mathbb{R})$ consisting of all orthogonal automorphisms of $\mathbb{R}^n$ and $SO(n) = O(n) \cap GL_+(n,\mathbb{R})$. We let $O_-(n) = O(n) \cap GL_-(n,\mathbb{R})$, $SO(n)$ and $O_-(n)$ are connected components of $O(n)$.

Since $GL_+(n,\mathbb{R})$ (resp., $SO(n)$) is a topological group there is a natural identification $\pi_k(GL_+(n,\mathbb{R})) = [S^k, GL_+(n,\mathbb{R})]$ (resp., $\pi_k(SO(n)) = [S^k, SO(n)]$, comp. [H], p. 212).

Moreover $GL_-(n,\mathbb{R})$ is homeomorphic to $GL_+(n,\mathbb{R})$ (resp., $O_-(n)$ is homeomorphic to $SO(n)$), therefore $\pi_k(GL_-(n,\mathbb{R})) = [S^k, GL_-(n,\mathbb{R})]$ (resp., $\pi_k(O_-(n)) = [S^k, O_-(n)]$).

In what follows we let

$$\pi_k(GL(n)) = \pi_k(GL_+(n)) \oplus \pi_k(GL_-(n,\mathbb{R})),$$

$$\pi_k(O(n)) = \pi_k(SO(n)) \oplus \pi_k(O_-(n)),$$

where $\oplus$ denotes the direct sum of abelian groups. Therefore $[S^k, GL(n,\mathbb{R})]$ (resp., $[S^k, O(n)]$) generates the abelian group $\pi_k(GL(n,\mathbb{R}))$ (resp., $\pi_k(O(n))$).

Finally, let $\phi : S^{k-1} \to O(n)$ be a continuous map and define $f : S^{k-1} \times S^{n-1} \to S^{n-1}$ by $f(x,y) = \phi(x)(y)$. The <u>classical J-homomorphism</u>

$J : \pi_{k-1}(O(n)) \to \pi_{n+k}(S^n)$ is defined by the formula $J([\phi]) = \chi[f]$, (see [H.]).

We shall need the following extension of the J-homomorphism. Let $\phi : S^{k-1} \to GL(n,\mathbb{R})$ be continuous. Define $\tilde{f} : S^{k-1} \times S^{n-1} \to S^{n-1}$ by $\tilde{f}(x,y) = \phi(x)y/|\phi(x)y|$ and let $\tilde{J}([\phi]) = \chi([\tilde{f}])$. If $i : O(n) \to GL(n,\mathbb{R})$ stands for the inclusion map, then the following diagram is obviously commutative

$$\begin{array}{ccc} \pi_{k-1}(O(n)) & & \\ & \searrow^{J} & \\ \downarrow i_* & & \pi_{n+k}(S^n) . \\ & \nearrow_{\tilde{J}} & \\ \pi_{k-1}(GL(n,\mathbb{R})) & & \end{array}$$

If for a given continuous map $\phi : S^{k-1} \to GL(n,\mathbb{R})$ we let $f : S^{k-1} \times S^{n-1} \to \mathbb{R}^n/\{0\}$, $f(x,y) = \phi(x)(y)$, then $\hat{J}([\phi]) = \hat{\chi}([f])$ defines $\hat{J} : \pi_{k-1}(GL(n,\mathbb{R})) \to \pi_{n+k}(S^{n+1})$.

From Proposition 3.1 we obtain the commutativity of the following diagram

$$\begin{array}{ccc} \pi_{k-1}(O(n)) & \xrightarrow{J} & \pi_{n+k-1}(S^n) \\ \downarrow i_* & & \downarrow \Sigma \\ \pi_{k-1}(GL(n,\mathbb{R})) & \xrightarrow{\hat{J}} & \pi_{n+k}(S^{n+1}) . \end{array}$$

The above defined $\hat{J}$-homomorphism will be used in the next section to derive bifurcation results (see Proposition 4.4).

4. SOME RESULTS ON BIFURCATION

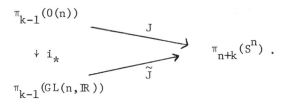

In this section we present some simple bifurcation results for maps defined on the whole space. More precisely, we will consider bifurcation problems for continuous maps $f : \mathbb{R}^k \times \mathbb{R}^n \to \mathbb{R}^n$, where $\mathbb{R}^k$ is the "parameter space". We consider globally defined maps in order to avoid some technicalities. However, the results of this section could be stated for maps $f : U \to \mathbb{R}^n$, where U is an open, not necessarily bounded, subset of $\mathbb{R}^k \times \mathbb{R}^n$ (cfr. [F.M.P.], [I.], [I.M.P.V.] and [R.]).

The results of this section are essentially not new and are stated not in their maximal generality. However, they are stated in accordance to the terminology of our generalized degree.

Assume that $f : \mathbb{R}^k \times \mathbb{R}^n \to \mathbb{R}^n$ is a continuous map satisfying

$$f(\lambda,0) = 0 \quad \text{for all } \lambda \in \mathbb{R}^k. \tag{4.1}$$

We will consider the following equation.

$$f(\lambda,x) = 0 \,. \tag{*}$$

Points of the form $(\lambda,0)$ are called <u>trivial solutions</u> of (*). All remaining solutions are called <u>non-trivial</u>.

<u>Definition 4.1</u>. A point $\lambda_0 \in \mathbb{R}^k$ is called a <u>bifurcation point</u> if any neighborhood of $(\lambda_0,0)$ contains a non-trivial solution of the equation (*).

We let $\Lambda = \Lambda(f)$ denote the set of all bifurcation points of (*).

Assume now that our map f satisfies (4.1) and the following condition

$$f(\lambda,x) \neq 0 \text{ for all } (\lambda,x) \in S^{k-1} \times (D^n/\{0\}). \tag{4.2}$$

If this is the case, then the restriction of f (still denoted by f)

$$f : S^{k-1} \times S^{n-1} \to \mathbb{R}^n/\{0\}$$

gives an element of $[S^{k-1} \times S^{n-1}, \mathbb{R}^n/\{0\}]$.

Let us set

$$\gamma(f) = \hat{\chi}[f] \in \pi_{n+k}(S^{n+1}),$$

where $\hat{\chi}$ denotes the map from $[S^{k-1} \times S^{n-1}, \mathbb{R}^n/\{0\}]$ into $\pi_{n+k}(S^{n+1})$, defined in Section 3.

We are now in a position of stating the following (local) result on the existence of bifurcation points.

<u>Proposition 4.2</u>. <u>Let</u> $f : \mathbb{R}^k \times \mathbb{R}^n \to \mathbb{R}^n$ <u>be a continuous map satisfying</u> (4.1) <u>and</u> (4.2). <u>Assume that</u> $\gamma(f) \neq 0$, <u>then there exists a bifurcation point</u> $\lambda_0 \in B^k$.

Proof. Let $\sigma : D^k \times D^n \to \mathbb{R}$ be the continuos map defined by $\sigma(\lambda,x) = ||\lambda||^2 - ||x||^2$. Clearly, the real-valued function σ satisfies (a) and (b) of Section 3 on $S^{k-1} \times B^n$ and $B^k \times S^{n-1}$ respectively. Therefore, the map $F : D^k \times D^n \to \mathbb{R}^{n+1}$ defined by

$$F(\lambda,x) = (f(\lambda,x), \sigma(\lambda,x)), \ (\lambda,x) \in D^k \times D^n$$

has the generalized degree different from zero, since $d(F, B^k \times B^n) = \gamma(f) \neq 0$.

Let $0 < r \leq 1$ and let $h_t : D^k \times D^n \to \mathbb{R}$, $t \in [0,1]$ be the family of maps defined by

$$h_t(\lambda,x) = (1 - t)||\lambda||^2 + tr^2 - ||x||^2, \ (\lambda,x) \in D^k \times D^n,$$

$$t \in [0,1].$$

Consider the homotopy $F_t : D^k \times D^n \to \mathbb{R}^{n+1}$, $t \in [0,1]$ define by

$$F_t(\lambda,x) = (f(\lambda,x), h_t(\lambda,x)), \ (\lambda,x) \in D^k \times D^n, \ t \in [0,1].$$

Clearly, F_t is an admissible homotopy and thus, by Proposition 2.4, we have that $d(F_1, B^k \times B^n) \neq 0$. Now, by Proposition 2.6, it follows that the map F_1 is essential on $D^k \times D^n$. In particular, the equation $f(\lambda,x) = 0$ has a solution for some $(\lambda,x) \in D^k \times D^n$ with $||x|| = r$. The statement now follows from the fact that $0 < r \leq 1$ is arbitrary. Q.E.D.

Remark 4.3. Under the assumptions of Proposition 4.2 much more can be said: a result of global nature is available. Namely, in the proof of Proposition 4.2 we evidenced the fact that the assumption $\gamma(f) \neq 0$ yields that the map $(f(\lambda,x), r^2 - ||x||^2)$ is essential in $D^k \times D^n$, $0 < r \leq 1$. This allows us to apply Theorem 4.2 of [I.M.P.V.] (the well-known result of [R.] if $k = 1$) in order to obtain the existence of a connected subset C of non-trivial solutions of (*) such that $\overline{C}$ has nonempty intersection with the set of bifurcation points in $D^k \times \{0\}$, which is either unbounded or intersects $\Lambda(f)$ outside $D^k \times \{0\}$.

The remaining of this section is devoted to the computation of $\gamma(f)$ under some smoothness assumptions on the map f.

Assume now that our map $f : \mathbb{R}^k \times \mathbb{R}^n \to \mathbb{R}^n$ is of class C^1 and let D_x denote the derivative with respect to $x \in \mathbb{R}^n$. Therefore, the map $D_x f(\lambda_0, x_0)$, $(\lambda_0, x_0) \in \mathbb{R}^k \times \mathbb{R}^n$ is linear. If $D_x f(\lambda_0, 0) \in GL(n, \mathbb{R})$, then the Inverse Function Theorem implies that λ_0 is not a bifurcation point. Consider the following condition

$$\{\lambda \in \mathbb{R}^k : Df(\lambda,0) \notin GL(n,\mathbb{R})\} \text{ \underline{is a discrete subset of} } \mathbb{R}^k. \tag{4.3}$$

Suppose now further that f satisfies (4.1), (4.3) and we are given $\lambda_0 \in \mathbb{R}^k$ such that $D_x f(\lambda_0, 0) \notin GL(n, \mathbb{R})$.

Under these assumptions on the map f, there exist $p > 0$ and $r > 0$ such that the map $f_1 : S^{k-1} \times S^{n-1} \to \mathbb{R}^n/\{0\}$ defined by $f_1(\lambda, x) = f(\lambda_0 + p\lambda, r \cdot x)$ satisfies (4.3). Moreover, if r is sufficiently small then the map f_1 is homotopic to the map $f_2 : S^{k-1} \times S^{n-1} \to \mathbb{R}^n/\{0\}$ defined by $f_2(\lambda, x) = D_x f(\lambda_0 + p\lambda, 0)(x)$.

On the basis of the results of Section 3 we readily obtain the following result.

<u>Proposition 4.4.</u> <u>Under the above assumptions on the map</u> f, <u>we have</u>

$$\gamma(f_1) = \hat{J}[f_2].$$

Thus, if $\hat{J}[f_2]$ is non-trivial, then λ_0 is a bifurcation point for the map f (we would like to point out that actually Proposition 4.4 allows global bifurcation results in the spirit of [A.], [A.F.], [A.Y.] and [I.]).

REFERENCES

[R.] Rabinowitz, P. H., *Some global results for nonlinear eigenvalue problems*, J. Funct. Anal., 7 (1971), 487-513.

[A.] Alexander, J. C., *Bifurcation of zeros of parametrized functions*, J. Funct. Anal., 29 (1978), 37-53.

[A.F.] Alexander, J. C. and P. M. Fitzpatrick, *The homotopy of certain spaces of nonlinear operators and its relation to global bifurcation of fixed points of parametrized condensing operators*, J. Funct. Anal., 34 (1979), 87-106.

[A.Y.] Alexander, J. C. and J. A. Yorke, *The implicit function Theorem and the global methods of cohomology*, J. Funct. Anal. 21 (1976), 330-339.

[A.W.] Amann, J., and S. Weiss, *On the uniqueness of the topological degree*, Math. Z, 130 (1973), 39-54.

[D.] Dancer, E. N., *On the existence of zeroes of perturbed operators*, Nonlinear Anal. T.M.A., 7 (1983), 717-727.

[F.M.P.] Fitzpatrick, P. M., Massabó, I. and J. Pejsachowicz, *Global several-parameter bifurcation and continuation theorems: a unified approach via complementing maps*, Math. Ann., 263 (1983), 61-73.

[G.] Geba, K., *Algebraic methods in the theory of compact fields in Banach spaces*, Fund. Math. 54 (1964), 168-209.

[G.G.] Geba, K., and A. Granas, *Infinite dimensional cohomology theories*, J. Math. Pures Appl., 52 (1973), 145-270.

[H.W.] Hurewicz, W., and H. Wallman, *Dimension theory*, Princeton University Press, 1948.

[H.] Husemoller, D., *Fibre bundles*, McGraw-Hill Inc., 1966.

[I.] Ize, J., *Introduction to bifurcation theory*, Differential Equations, Springer-Verlag Lecture Notes in Mathematics, 957 (1982), 145-203.

[I.M.P.V.] Ize, J., Massabo, I., Pejsachowicz, J., and A. Vignoli, *Structure and dimension of global branches of solutions to multiparameter nonlinear equations*, Trans. of A.M.S., 291(1985), 383-435.

[N.] Nirenberg, L., *An application of generalized degree to a class of nonlinear operators*, III me Coll. d'Analyse Fonct., Liege, (1971), 57-74.

GLOBAL RESULTS ON CONTINUATION AND BIFURCATION FOR EQUIVARIANT MAPS

Jorge Ize
IIMAS-UNAM
Apdo. Postal 20-726
Admón 20
Deleg. Alvaro Obregón
01000 México, D. F.
MEXICO

Ivar Massabó
Universitá della Calabria
Dipartimento di Matematica
87030 Arcavacata di Rende (CS),
ITALY

Alfonso Vignoli
II Universitá di Roma (Tor Vergata)
Dipartimento di Matematica
Via Orazio Raimondo (La Romanina)
00173 Roma, ITALY

ABSTRACT. This paper gives an extension for equivariant maps of our previous results concerning the study of global branching phenomenae for parameter dependent maps. Using elementary point set topology, we study the class of Γ - epi maps and give a detailed description of the structure, local dimension and global behaviour of the set of solutions for equations involving equivariant maps. A new degree theory for S^1-maps is sketched and our results are applied to global continuation and global bifurcation in presence of symmetries.

0. INTRODUCTION

In this paper we extend to the context of equivariant maps the results contained in [I.M.P.V,1], concerning the study of global branching phenomenae for parameter dependent maps.

The general abstract setting may be visualized as follows (for details take a look at Section 1).

Let E, F be two Banach spaces and let Γ be a compact Lie group acting linearly on E and F with isometric representations ρ and ρ' respectively. Let Λ be the parameter space (for simplicity you may think of Λ as being $\mathbb{R}^n$, for some $n \geqslant 1$). Let U be an open invariant subset of E and let $f\colon U \times \Lambda \longrightarrow F$ be a map such that $f(\rho(\gamma)x,\lambda) = \rho'(\gamma)f(x,\lambda)$, $\gamma \in \Gamma$, i. e., the map f is <u>equivariant</u> with respect to the first variable.

We are interested in describing the global behavior of the solutions of the equation

$$f(x,\lambda) = 0, \quad x \in U$$

as λ varies over Λ . Typical examples of this type are, of course,

S. P. Singh (ed.), Nonlinear Functional Analysis and Its Applications, 75–111.

global continuation and bifurcation problems in presence of symmetries (for example Hopf continuation and bifurcation).

The general philosophy of this paper is essentially the same as that of [I.M.P.V,1]. Namely, we try to obtain very general results with very elementary tools. It is indeed surprising, in our opinion, how one can get so general results with so little mathematical background. This is the reason why we start the paper with a list of elementary facts such as the construction of invariant neighborhoods, invariant Urysohn functions, the Dugundji-Gleason theorem and the like. This is solid elementary ground on which we build up our theory of Γ - epi maps. We would like to stress that this by no means implies that one cannot use fancy and refined mathematical instruments. We think just the way around: once you have a general theorem what you usually do is to try to show that it contains as many different particular results as possible. This is exactly the point where craftsmanship and knowledge come in. This point will be clearly evidenced in some parts of the present paper which is structured as follows.

We already mentioned that the first thing we do is to give a very short list of elementary facts (which is actually all we need). We proceed then by giving the definition of Γ - epi maps and proving the more elementary (but useful) facts for this class of maps. This first section closes with a result concerning the structure of the set of zeroes of Γ - epi maps that will be useful in the sequel (see Theorem 1.1).

Section 2 contains some examples of Γ - epi maps. In particular, this class of maps contains that for which the Fuller index is defined (see [F]) and nontrivial. It also includes the degree defined by Dancer for S^1 - gradient maps (see [Da]). This is done by defining a degree for S^1 - maps in the spirit of the rest of this paper, that is by looking at S^1 - extensions of maps and hence at equivariant obstruction. Due to the length of this paper we have postponed for a future publication the definition and study of a general degree for equivariant maps and a general Lie group. Here we will sketch only a very particular case.

Section 3 contains the main result of this paper (see Theorem 3.1). It gives a detailed description of the structure of the zero set of Γ - epi maps. It describes not only the behavior of global branches of solutions but also its covering dimension at each point of the branch itself.

In the last part of this paper we apply the general result of the previous section to global continuation and bifurcation problems for Γ - epi maps. We study in particular the Hopf bifurcation problem in its full generality, including the case of possible resonance on the stationary solutions.

ACKNOWLEDGMENT

The authors feel indebted to their common friend Jacobo Pejsachowicz, whose helpful observations and comments on the subject matter we are dealing with here, have greatly improved the content of the present work. Research done during visits of the first author to the Università della Calabria.

1. PRELIMINARY RESULTS

Let E, G be real Banach spaces and let Γ be a compact Lie group acting on E and G with representations $\rho:\Gamma \to GL(E)$ and $\rho':\Gamma \to GL(G)$. Without loss of generality we will assume that both ρ and ρ' are isometries since, if this is not the case, one may renorm the space by setting

$$|||x||| = \int_\Gamma \|\rho(\gamma)x\| \, d\gamma$$

where the integral is the normalized Haar integral on Γ . Under this new norm ρ (and analogously ρ') becomes an isometry since the Haar integral is invariant (see [B, page 11]).

Definition 1.1. Let U be an open invariant (i.e., $\Gamma U \subset U$) subset of E. A map $g: U \to G$ is called equivariant if

$$g(\rho(\gamma)x) = \rho'(\gamma)g(x) \quad , \quad \gamma \in \Gamma \quad , x \in U .$$

In what follows we shall use the simplified notation $\rho(\gamma) = \gamma$, $\rho'(\gamma) = \gamma'$. We will also frequently apply the following well-known facts.

1. Invariant Urysohn functions. If X is a normal space, Γ is a compact Lie group acting on X and A and B are closed invariant subsets of X with $A \cap B = \phi$, then there exists a continuous function $\varphi:X \to [0,1]$ which is invariant, i. e., $\varphi(\gamma x) = \varphi(x)$, $\gamma \in \Gamma$, $x \in X$, such that $\varphi(x) = 0$ if $x \in A$ and $\varphi(x) = 1$ if $x \in B$. In fact, if $\tilde{\varphi}: X \to [0,1]$ is any Urysohn function relative to A and B, then

$$\varphi(x) = \int_\Gamma \tilde{\varphi}(\gamma x) d\gamma$$

is the required invariant Urysohn function.

2. Construction of invariant neighborhoods. Let X and Γ be as above and let $A \subset X$ be an invariant closed subset and $U \subset X$ be open invariant such that $A \subset U$. Then there exists an open invariant subset V such that $A \subset V \subset \bar{V} \subset U$. To see this let $\varphi: X \to [0,1]$ be an invariant Urysohn function such that $\varphi(x) = 0$ on A and $\varphi(x) = 1$ on U^c. The desired set is for example $V = \varphi^{-1}([0,1/2))$.

3. Dugundji-Gleason Theorem. Let A and B be closed invariant subsets of E such that $A \subset B$. Let $\bar{g}: A \to G$ be an equivariant and compact map. Then there exists an equivariant and compact extension $\tilde{g}: B \to G$ of g. This follows by letting $\bar{g}: B \to G$ to be any Dugundji extension of g(see [D]) and by setting

$$\tilde{g}(x) = \int_{\Gamma} \gamma^{-1}\, \bar{g}(\gamma x)d\gamma \quad , \quad \gamma \in \Gamma \; , \; x \in B \; .$$

The compactness of $\tilde{g}$ follows by an argument given in [I, Lemma III-2.1].

Definition 1.2. Let $S \subset E$ be an arbitrary invariant set. A continuous equivariant map $g:U \to G$ is said to be admissible on $S \cap U$ if there exists an open bounded and invariant set V_0 such that $g^{-1}(0) \cap S \subset V_0 \subset \bar{V}_0 \subset U$.

We are now ready to introduce the class of maps that will be the object of our studies in this paper. In order to speed up our exposition we will use the following notation.

$$\Gamma K(U,G) = \{h : E \to G : h \text{ is compact equivariant with bounded support contained in } U\}.$$

The support of a map h will be denoted by supp h.

Definition 1.3. Let $g: U \to G$ be an admissible equivariant map. The map g is said to be Γ - epi on $S \cap U$ if the equation $f(x) = h(x)$ has a solution $x \in S \cap U$ for any $h \in \Gamma K(U,G)$.

The class of Γ - epi maps turns out to be equivalent to the following class of maps as shown in Propostion 1.1 below.

Definition 1.4. Let $g: U \to G$ be an admissible equivariant map. The map g is called Γ - essential on $S \cap U$ if, for any open bounded and invariant set V such that $g^{-1}(0) \cap S \subset V \subset \bar{V} \subset U$, any continuous equivariant extension $\bar{g}: \bar{V} \to G$ of $g:\partial V \to G$ with $g-\bar{g}$ compact on $\bar{V}$, has a zero on $S \cap V$.

Proposition 1.1. The map $g: U \to G$ is Γ - epi on $S \cap U$ if and only if g is Γ - essential on $S \cap U$.

Proof. (If). Let $h \in \Gamma K(U,G)$ and consider the set $V = \{x \in E : h(x) \neq 0\} \cup V_0$, where V_0 is as in Definition 1.2. Clearly, V is open bounded invariant and $\bar{V} = \text{supp } h \cup \bar{V}_0$. Moreover, V satisfies the properties of Definition 1.4. Since h vanishes on ∂V then $\bar{g} = (g-h)|_{\bar{V}}$ is an equivariant extension of $g|_{\partial V}$ satisfying the requirements of Definition 1.4 and, as such, $\bar{g}$ has a zero on $S \cap U$.

(Only if) Let V and $\bar{g}$ be as in Definition 1.4. Define $h: E \to G$ as $g-\bar{g}$ on $\bar{V}$ and zero outside. Clearly, supp $h \subset \bar{V}$ and h is continuous compact and equivariant (notice that the identically zero map is equivariant). Hence the equation $g(x) = h(x)$ has a solution $x \in S \cap U$. Since x cannot lie in $E \setminus \bar{V}$, then $x \in S \cap V$.

Q. E. D.

We list now some elementary properties of Γ - epi maps.

Property 1.1. (Existence) Let $g: U \to G$ be Γ - epi on $S \cap U$. Then $g^{-1}(0) \cap S$ is non empty (and invariant).

This follows immediately by taking $h \in \Gamma K(U, G)$ to be the identically zero map.

Property 1.2. (Localization). Let $g: U \to G$ be Γ - epi on $S \cap U$. Then g is Γ - epi on $S \cap V$ for any open invariant set V such that $g^{-1}(0) \cap S \subset V_0 \subset \bar{V}_0 \subset V \subset U$.

If H is a closed subgroup of Γ, then we put $E^H = \{x \in E : \gamma x = x$, for all $\gamma \in H\}$ - the fixed point set of H, which is clearly a closed subspace of E, being the action continuous.

We shall use the notation $U^H = E^H \cap U$. Note that an equivariant map $g: U \to G$ sends U^H into G^H.

Property 1.3. Let $g: U \to G$ be equivariant and let H be a closed subgroup of Γ. Then

(a) If g is H - epi on $S \cap U$ then g is Γ - epi on $S \cap U$.

(b) If the restriction $g|_{U^H}$ into G^H is Γ - epi on $S^H \cap U^H$, then g is Γ - epi on $S \cap U$ (here we assume tacitly that both S^H and U^H are Γ - invariant).

(c) If the restriction $g|_{U^H}$ into G^H is zero-epi (see Remark 1.1 below) on $S^H \cap U^H$, then g is Γ - epi on $S \cap U$.

Proof. (a) Follows from the fact that if $h: E \to G$ is Γ-equivariant then it is also H-equivariant (as a matter of fact $\Gamma K(U,G) \subset HK(U,G)$).

(b) If $h \in \Gamma K(U,G)$, then $h: U^H \to G^H$ is equivariant with respect to Γ and the following inclusion holds $\mathrm{supp}\,(h|_{U^H}) \subset (\mathrm{supp} h) \cap U^H$. Therefore, the equation $g(x) = h(x)$ has a solution $x \in S^H \cap U^H \subset S \cap U$.

(c) Take $h \in \Gamma K(U,G) \subset HK(U,G)$. Then the equation $g(x) = h(x)$ has a solution in $S^H \cap U^H \subset S \cap U$.

Q. E. D.

Remark 1.1. In the case when $H = \{1\}$ the notion of H - epi map on $S \cap U$ reduces to the concept of zero-epi map on $S \cap U$ introduced in [I.M.P.V.,1]. The latter definition reads, consequently, as follows. Given an admissible map $g: U \to G$ (i.e., such that there exists an open and bounded subset V_0 with the property $g^{-1}(0) \cap S \subset V_0 \subset \bar{V}_0 \subset U$), we say that g is zero-epi on $S \cap U$ if the equation $g(x) = h(x)$

has a solution $x \in S \cap U$ for any compact map $h: E \to G$ having bounded support contained in U.

The following examples show that the inverse implications of Property 1.3 (a), (b) and (c) do not hold in general.

Example 1.1. The converse of (a) does not hold. In fact, let $E = \mathbb{C}^3$, $G = \mathbb{C}^2 \times \mathbb{R}$, with S^1 - actions of the form

$$e^{i\varphi}(z_1, z_2, \lambda) = (e^{i\varphi}z_1, e^{i\varphi}z_2, \lambda), \text{ for } (z_1, z_2, \lambda) \in E,$$

$$e^{i\varphi}(z_1', z_2', x) = (e^{i\varphi}z_1', e^{i\varphi}z_2', x), \text{ for } (z_1', z_2', x) \in G.$$

Let $U = \{ (z_1, z_2, \lambda) \in E : \frac{\varepsilon^2}{4} < |z_1|^2 + |z_2|^2 \}$, $\varepsilon > 0$,

and let $S = E$. Consider the map $g: U \to G$ defined by

$$g(z_1, z_2, \lambda) = (\lambda z_1, \lambda z_2, |z_1|^2 + |z_2|^2 - 4\varepsilon^2).$$

Clearly, g is equivariant and admissible since

$$g^{-1}(0) = \{(z_1, z_2, \lambda) \in E : \lambda = 0 \quad |z_1|^2 + |z_2|^2 = 4\varepsilon^2\} \subset U.$$

For a proof of the fact that g is S^1 - epi on U see [I, pag.787]. Now, g is not H - epi, where $H = \{1\}$, being the action free, is the only isotropy subgroup of S^1. In fact, let $\varphi : \mathbb{R}^+ \to \mathbb{R}^+$ be defined by

$$\varphi(r) = \begin{cases} 0, \text{ if } r < \varepsilon \\ \frac{1}{\varepsilon}(r-\varepsilon), \text{ if } \varepsilon \leq r \leq 2\varepsilon \\ -\frac{1}{\varepsilon}(r-3\varepsilon), \text{ if } 2\varepsilon \leq r \leq 3\varepsilon \\ 0, \text{ if } r \geq 3\varepsilon. \end{cases}$$

Consider the equations

$$\lambda z_1 - \bar{z}_2\, \varphi((|z_1|^2 + |z_2|^2 + |\lambda|^2)^{1/2}) = 0$$

$$\lambda z_2 + \bar{z}_1\, \varphi((|z_1|^2 + |z_2|^2 + |\lambda|^2)^{1/2}) = 0$$

$$|z_1|^2 + |z_2|^2 = 4\varepsilon^2 .$$

Clearly, the above system has no solutions. Note that the perturbation has bounded support (and obviously is not equivariant). Hence, g is not H - epi.

Example 1.2. The converse of (b) is not true in general (unless $H=\{1\}$). Let E, G, U and S be as in Example 1.1 together with actions

$e^{i\varphi}(z_1, z_2,\lambda) = (e^{i\varphi} z_1, e^{i2\varphi} z_2,\lambda)$ in E and

$e^{i\varphi}(z_1', z_2',\lambda) = (e^{i\varphi} z_1', e^{i2\varphi} z_2',\lambda)$ in G. Consider the map $g: U \to G$

defined by $g(z_1, z_2,\lambda) = (\lambda z_1, |\lambda| z_2, |z_1|^2+|z_2|^2 - 4\varepsilon^2)$. As above g is S^1 - epi on U. On the other hand, if $H = \{-1,+1\}$ then g is not S^1 - epi on U^H. Indeed, $E^H = \{(0,z_2,\lambda) \in \mathbb{C}^3\}$, U^H is S^1 - invariant, $G^H = \{(0, z_2',x) \in \mathbb{C}^2\times \mathbb{R}\}$. The restriction $g|_{U^H}$ is not S^1 - epi on U^H since the system

$$|\lambda| z_2 + z_2\, \varphi\, ((|z_2|^2+|\lambda|^2)^{1/2}) = 0$$

$$|z_2|^2 = 4\, \varepsilon^2$$

where φ is as in Example 1.1 , has no solutions in U^H .

Example 1.3. To show that the converse of (c) does not hold it suffices to consider Example 1.2 above and notice that the map g is not zero-epi on U^H.

In order to obtain further properties for Γ - epi maps we shall assume in what follows that $S \cap U$ is closed in U.

Property 1.4. (Normalization) Let U be such that $0 \notin \partial U$. Then the inclusion $i: U \to E$ is Γ - epi on $S \cap U$ if and only if $U_0^\Gamma \subset S$, where U_0^Γ is the component of zero in $U^\Gamma = E^\Gamma \cap U$.

Proof. (Only if). If $E^\Gamma = \{0\}$ then $0 \in S \cap U$ since the inclusion is Γ - epi on$_\Gamma$ $S \cap U$. Hence $U_0^\Gamma = \{0\} \subset S$. Assume that $E^\Gamma \neq \{0\}$ and let $x_0 \in U_0^\Gamma$. Since U_0^Γ is open in the Banach space E^Γ then U_0^Γ is path connected. Let $\sigma : [0,1] \to U_0^\Gamma$ be a path (of invariant points) such that $\sigma(0) = 0$ and $\sigma(1) = x_0$. Clearly, there exists an ε-neighborhood of the path which is invariant since the path is invariant and the action is an isometry. Let φ be an Urysohn function taking value 1 on the path and vanishing outside the ε-neighborhood. Let $h(x) = \sigma(\varphi(x))$. Notice that h, φ are invariant and supp h is contained in the ε-neighborhood of the path since $\sigma(0) = 0$. Then the equation $x - \sigma(\varphi(x))$ has a solution $\bar{x} \in S \cap U$. This implies that $\bar{x}$ belongs to the path σ and so $\varphi(\bar{x}) = 1$, i.e., $\bar{x} = \sigma(1) = x_0 \in S$.

(If). By Property 2.7 of [I.M.P.V.,1] we have that U_0^Γ is contained in S^Γ if and only if the restriction $i : U^\Gamma \to G^\Gamma$ is zero-epi on $S^\Gamma \cap U^\Gamma$. Hence, by Property 1.3, $i : U \to G$ is Γ - epi on $S \cap U$.

Property 1.5. Let G_i , $i = 1,2$ be Banach spaces with actions ρ_i , $i = 1,2$ and let $g_i : U \to G_i$, $i = 1,2$ be continuous equivariant maps. Define $g : U \to G_1 \times G_2$ by $g(x) = (g_1(x), g_2(x))$. Assume that g is Γ - epi on $S \cap U$ (taking on $G_1 \times G_2$ the product action $\rho = \rho_1 \times \rho_2$). Then $g_2 : U \to G_2$ is Γ - epi on $g_1^{-1}(0) \cap S \cap U$.

Proof. If $h_2 \in \Gamma\, K(U,G_2)$, then the map $h = (0, h_2)$ belongs to $\Gamma\, K(U,G_1 \times G_2)$ so that the equations $g_2(x) = h_2(x)$ and $g_1(x) = 0$ are solvable in $S \cap U$.

Q. E. D.

Property 1.6. (*Homotopy property*). *Let* $g: U \to G$ *be* Γ - *epi on* $S \cap U$ *and let* $h : U \times [0,1] \to G$ *be compact*, $h(x,0) = 0$ *for all* $x \in U$ *and* $h_t : U \to G$ *equivariant for any* $t \in [0,1]$. *Assume that there exists an open, bounded and invariant set* V_0 *such that* $A_0 \subset V_0 \subset \bar{V}_0 \subset U$, *where* $A_0 = \{x \in S \cap U : g(x) = h(x,t)$ *for some* $t \in [0,1]\}$. *Then* $g(\cdot)-h(\cdot,1)$ *is* Γ - *epi on* $S \cap U$.

Proof. Let $k \in \Gamma\, K(U,G)$ and consider the set $A = \{ x \in S \cap U : g(x) - h(x,t) = k(x)$ for some $t \in [0,1]\}$. The set A is invariant and bounded (the latter follows from the fact that if $x \in A \cap (\overline{\text{supp}k})^c$, then $k(x) = 0$ and $x \in A_0$, thus $A \subset \text{supp}\, k \cup A_0 \subset \text{supp}\, k \cup V_0 \subset U$). Moreover, the set A is closed and disjoint from the set $(\overline{\text{supp}\, k})^c \cap V_0^c$ (if $x \in (\overline{\text{supp}k})^c \cap V_0^c \cap A$, then $x \in V_0^c$ and $k(x) = 0$; so that $x \in A_0 \subset V_0$). Let φ be an invariant Urysohn function such that $\varphi(A) = 1$ and $\varphi((\overline{\text{supp}\, k})^c \cap V_0^c) = 0$. Now, the map $k(x) + h(x,\varphi(x))$ may be considered compact on E since the assumption $h(x,0) = 0$ for all $x \in U$ allows us to extend h to the whole space E by setting it identically zero outside $\text{supp}k \cup V_0$. Clearly, since φ is invariant, the map $k(\cdot) + h(\cdot, \varphi(\cdot)) \in \Gamma\, K(U,G)$. Thus, the equation $g(x)=k(x)+h(x,\varphi(x))$ has a solution $\bar{x} \in S \cap U$. Hence, $\bar{x} \in A$ and $\varphi(\bar{x}) = 1$, i.e., $g(\bar{x}) - h(\bar{x},1) = k(\bar{x})$.

Q. E. D.

Remark 1.2. The above homotopy principle can be extended to a much broader class of perturbations as it has been done in Theorem 2.2 of [I. M. P. V,1]. Also g need not be continuous (see Proposition 2.5 of [I.M.P.V,1] and the observations following it).

The following property of Γ - epi maps will be used in the last section of this paper which will be devoted to applications.

Property 1.7. (*Scaling property*). *Let* $E = E_1 \times E_2$, $G = G_1 \times G_2$ *be endowed with the maximum norm. Let* $S = E$ *and let* Γ *be a compact Lie group acting on* E_i *and* G_i, $i = 1,2$. *Let the equivariant map* $g : E \to G$ *be written as* $g = (g_1, g_2)$, *where* $g_i(x_1,x_2) = L_i x_i - k_i(x_1,x_2)$ *is such that* L_i *is a bounded linear operator and* k_i *is compact*, $i = 1,2$. *Assume that* $g^{-1}(0) \subset (B_1^\varepsilon \times B_2^\varepsilon) \cup A$, $g^{-1}(0) \cap ((B_1 \times B_2) \setminus A))$ *is closed, where* $B_i = \{x_i \in E_i : \|x_i\| < r_i\}$, $B_i^\varepsilon = \{x_i \in E_i : \|x_i\| < r_i - \varepsilon\}$, $\varepsilon > 0, i=1,2$ *and* A *is a closed invariant set such that if* $(x_1,x_2) \in A$, *then* $(t\, x_1,x_2) \in A$ *and* $(x_1, t\, x_2) \in A$ *for all* $t \geqslant 1$. *Then, the map* g *is* Γ - *epi on* $E \setminus A$ *if and only if* g *is* Γ - *epi on* $(B_1 \times B_2) \setminus A$.

Proof. (Only if). Since the set $g^{-1}(0) \cap (E \setminus A) = g^{-1}(0) \cap ((B_1^\varepsilon \times B_2^\varepsilon) \setminus A))$ is closed invariant and bounded, then, by normality, the map g is admissible on $(B_1 \times B_2)\backslash A$. The implication now follows from Property 1.2.

(If). Assume that g is not Γ - epi on $E\backslash A$. Then, from Proposition 1.1 and Definition 1.4, there is an open, bounded and invariant set V such that $g^{-1}(0) \cap (E\backslash A) \subset V \subset \bar{V} \subset E\backslash A$, and a continuous equivariant map $\bar{g} : \bar{V} \to G\backslash\{0\}$, with $g - \bar{g}$ compact on $\bar{V}$ and $g - \bar{g} = 0$ on ∂V. Clearly, $\bar{V} \subset \tilde{B}_1 \times \tilde{B}_2$, where $\tilde{B}_i = \{x_i \in E_i : \|x_i\| < R_i\}$, $R_i > r_i, i=1,2$. Now, the map $\tilde{g} : E \to G$ defined by

$$\tilde{g}(x) = \begin{cases} \bar{g}(x) & , \text{ if } x \in \bar{V} \\ g(x) & , \text{ if } x \in E\backslash V \end{cases}$$

has no zeros on $E\backslash A$. Thus, g is not Γ - epi on $(\tilde{B}_1 \times \tilde{B}_2)\backslash A$. Consider now the following equivariant scaling on B_1 defined by $s(t,x_1) = \alpha(t,x_1)x_1$, where

$$\alpha(t,x_1) = \begin{cases} 1, & \text{if } \|x_1\| \leqslant r_1 - \varepsilon \\ \left(1 + \dfrac{t(R_1 - r_1)(\|x_1\| - r_1 + \varepsilon)}{\varepsilon\, r_1}\right), & \text{if } r_1 - \varepsilon \leqslant \|x_1\| \leqslant r_1 \end{cases} .$$

For any $t \in [0,1]$ the scaling is an equivariant homeomorphism from B_1 into $\tilde{B}_1$, leaving fixed B_1^{ε} and $s_1(B_1) = \tilde{B}_1$, where $s_t(\cdot) = s(t,\cdot)$. Clearly the map g is Γ - epi on $(\tilde{B}_1 \times \tilde{B}_2)\backslash A$ if and only if $g(s_1(\cdot),\cdot)$ is Γ - epi on the set $(B_1 \times \tilde{B}_2)\backslash s_1^{-1}(A)$. Consider on $(B_1 \times \tilde{B}_2)\backslash s_1^{-1}(A)$ the homotopy

$$\left(\frac{g_1(s(t,x_1),\, x_2)}{\alpha(t,x_1)} ,\quad g_2(s(t,x_1),x_2)\right) =$$

$$= \left(L_1\, x_1 - \frac{k_1(s(t,x_1),x_2)}{\alpha(t,x_1)} ,\ L_2\, x_2 - k_2(s(t,x_1),x_2)\right).$$

This homotopy is equivariant and its set of zeros is that of g for $\|x_1\| \leqslant r_1 - \varepsilon$ and is contained in $s_1^{-1}(A)$ if $r_1 - \varepsilon \leqslant \|x_1\|$. This follows from the structure of A and the inclusion $s_1^{-1}(A) \subset s_t^{-1}(A)$ (recall that $\alpha(t,x)$ is nondecreasing in t). Hence, on $(B_1 \times \tilde{B}_2)\backslash s_1^{-1}(A)$ the zeros of the map g are kept fixed under the homotopy. Thus, g is not Γ - epi on $(B_1 \times \tilde{B}_2)\backslash s_1^{-1}(A)$ and from Property 1.2 we have that g is not Γ - epi on $(B_1 \times \tilde{B}_2)\backslash A$ (since $g^{-1}(0) \cap (E\backslash s_1^{-1}(A)^c) = g^{-1}(0) \cap (E\backslash A)$). Finally, the scaling in the x_2 - direction will contradict the hypothesis.

Q. E. D.

<u>Remark 1.3.</u> (a) Property 1.7 above can be regarded as a version of the excision property holding in the context of classical degree theories.

It can be stated in much more general terms. In particular the domain of g may be replaced by any open subset $U \subset E$ and g may admit a more general decomposition in the spirit of Proposition 4.1 of [I.M.P.V,1]

(b) In our applications we will deal mainly with the case when $E_2=E^\Gamma$ and $A \supset \{(0,x_2): x_2 \in E_2\}$ since we will be looking for nontrivial solutions, i. e., solutions with $x_1 \neq 0$.

The following is an easy but useful consequence of the homotopy principle.

Proposition 1.2. Let $g : U \to G$ be Γ - epi on $S \cap U$. Then
(a) either, $S \cap \partial V \neq \phi$, or $g(S \cap \bar{V}) \supset G^\Gamma$ for any open, bounded and invariant set V such that $g^{-1}(0) \cap S \subset V \subset \bar{V} \subset U$. In particular, if $G^\Gamma \neq \{0\}$ and the projection g_1 of g onto G^Γ sends bounded, closed (in E) invariant subsets of $S \cap U$ into bounded subsets of G^Γ, then $S \cap \partial V \neq \phi$.
(b) either, $S \cap U$ is unbounded, or $(\overline{S \cap U}) \cap \partial U \neq \phi$, or there exists V as above such that $g(S \cap \bar{V}) \supset G^\Gamma$.

Proof. (a). Assume there is such a V for which $S \cap \partial V \neq \phi$. By Property 1.2, the map g is Γ - epi on $S \cap V$. Let $p \in G^\Gamma$ and consider the compact map $h: U \times [0,1] \to G$ defined by $h(x,t)= t\,p$. By the choice of p the map h_t is equivariant. Now, the set $A = \{ x \in S \cap V : g(x) = h(x,t)$ for some $t \in [0,1]\}$ is invariant and properly contained in V since $S \cap \partial V = \phi$. By the homotopy principle $g-p$ is Γ - epi on $S \cap V$ so that $p \in g(S \cap V)$.
(b). If $(\overline{S \cap U}) \cap \partial U = \phi$ and $S \cap U$ is bounded, we may construct an invariant bounded open set V such that $\overline{S \cap U} \subset V \subset \bar{V} \subset U$. Since $S \cap \partial V = \phi$, then, from part (a) we obtain $g(S \cap \bar{V}) \supset G^\Gamma$.

Q. E. D.

Remark 1.4. If $G^\Gamma = \{0\}$ and $g : U \to G$ is equivariant, then g is Γ - epi provided that $0 \in S \cap U$ (as a matter of fact, in this case, any equivariant map g satisfies $g(0)=0$). Hence, under the above assumptions, we may have $S \cap U$ bounded and $S \cap \partial U = \phi$ (for example one can take $\{0\} = S$).

Imposing further (rather mild) assumptions on the map g we obtain the following refinement of Proposition 1.2.

We recall that a map g is bounded if it sends bounded sets into bounded sets and proper if the inverse image under g of any compact set is a compact set.

Theorem 1.1. Let $g : U \to G$ be Γ - epi on $S \cap U$. Assume that g is bounded and proper on bounded and closed (in E) subsets of $S \cap U$. Then, there exists an invariant set $\Sigma \subset S \cap U$ which is minimal closed (in U) and such that
(a) The map g is Γ - epi on $\Sigma \cap U = \Sigma$ (this implies in particular that $g^{-1}(0) \cap \Sigma \neq \phi$).
(b) If $\Sigma = \Sigma_1 \cup \Sigma_2$, where Σ_1 and Σ_2 are proper, closed and invariant subsets with $\Sigma_1 \cap \Sigma_2 = \phi$, then, either $\Sigma_1 = \phi$, or $\Sigma_2 = \phi$

(c) The set Σ is minimal for any map g_1 homotopic to g.
(d) If, moreover, $G^{\Gamma} \neq \{0\}$, then Σ is either unbounded, or $\Sigma \cap \partial U \neq \phi$.

Proof. (a) Note that, since $g^{-1}(0) \cap S$ is bounded, then it is compact. Now, let

$$\mathcal{C} = \{ C \subset S \text{ , } C \text{ closed (in } S \cap U) \text{ invariant : } g \text{ is } \Gamma - \text{epi on } S \cap U\}.$$

The family $\mathcal{C}$ is non-empty since $S \cap U \in \mathcal{C}$. Define an order in $\mathcal{C}$ by inclusion of sets and let $\mathcal{C}'$ be a chain in $\mathcal{C}$. Consider $\Sigma = \bigcap_{C \in \mathcal{C}'} C$.

Since $g^{-1}(0) \cap C$ is a descending family of invariant and non-empty compact sets, then $g^{-1}(0) \cap \Sigma$ is invariant and compact (non-empty). Let $h \in \Gamma K(U,G)$ and let V_0 be an open, bounded and invariant set such that $g^{-1}(0) \cap S \subset V_0 \subset \bar{V}_0 \subset U$. Set $V = V_0 \cup \{x \in E : h(x) \neq 0\}$ which is open, bounded, invariant and $\bar{V} \subset U$. Since g is proper on the bounded and closed (in E) subset $C \cap \bar{V}$, then $(g-h)^{-1}(0) \cap C \cap \bar{V}$ is a descending family of non-empty compact sets. Hence, $(g-h)^{-1}(0) \cap \Sigma \cap \bar{V} \neq \phi$, i.e., g is Γ-epi on $\Sigma \cap U$. Thus, $\Sigma \in \mathcal{C}$. By Zorn's lemma $\mathcal{C}$ has a minimal element (also denoted by Σ). Since g is Γ - epi on $\Sigma \cap U$, then, by the minimality of Σ, we have $\Sigma \subset S \cap U$ (otherwise $\Sigma \cap U$ would be a proper subset of Σ, contradicting its minimality).
(b) Let $\Sigma = \Sigma_1 \cup \Sigma_2$ with the properties listed in the above statement. If both Σ_1 and Σ_2 are non-empty then g is not Γ - epi on $\Sigma_i \cap U$, $i = 1,2$. Thus, there exist $h_i \in \Gamma K(U,G)$, $i = 1,2$, such that $g(x) \neq h_i(x)$ on $\Sigma_i \cap U$, $i = 1,2$. Since $\Sigma_1 \cap \Sigma_2 = \phi$, then there exist open and invariant subsets U_1, U_2 such that $\Sigma_1 \subset U_1$, $\Sigma_2 \subset U_2$ and $U_1 \cap U_2 = \phi$. Let φ_i be an invariant Urysohn function such that $\varphi_i(x) = 1$ on Σ_i and zero outside U_i, $i = 1, 2$. Define $h: E \to G$ by

$$h(x) = \begin{cases} \varphi_1(x) \; h_1(x) & \text{, if } x \in U_1 \\ \varphi_2(x) \; h_2(x) & \text{, if } x \in U_2 \\ 0 & \text{elsewhere.} \end{cases}$$

Clearly, $h \in \Gamma K(U,G)$. Hence the equation $g(x) = h(x)$ has a solution $\bar{x} \in \Sigma \cap U$. If $\bar{x} \in \Sigma_1$ then $\varphi_2(\bar{x}) = 0$ and $\varphi_1(\bar{x}) = 1$, so that $g(\bar{x}) = h_1(\bar{x})$ with $\bar{x} \in \Sigma_1 \cap U$, a contradiction.
(c) Let $g(\cdot)-h(\cdot,t)$ be the homotopy joining g with $g_1 = g - h_1$. Then, by the homotopy principle g_1 is Γ - epi on $\Sigma \cap U$. Thus, from (a) there exists a minimal subset Σ_1 of Σ on which g_1 is Γ-epi. Since the homotopy is reversible, then g is Γ - epi on Σ_1 and hence $\Sigma_1 = \Sigma$.
(d) Follows from Proposition 1.2.

Remark 1.5. Property (b) of Theorem 1.1 is equivalent to say that the quotient set Σ/Γ is connected. This does not imply that the set Σ itself is connected unless Γ is connected.

2. SOME CLASSES OF Γ - EPI MAPS.

It is clear that the class of Γ - epi maps contains any class of equivariant maps for which a generalized equivariant degree is defined, with the properties that:

1) If this degree is non-trivial then one has a zero of the map in the set U.

2) This degree is invariant under compact perturbations with bounded support contained in the set U.

However no general theory of this sort is yet available, except for certain classes of S^1 - problems such as the degree defined by Dancer, in [Da], for gradient maps and the degree defined by Fuller, in [F] for periodic solutions of autonomous differential equations. It is, in fact, possible to define a general degree: it is given by the class of a map, related to g, in the "stable" equivariant homotopy group of spheres (for any bounded open set U in a Banach space). However several reasons have prevented us to present this generalized degree in this paper:

i) The extra length due to the proofs of the properties of the degree: so independence of the approximations, homotopy property, excision and addition (up to one suspension), stability property and so on...

ii) Equivariant homotopy groups are not as well studied as the ordinary groups. This fact has implied the use of much heavier machinery from algebraic topology and group theory, contrary to the spirit of the present paper.

We shall thus restrict our attention to the case of S^1 - maps, for which most of the tools from algebraic topology have already been used in [I]. Furthermore our S^1-degree will be defined only for spheres (this is not a real restriction since the general degree will reduce the case of any open set U to the case of a sphere).

2.1 <u>Definition of the S^1 - degree in finite dimensional spaces</u>. If E and G are finite dimensional, let $E^{S^1} \cong \mathbb{R}^k$, $G^{S^1} \cong \mathbb{R}^\ell$, be the fixed point subspaces. Then the action of S^1 on the orthogonal complements will give these spaces a complex structure. Thus $E \cong \mathbb{C}^n \times \mathbb{R}^k$, $G \cong \mathbb{C}^m \times \mathbb{R}^\ell$ and the action will be represented by

$$e^{i\varphi}(Z_1,\dots,Z_n) \equiv (e^{in_1\varphi}Z_1,\dots,e^{in_n\varphi}Z_n) \equiv e^{i\varphi}Z$$

$$e^{i\varphi}(z_1,\dots,z_m) \equiv (e^{im_1\varphi}z_1,\dots,e^{im_m\varphi}z_m)$$

where z_i, Z_i are complex numbers, n_i, m_i are non-zero integers. Any element of E will be written as (Z,X) and any equivariant map g from E into G as $g(Z,X) = (F(Z,X), F_0(Z,X))$. Note that $F(0,X) = 0$.

Assume g is non-zero on the boundary of the ball $B \times B_0$, with $B = \{Z/ \|Z\| < R\}$, $B_0 = \{X/\|X\| < R_0\}$. To say that g is S^1 - epi on $B \times B_0$ means that g is not equivariantly deformable on $\partial(B \times B_0)$ to a map without zeros in the ball. Now if $F_0(0,X)$ defines a non-trivial element in $\Pi_{k-1}(S^{\ell-1})$, then F_0 is 0-epi on B_0 and, from Property 1.3.c, g is S^1 - epi on $B \times B_0$. This gives the existence of a stationary solution, i. e. Z = 0.

Remark 2.1. The question of knowing if the fact that F_0 is non-trivial implies that g itself has no non-zero extension to $B \times B_0$ (non-equivariant extensions included hence a non-trivial element in $\Pi_{k+2n-1}(S^{\ell+2m-1})$) is a Borsuk - Ulam problem. This is known in some special cases, see [N] and Professor Fadell's lecture in these proceedings. In general one expects a rigid relationship between the two classes and equality of n and m, due to the group action. Thus, this relationship is useful for problems with different dimensions and the application to the indices of sets of zeros in variational problems. Such a result will be given in Proposition 2.1 below.

It is then natural to suppose that $F_0(0,X)|_{\partial B_0}$ has a continuous non-zero extension $\tilde{F}_0(X)$ to B_0. Let $\tilde{F}_0(Z,X)$ be an invariant extension of $\tilde{F}_0(X)$ and of $F_0(Z,X)|_{\partial B_0}$ to $B \times B_0$. Define $\tilde{g}(Z,X) = (F(Z,X), \varphi(|Z|)F_0(Z,X) + (1-\varphi(|Z|))\tilde{F}_0(Z,X))$ where $\varphi(r)$ has the value 1 for r = R, and 0 for $0 \leqslant r \leqslant \varepsilon$. ($\varepsilon$ is chosen so small that $\tilde{F}_0(Z,X) \neq 0$ for $|Z| \leqslant \varepsilon, |X| \leqslant R_0$). Thus one may replace g by the map $\tilde{g}$, with $\tilde{g}(Z,X) \neq 0$ for $|Z| = \varepsilon$, and g coincides with $\tilde{g}$ on $\partial(B \times B_0)$. The hypothesis that $F_0(0,X) \neq 0$ is also present in Dancer's and Fuller's degree. However we shall make a weaker hypothesis: (H) Assume $g(Z,X) \neq 0$ for $|Z| = \varepsilon$, $|X| \leqslant R_0$. (In some of our applications we shall leave open the triviality of the class of $F_0(0,X)$, in particular in bifurcation problems). We are thus interested in giving sufficient conditions for the existence of zeros of any equivariant extension of g from the boundary of the set $\{\varepsilon \leqslant |Z| \leqslant R\} \times \{|X| \leqslant R_0\}$ to itself. Necessary conditions require a very close study of the classes of the restriction of g to the fixed point subspaces of the different isotropy subgroups of S^1. To do so in the present paper would take us very far from global problems; thus the study of the generalized Hopf extension problem will undertaken in another publication.

Let $\{(\eta_1,\ldots,\eta_n) / \Sigma |\eta_i|^2 = 1\} = S^{2n-1}$ be the unit sphere in $\mathbb{C}^n$. with the standard action

$$e^{i\varphi}(\eta_1,\ldots,\eta_n) = (e^{i\varphi}\eta_1,\ldots,e^{i\varphi}\eta_n)$$

and consider the map $(\mathcal{F},\mathcal{F}_0)(\eta,r,X)$, from $S^{2n-1} \times [\varepsilon,R] \times B_0$ into $\mathbb{C}^m \times \mathbb{R}^\ell$, defined as

$$\mathcal{F}_j(\eta,r,X) = [F_j(\tilde{r}\,\eta_1^{n_1^*},\ldots,\tilde{r}\,\eta_n^{n_n^*}, X)]^{|M|/m_j^*} \qquad j = 1,\ldots,n,$$

$$\mathcal{F}_0(\eta,r,X) = F_0(\tilde{r}\,\eta_1^{n_1^*},\ldots,\tilde{r}\,\eta_n^{n_n^*},X)$$

$$\text{where} \qquad Z^{n*} = \begin{cases} Z^n & \text{if } n > 0 \\ \bar{Z}^{|n|} & \text{if } n < 0 \end{cases}$$

$$M = \Pi m_j \ , \quad N = \Pi n_j \ , \ \tilde{r} = r\,(\Sigma\, |\eta_j|^{2|n_j|})^{-1/2}$$

Remark 2.2. The reason for using η instead of Z is because the action on S^{2n-1} is free. The reason for taking powers for F_j is for computational purposes, mainly for bifurcation. Note that many of the ideas behind our construction and most of the technical results needed in this part of the paper are taken from [I].

The map (F,F_0) is thus the composition of the maps:

1) $S^{2n-1} \times [\varepsilon,R] \times B_0 \to B \times B_0 \setminus \{|Z| < \varepsilon\}$ given by

$$(\eta,r,X) \to (\tilde{r}\,\eta_1^{n_1^*},\ldots,\tilde{r}\,\eta_n^{n_n^*},X)$$

2) $B \times B_0 \to \mathbb{C}^m \times \mathbb{R}^\ell$ given by (F,F_0)

3) $\mathbb{C}^m \times \mathbb{R}^\ell \to \mathbb{C}^m \times \mathbb{R}^\ell$ given by $(z_1,\ldots,z_m, X) \to (z_1^{|M|/m_1^*},\ldots,z_m^{|M|/m_m^*},X)$

Each of these maps is equivariant, but the actions of S^1 are different (on the last space the action is quasi-free and given by $e^{i\varphi}(z_1,\ldots,z_m) = (e^{i|M|\varphi}z_1,\ldots,e^{i|M|\varphi}z_m)$. The first and last applications are clearly not one to one, but they are onto. They have an ordinary degree (index at 0) N and $|M|^m/M$, respectively.

It is easy to see that (F,F_0) is non-zero on $\partial(B_\varepsilon \times B_0)$ if and only if (F,F_0) is non zero on $S^{2n-1} \times \partial B^{k+1}$, where $B_\varepsilon = \{ Z/\varepsilon < |Z| < R \}$, $B^{k+1} = \{(r,X)/\varepsilon < r < R,\ |X| < R_0\}$. Furthermore if (F,F_0) has an equivariant (with respect to the given actions) extension to $B_\varepsilon \times B_0$, then (F,F_0) has an equivariant extension to $S^{2n-1}\times B^{k+1}$, that is if (F,F_0) is S^1 - epi on $S^{2n-1} \times B^{k+1}$ then (F,F_0) is S^1 - epi on $B_\varepsilon \times B_0$ (not conversely).

Since the action on η is free, $(F,F_0)|_{S^{2n-1}\times S^k}$ defines a cross-section of the bundle

$$S^{2m+\ell-1} \to S^k \times (S^{2n-1} \times_{S^1} S^{2m+\ell-1}) \xrightarrow{p} S^k \times \mathbb{C}P^{n-1}$$

where $p(r,X,[\eta, Y, Y_0]) = (r,X,[\eta])$ and $[\eta,Y,Y_0]$ is identified with $[e^{i\varphi}\eta, e^{i|M|\varphi}Y,Y_0]$. The cross-section is just the map $(r,X,[\eta]) \to (r,X,[\eta,F,F_0])$. (see [B, chapter II, 2.6] and [I] for details).

Now (F,F_0) is not S^1 - epi on $S^{2n-1}\times B^{k+1}$ if and only if the above cross-section extends to a cross-section to the sphere bundle with S^k replaced by B^{k+1}; this will happen if and only if the obstruction sets, for extension on the $q+1$ skele of $B^{k+1} \times \mathbb{C}P^{n-1}$, contain 0, for all q's between 0 and the top dimension $k+2n-1$. These obstruction sets are subsets of $H^{q+1}(B^{k+1} \times \mathbb{C}P^{n-1}, S^k \times \mathbb{C}P^{n-1};\ \Pi_q(S^{2m+\ell-1}))$. These cohomology groups vanish for all q's, except if $q=k+2p$, $p=0,\ldots,n-1$, in which case they are equal to the coefficient group $\Pi_q(S^{2m+\ell-1})$. The first non-vanishing coefficient group is for $q=2m+\ell-1$. This gives the primary obstruction, an integer, which is unique (see [I]). The obstructions are defined as follows: $\mathbb{C}P^{n-1}$ is given the cell structure $B^0 \times B^2 \times\ldots\times B^{2p}\times\ldots\times B^{2n-2}$, with $B^{2p}=\{\eta_1,\ldots,\eta_p,0,\ldots,0\}$. If one has an equivariant extension $\tilde{F}_p(\eta_1,\ldots,\eta_p,X)$ to $S^{2p-1} \times B^{k+1}$, then the obstruction for an extension to $B^{2p} \times B^{k+1}$ is the class, in $\Pi_{2p+k}(S^{2m+\ell-1})$, of the map defined as $\tilde{F}_p$ on $S^{2p-1} \times B^{k+1}$ and

$(F,F_0)(\eta_1,\ldots,\eta_p, (1 - \Sigma_{i=1}^{p}|\eta_i|^2)^{1/2},0,\ldots,0,r,X)$ on $B^{2p} \times S^k$. (see [I, Remark II-4-b] for this step by step construction). One could then define the S^1-degree as the set of all the obstructions. However, as it is noted in [I], secondary obstructions are not easy to compute and their interpretation is not clear. For the purpose of this paper, we shall restrict our attention to the primary obstruction and thus, in order to get a non-trivial invariant, to the case where k and ℓ have different parities, $2p = 2m+\ell-1-k$.

Definition 2.1. The S^1 - degree of g is the primary obstruction for (F,F_0), with the notation $\deg_{S^1}(g)$.

Remark 2.3. In order to extend this degree to infinite dimensional spaces, we shall also define the normalized degree of g as the rational number given by the primary obstruction divided by $N|M|^m M^{-1}$, i.e.,

$$\deg^n_{S^1}(g) = \deg_{S^1}(g) \;/\; (N|M|^m M^{-1})$$

Suppose then that another equation is added to g, of the form: z with action $e^{ik\varphi}z$. Then n and m are replaced by $n+1$ and $m+1$, the primary obstruction occurs for $2(p+1)$ and is the degree of the map : $(\eta_1\eta_1,\ldots,\eta_p,r,X) \longrightarrow (\eta^{|k\,M|}, F_j^{|kM|/m_j*},F_0)$ as a mapping from $\mathbb{R}^{2(m+1)+\ell}$ into itself, (clearly the previous extension with the new powers is still a valid extension). From the product and composition properties of the ordinary degree, the degree of the suspended map is $|kM|\,|k|^m$ times the old degree. Since the new normalizing factor is $kN\,|kM|^{m+1} k^{-1} M^{-1}$, both normalized degrees are equal. Note also that a suspension in X will not alter the degree.

2.2. Extension to infinite dimensional spaces. Let $E = G$ be a real Banach space, with S^1 acting by isometries, and assume that $g(Z,X)$ has the form of Identity-Compact (implicitely we are supposing that E^{S^1} has an invariant complement). By averaging the finite dimensional approximations of the compact part of g, we obtain, as in [I, Remark III.2.1], equivariant finite dimensional approximations of the form $(Z,X) - (F_n(Z,X), F_{n0}(Z,X))$. By considering the projection of this equation on the finite dimensional subspace, one gets a normalized degree. Due to Remark 2.3, the degree is independent of the approximation used.

Note that, since our degree is "globally" defined, the homotopy and normality properties follow at once (the fact that if the degree is non-zero, then one has a zero of g is a standard compactness argument). Since we are just considering one set, no additivity can be proved for the moment (see the introduction to this part), but the relationship with a local index will be given in section 2.4.

2.3. The (ordinary) degree of S^1 - maps. In this note we shall illustrate how the idea of equivariant extension may be used to compute the degree of an S^1 - map. This is based on the fact that if $2m+\ell > 2n + k - 1$, then there is always an extension ($2n + k - 1$ is the top dimension and $q + 1 \geq 2m + \ell$). Assume thus that $m = n$, $k = \ell$, and that $g(Z,X)$ is an equivariant map defined on $\mathbb{C}^n \times \mathbb{R}^k$, which is non-zero on the boundary of a bounded invariant neighborhood Ω of the origin. One has then a slight generalization of a result of Nirenberg [N] (he assumes $F_0(0,X) = X$).

Proposition 2.1. $\deg(F(Z,X), F_0(Z,X); \Omega, 0) = (M/N)\deg(F_0(0,X), \Omega^{S^1}, 0)$

Proof. Replace first Z by (r,η) and (F,F_0) by $(\mathcal{F},\mathcal{F}_0)$ as above. By composition of the degree, $\deg(F,F_0) = N^{-1}|M|^{-n} M \deg(\mathcal{F},\mathcal{F}_0,\Omega',0)$, where Ω' is the inverse image of Ω. Choose ε so small that on $(\partial\Omega')^{S^1} \times \{r \leq 2\varepsilon\}$, $\mathcal{F}_0(r,\eta,X) \neq 0$. Let $\varphi(r,\eta,X)$ be an invariant Uryshon function with values 1 on $(\partial\Omega') \cap \{r \geq 2\varepsilon\}$ and 0 on $\bar{\Omega}' \cap \{r \leq \varepsilon\}$. Clearly the map $(\varphi \mathcal{F}(r,\eta,X) + (1-\varphi)(\tilde{r}\,\eta_1^{|M|}, \ldots, \tilde{r}\,\eta_n^{|M|}), \mathcal{F}_0(r,\eta,X))$ has the same degree as $(\mathcal{F},\mathcal{F}_0)$. Since the action on (r,η,X), for $r \geq \varepsilon$, is free and from [B p. 90 and 105] (that is from the dimension considerations), the above map, when restricted to $\partial(\Omega' \cap \{r > \varepsilon\})$ has a non-zero equivariant extension to $\Omega' \cap \{r > \varepsilon\}$. Thus $\deg(\mathcal{F},\mathcal{F}_0, \Omega', 0) = \deg(\tilde{r}\,\eta_1^{|M|}, \ldots, \tilde{r}\,\eta_n^{|M|}, \mathcal{F}_0(r,\eta,X); \Omega' \cap \{r < \varepsilon\}, 0)$. Clearly (r,η) can be deformed to 0 in $\mathcal{F}_0$, $\Omega' \cap \{r < \varepsilon\}$ may be replaced by $\Omega'^{S^1} \times \{r < \varepsilon\}$ and, from the product property, the degree is $|M|^n \deg(\mathcal{F}_0(0,X); \Omega'^{S^1}; 0)$. Since $\Omega'^{S^1} = \Omega^{S^1}$, one gets the result.

Q. E. D.

2.4. Finite number of orbits: a Poincaré map. The next easiest case is for $m = n$, $k = \ell + 1$, which is the case of our applications. One has then just the primary obstruction. From these dimension hypotheses, it is not difficult to see that one may approximate (F, F_0) by a smooth map, which vanishes only on a finite number of orbits (either by using Sard's lemma on $\mathbb{C}P^{n-1}$, or at the level of $B \times B_0$ as in [D] or [N]). One gets then a finite number of points in $\mathbb{C}P^{n-1} \times B^{k+1}$, where $(\mathcal{F},\mathcal{F}_0)$ is 0. Make then a small rotation P in $\mathbb{C}P^{n-1}$, such that, if $[\eta] = [P\xi]$, then for each of these points one has $\xi_n \neq 0$ (this is an open set). Since P is equivariantly deformable to I, or from the fact that the primary obstruction is defined independently of the cell decomposition (see [I, p. 765]), then the S^1 degree is unchanged. Now on $\partial(B^{k+1} \times B^{2n-2})$, the map $(\mathcal{F},\mathcal{F}_0)(r,X,\xi_1,\ldots,\xi_{n-1},\sqrt{1-\Sigma|\xi_i|^2})$ is non zero (on $S^k \times B^{2n-2}$ by hypothesis and, for $\Sigma_{i=1}^{n-1}|\xi_i|^2 = 1$, from the fact that $\xi_n \neq 0$ on the zeros of $g(Z,X)$). Hence the above map is a good extension to the $k+1+2n-3$ skeleton of $B^{k+1} \times \mathbb{C}P^{n-2}$, and the primary obstruction is thus the ordinary degree of $(r,X,\xi_1,\ldots,\xi_{n-1}) \rightarrow (\mathcal{F},\mathcal{F}_0)(r,X,\xi_1,\ldots,\xi_{n-1},\sqrt{1-\Sigma|\xi_i|^2})$. Thus this degree is the sum of the local indices at each zero. Now if (Z_0,X_0) is a point on an orbit, then from the relation $Z_j = \tilde{r}\,\eta_j^{n_j^*}$, one gets $|n_j|$ pre-images, if $Z_j \neq 0$. Hence one gets $\prod_J |n_j|$ preimages in $\mathbb{C}^n$, if $J = \{j/Z_j \neq 0\}$. But two of these preimages give the same point in

$\mathbb{C}P^{n-1}$ if $\eta = e^{i\varphi}\tilde{\eta}$, that is if $Z_0 = e^{i\varphi}(\tilde{Z}_0)$, thus if and only if $e^{i\varphi}$ belongs to $S^1{}_{Z_0}$, the isotropy subgroup of Z_0, of order n_0 which is the greatest common divisor of $|n_j|$, for j in J. Thus each orbit generates $\Pi_J |n_j| / n_0$ points in $\mathbb{C}P^{n-1}$ and on $B^{k+1}\times B^{2n-2}$. By deforming P to I and letting the group act on Z, the local index of the map at $(r_0,X_0,\xi_1^0,\ldots,\xi_{n-1}^0)$ will be the index of the composition of the maps:

1) $(r,X,\xi_1,\ldots,\xi_{n-1}) \longrightarrow (r,X,\eta_1,\ldots,\eta_{j-1},\eta_{j+1},\ldots,\eta_n)$, where $\eta_1^0 = \eta_{j-1}^0 = 0$, $j,\ldots,n$ are the elements of J, and the group is left to act until $\xi_n = \sqrt{1-|\xi|^2}$ gives $\eta_j = \sqrt{1-|\eta|^2}$ (this map has degree 1).

2) $(r,X,\eta_1,\ldots\eta_{j-1},\eta_{j+1},\ldots,\eta_n) \to (X,\tilde{r}\eta_1^{n_1^*},\ldots,\tilde{r}\eta_{j-1}^{n_{j-1}^*}, \tilde{r}(1-|\eta|^2)^{n_j^*/2},$

$\tilde{r}\eta_{j+1}^{n_{j+1}^*},..,\tilde{r}\eta_n^{n_n^*})$ with index at $(r_0,X_0,0,\ldots,0,\eta_{j+1}^0,\ldots,\eta_n^0)$ equal to

the degree of

$$(X-X_0,\tilde{r}\eta_1^{n_1^*},\ldots,\tilde{r}\eta_{j-1}^{n_{j-1}^*},\tilde{r}(1-|\eta|^2)^{n_j^*/2} - \tilde{r}_0(1-|\eta_0|^2)^{n_j^*/2},\ldots\tilde{r}\eta_n^{n_n^*} - \tilde{r}_0\eta_n^{0\;n_n^*}).$$

Replacing $\tilde{r}$ by $t\tilde{r} + (1-t)\tilde{r}_0$, except in the j-component and, in that component $|\eta|$ by $t|\eta| + (1-t)\,|\eta^0|$, then, if $\tilde{r} < \tilde{r}_0$ one has $(t\tilde{r} + (1-t)\tilde{r}_0)/\tilde{r}_0 \equiv \alpha < 1$. At a zero of the map, one needs $|\eta_k^0| < |\eta_k|$, $k = j+1,\ldots,n$; thus the j-component is negative. This implies that $\alpha = 1$, $\tilde{r} = \tilde{r}_0$, $|\eta| = |\eta^0|$, and hence one has an isolated zero, with index equal to the degree of

$$(X-X_0,\eta_1^{n_1^*},\ldots,\eta_{j-1}^{n_{j-1}^*},\; r-r_0,\eta_{j+1}^{n_{j+1}^*/|n_{j+1}|} - \eta_{j+1}^{0\;\;n_{j+1}^*/|n_{j+1}|},\ldots,$$

$$\eta_n^{n_n^*|n_n|} - \eta_n^{0\;\;n_n^*/|n_n|})$$

$(\tilde{r}_0,(1-|\eta|^2)$ are deformable to 1, $\eta^{k^*}-\eta^{0k^*}$ to $\eta^{k^*/|k|}-\eta^{0k^*/|k|}$, near η^0).

Thus the degree of this map is $\prod_{k<j} n_k \prod_{k>j} \operatorname{sign} n_k$.

3) $(X,Z_1,\ldots,Z_{j-1},\alpha_j,Z_{j+1},\ldots,Z_n) \to (F,F_0)\;(X,Z_1,\ldots,\alpha_j,\ldots,Z_n)$

which is the map defined as a transversal section to the orbit by taking Z_j real and near Re $Z_j^0 > 0$.

4) $(z_1,\ldots,z_n,x) \to (z_1^{|M|/m_1^*},\ldots,z_n^{|M|/m_n^*},x)$ with index at 0 equal to $|M|^n/M$.

Thus the local index at $(r_0,X_0,\xi_1^0,\ldots,\xi_{n-1}^0)$ is

$$\prod_{k<j} n_k \prod_{k>j} \operatorname{sign} n_k \;|M|^n/M \;\text{Index}\,(F,F_0;X,Z_1,\ldots,\alpha_j,\ldots,Z_n).$$

Since all the preimages contribute this index, the orbit will give a total degree

$N\ |M|^{n}/M\ \text{sign}(n_j)\ \text{Index}\ (F,F_0;X,Z_1,\ldots,\alpha_j,\ldots,Z_n)/n_0$

Definition 2.2. We shall call

$$\text{sign}(n_j)\ \text{Index}\ (F,F_0;X,Z_1,\ldots,\alpha_j,\ldots Z_n)/n_0 \equiv \text{In}^n_{S^1}\ (g;X_0,Z_0)$$

the normalized S^1 - index of the orbit (X_0,Z_0). n_0 is the order of isotropy subgroup of the orbit, $\alpha_j = \text{Re}Z_j > 0$. The sign of n_j reflects the action of the group near the transversal section at $(X_0,0,\ldots,0,\alpha_j,Z^0_{j+1},\ldots Z^0_n)$, $Z^0_k \neq 0$, $k = j,\ldots,n$.

Remark 2.4. Note that from our definition of S^1 - index of orbits, one may define a degree for a map g , by genericity arguments, with the task of proving its independence from the section chosen. In our case this is already done. We have thus proved the following:

Proposition 2.2. If $m = n$, $k = \ell + 1$ and one has a finite number of orbits, then: $\deg^n_{S^1}(g) = \Sigma\ \text{In}^n_{S^1}(g;X_0,Z_0)$

In the last two sections of this part, we shall relate our degree, with two known approaches.

2.5. Gradient maps: Dancer's degree. In [D], Dancer has considered gradients of real-valued maps $f(Z,X)$, with the property that $f(e^{i\varphi}Z,X) = f(Z,X)$, Z in $\mathbb{C}^n$, X in $\mathbb{R}^{\ell}$. It is easy to see that if $f(T_g x) = f(x)$, then $(D f(T_g x),T_g h) = (D f(x),h) = (T_g\ D f(x),T_g h)$ (Since $T^T_g = T_{g^{-1}}$). Thus $g(x) = D f(x)$ is equivariant. Writing $Z_j = x_j + iy_j$, $g_j = f_{x_j} + i f_{y_j}$ and taking the φ-derivative, at $\varphi = 0$, of $f(T_\varphi(x,y),X) = f(x,y,X)$, one obtains the relation

$$\text{Re}\ (g(Z,X),AZ) = 0 \qquad (1)$$

where the scalar product is complex and A is the diagonal matrix with components $in_1,\ldots,in_n$. $g = (F,F_0)$.

If one considers the problem

$$\nu\ A\ Z + F(Z,X) = 0 \qquad (2)$$

$$F_0(Z,X) = 0$$

taking the scalar product with AZ, from (1), one gets $\nu\|AZ\|^2 = 0$. Thus either $Z = 0$ and one has a stationary solution (this is ruled out in Dancer's and our hypothesis), or $\nu = 0$, giving a zero of $g(Z,X)$.

Dancer defines an S^1 - degree for $g(Z,X)$, via the same genericity arguments, through a local index at an orbit, which is, in our notation, the $\text{Index}(F_1,\ldots,F_{j-1},\text{Re}F_j,F_{j+1},\ldots,F_n,F_0)/n_0$ at $(X,Z_1,\ldots,\alpha_j,\ldots,Z_n)$,

as a mapping from $\mathbb{R}^{2n-1} \times \mathbb{R}^{\ell}$ into itself.

Proposition 2.3. Dancer's degree and $\deg^n_{S^1}((2))$ coincide.

Proof. It is enough to check the equality at the level of the local indices. We have another variable, ν, and another equation, $\operatorname{Im} F_j = 0$. Write (2) as (for $t = 1$)

$$(i\, \nu t n_k Z_k + F_k,\ k \neq j,\ \ \operatorname{Re} F_j + i\, \nu n_j \alpha_j + i \operatorname{Im} F_j,\ F_0)$$

At a zero, multiply by AZ and take the real part, to get

$$\operatorname{Re}(F,AZ) + \nu t \|AZ\|^2 + \nu(1 - t) n_j^2\, \alpha_j^2 = 0.$$

Thus, from (1), $\nu = 0$ and the deformation is valid on the boundary of a neighborhood of $(\nu = 0, X_0, Z_0, Z_j = \alpha_j > 0)$. Using $it\operatorname{Im} F_j$, at a zero one has $F_k = 0$, $k \neq j$, and, from $(1)^j$, $\alpha_j n_j \operatorname{Im} F_j = 0$, a valid deformation. Deforming $\nu n_j \alpha_j$ to $\nu\ \operatorname{sign}(n_j)$ and taking the real map, then $\operatorname{In}^n_{S^1}\ (\nu\, AZ + g(Z,X)^j; 0, X_0, Z_0) = -$ Dancer's index. (the minus sign comes from the orientation chosen).

Q. E. D.

2.6. Autonomous differential equations : Fuller's index. Consider the differential equation

$$\frac{dX}{dt} = f(X) \qquad , \quad X \in \mathbb{R}^M \tag{3}$$

By letting $\tau = \nu t$, $2\pi/\nu$ - periodic solutions of (3) correspond to 2π-periodic solutions of

$$\nu\, \dot{X} = f(X) \tag{4}$$

The spaces $W^{1,2}(S^1)$, $L^2(S^1)$ of real valued functions are identified with the space of Fourier series $\Sigma\, x_n e^{int}$, with $x_{-n} = \bar{x}_n$ and $\Sigma(1 + \varepsilon n^2)|x_n|^2 < \infty$, $\varepsilon = 1, 0$.

Let E be $W^{1,2}(S^1)^M$, $G = L^2(S^1)^M$. Then the map $\nu \dot{X} - f(X)$ defines a C^1 Fredholm map from E into G.

Furthermore if $K : E \to E$ is the compact operator defined by

$$K(Y_0 + \Sigma_{n\neq 0}\ Y_n e^{int}) = Y_0 + \Sigma_{n\neq 0}\ Y_n\, e^{int}/in$$

then (4) is equivalent to the equation

$$g(Z, X_0, \nu) \equiv Z - K\, f(Z, X_0)/\nu = 0 \tag{5}$$

where $X(t) = X_0 + Z = \Sigma\, X_n\, e^{int}$, $X_{-n} = \bar{X}_n \in \mathbb{C}^M$.

Componentwise this is just the same as

$$i\, \nu\, n\, Z_n - f_n(Z, X_0) = 0 \qquad n \geqslant 0 \tag{6}$$

The action, by translation on t, is given by $e^{in\varphi}Z_n$ and (6) is equivariant. See [I pp 761, 781] for details.

Assume that, (5) has no zeros on

$$\partial(\{\varepsilon < \|Z\| < R\} \times \{|X_0| < R_0\} \times \{0 < \nu_1 < \nu < \nu_2\})$$

(true if the solutions in that range of frequencies ν are bounded and there are no stationary solutions for $|X_0| < R_0$: a hypothesis used for defining Fuller's degree).

Then, since $K f$ is compact, $g(Z,X,\nu)$ has a well-defined normalized S^1 - degree, by § 2.2.

Recall that the Fuller degree is defined generically for the field $f(X)$ and a bounded open set Ω, in $\mathbb{R}^M \times \mathbb{R}^+$, of orbits and frequencies, by considering first the case where Ω contains finitely many periodic orbits of hyperbolic type, as $d(\Omega,f) = \Sigma\, i(\Gamma)$ where $(\Gamma,\nu) \subset \Omega$ is a periodic orbit, corresponding to ν and $i(\Gamma)$ is the Fuller index of (Γ,ν) which is $(-1)^{\sigma}/n_0$. n_0 is the order of the isotropy subgroup of Γ or, equivalently, the integer such that the minimum period T of Γ is $2\pi/(\nu n_0)$. σ is the number of Floquet multipliers for the $n_0{}^{th}$ iterate of the return map, (eigenvalues of $(\partial X/\partial X_0)(n_0 T, X_0)$, where $X(t,X_0)$ is the solution of (3) with $X(0,X_0) = X_0$), on the interval $]1,\infty[$. $i(\Gamma)$ is also $\mathrm{ind}(\Pi^{n_0})/n_0$, where <u>ind</u> denotes the fixed point index and Π^{n_0} is the $n_0{}^{th}$ iterate of the Poincaré map for Γ. See [C.M.Y] and [V] for details.

<u>Proposition 2.4.</u> $\deg^n_{S^1}(g)$ <u>and</u> $d(\Omega,f)$ <u>coincide.</u>

<u>Proof</u>: By using generic arguments it is enough to check that the local indices agree for hyperbolic orbits (i. e. 1 is a simple Floquet multiplier and all the other multipliers are off the unit circle).

Assume Γ has the representation

$$X^0(t) = X^0_0 + \Sigma_{|n|\geq j}\ X^0_n e^{int} = X^0_0 + Z^0(t) \quad \text{where, after a rotation}$$

in $\mathbb{R}^M$ and a choice of the time origin, $X^0_j \equiv \vec{\alpha}_0 = (z^0_1,\ldots,z^0_{M-1},\alpha_0)$, with $\alpha_0 > 0$. j and the n's appearing in the sum are multiples of n_0. Then if P_N is the projection in E given by the truncation on the N first modes, then

$$\mathrm{In}^n_{S^1}(g,X^0) = \mathrm{Index}\ (F,F_0;X_0,\nu,Z_1,\ldots,\vec{\alpha},\ldots,Z_N)/n_0$$

where $(F,F_0) = (Z_n - f_n(P_N Z,X_0)/(i\nu n), n = 1,\ldots,N; -f_0(P_N Z,X_0))$ and $\vec{\alpha} = (z_1,\ldots,z_{M-1},\alpha)$, α in $\mathbb{R}$. See [I p.783].

Now from the definition of E and § 2.2, this index is the index of $g(Z,X_0,\nu)$ when restricted to the subspace E^0 of functions $X(t)$, with $X_j = \vec{\alpha}$, or else $X(t)$ L^2 orthogonal to $\vec{M}\sin jt$, $\vec{M} \equiv (0,\ldots,0,1)$ (this is easily seen by undoing the path covered in § 2.2 and 2.4). Note that the index above is the Leray-Schauder index at $(X^0(t),\nu)$ on the subspace $E^0 \times \mathbb{R} \cong E$. This index is the degree of $Z-Z^0-K(f(X)/\nu - f(X^0)/\nu_0) = Z-Z^0 - K\,A(t)(X-X^0)/\nu_0 + K\,f(X^0)(\nu-\nu_0)/\nu_0^2 + K\,G(X,\nu)$ where

$A(t) = D\,f(X^0(t))$ and $K\,G(X,\nu) = o(\|X-X^0\| + |\nu-\nu_0|)$. Thus one has to compute the degree on $E^0 \times \mathbb{R}$, at (X^0,ν_0), of $Z-Z^0- K\,A(t)(X-X^0)/\nu_0 + Z^0(\nu-\nu_0)/\nu_0$, or by translation and scaling, the index at $(U,V,\lambda) = (0,0,0)$ of

$$U - K\,A\,(U + V)/\nu_0 + \lambda\,Z^0 \qquad (7)$$

Now, zeros of (7) are periodic solutions of the equation

$$(U + V)' - A(t)\,(U + V)/\nu_0 + \lambda\,X^{0\prime} = 0 \quad (8)$$

with V constant and $U(t)$ orthogonal to $\vec{M}$ sin j t.

However, one may see that $X^{0\prime\prime} = A(t)X^{0\prime}/\nu_0$. Thus, by the hyperbolicity assumption, $X^{0\prime}$ is the only 2π-periodic solution of the homogeneous part of (8) (see[V p.240]) :
It is easy to verify that

$$\mathrm{Ker}\;(X' -AX/\nu_0 + \mu X) = \{X \in E/X(t) = e^{-\mu t}\Phi(t)W\;,$$

$$W \in \ker\;(\Phi(2\pi) - e^{2\pi\mu}\,I)\;\}\quad \text{where}\quad \Phi(0) = I\;,\quad \Phi(t)\ \text{is}$$

a fundamental matrix for $X' - A\,X/\nu_0$, $e^{2\pi\mu}$ are the Floquet multipliers of the $n_0{}^{th}$ iterate of the return map. Furthermore, if $X(t)$ belongs to $\ker(d/dt -(A/\nu_0-\mu)I)^{\alpha}$, by letting $Y(t)= e^{\mu t}\,\Phi^{-1}(t)\;X(t)$, it is easy to see that $(d/dt - (A/\nu_0-\mu))^k X = e^{-\mu t}\Phi(t)\;Y^{(k)}(t)$. (We are assuming here that $A(t)$ is smooth; this can be done by approximation, then the needed properties of the spectra are unchanged). Thus $Y(t) = \sum_0^{\alpha-1} W_k\; t^k/k!$. By induction on α, one may then prove that

$$\ker(d/dt - (A/\nu_0-\mu))^{\alpha} = \{\;X(t) \in E/X(t)=e^{-\mu t}\Phi(t)\sum_0^{\alpha-1} W_k\;t^k/k!\;/$$

$$W_k \in \ker((\Phi(2\pi)-e^{2\pi\mu}\;I)^{\alpha-k}),\, W_k \text{ uniquely defined by } W_0.\}$$

Thus, the Fredholm operator $d/dt - A/\nu_0$ has the same algebraic multiplicity at μ as $\Phi(2\pi)$ at the Floquet multiplier $e^{2\pi\mu}$.

Now (8) has a non-trivial solution if and only if $\lambda X^{0\prime}$ belongs to the range of $d/dt - A/\nu_0$. But, from the fact that 1 is a simple eigenvalue of $\Phi(2\pi)$, the kernel of $d/dt - A/\nu_0$, which is generated by $X^{0\prime}$, complements this range, hence $\lambda = 0$. Since $U + V$ must be L^2 - orthogonal to $\vec{M}$ sinjt, but $X^{0\prime}$ is not, we must have $U + V = 0$, i. e. on $E^0 \times \mathbb{R}$ the linear operator defined by (7) is invertible. We shall compute the index of the zero solution by adding to (7) the compact perturbation $\mu\,K(U + V)$, $\mu \geqslant 0$. This corresponds to adding to (8) the term $\mu(U + V)$. Since the index will remain constant as long as

$$(d/dt - (A/\nu_0 - \mu))\;X + \lambda\;X^{0\prime} \qquad (9)$$

has, in $E^0 \times \mathbb{R}$, only the (0,0) solution, we shall study the zeros of (9). If $Y = X + \lambda\;X^{0\prime}/\mu$, then

$$(d/dt - (A/\nu_0 - \mu))\ Y = 0 \qquad (10)$$

This equation has a non-trivial solution only if $e^{2\pi\mu}$ is a Floquet multiplier. If $Y = 0$ and X is in E^0, then $\lambda(X^{0'}, \vec{M}\ \sin jt) = 0$, thus $\lambda = 0$, and $X = 0$. Now, at such an eigenvalue the change of index of (10) (and thus of (9) and (8), as μ increases is $(-1)^n$, n the dimension of the generalized kernel (see [I_0 p. 79]). One has Index (8) = $(-1)^\sigma$ Index (10), μ large), σ as above. Now, since $Y^TAY = Y^T(A+A^T)Y/2$, multiplying (10) by Y^T and integrating over 2π, the equation

$$Y' + \tau\ (\mu - A/\nu_0)\ Y = 0$$

has only the 0 solution for μ large and $\tau > 0$. Thus, at the level of the integral operator, one may deform the corresponding operator to I. Index (8) = $(-1)^\sigma$.
(Some linear algebra is needed for the multiplicity result, see appendix)

Q. E. D.

3. COVERING DIMENSION AND MAIN RESULT

In this section we give the main result of this paper (see Theorem 3.1 below). This result tells us that if a given map g is Γ-epi on $S \cap U$ then, under some extra assumptions on the map g, we obtain the existence of a minimal subset of the orbit space having certain nice properties related not only to its global structure, but, also, to its covering dimension at each point.

It is well-known that the covering dimension of a normal space X can be characterized as follows (see [P, page 123]).

Definition 3.1. Let X be a normal space. Then the covering dimension of X, denoted $\dim X$, is less than or equal to $n \in \mathbb{N}$ if and only if for each closed subset A of X any map $g: A \to S^n \subset \mathbb{R}^{n+1}$ has a continuous extension over X.

We have chosen this definition of the covering dimension because it is closer in spirit to the techniques and ideas of this paper.

In the sequel we shall use the following notation. For a given invariant map $g: U \to G$ i.e. $g(\gamma x) = g(x), \gamma \in \Gamma$, denote by $\tilde{g}: U/\Gamma \to G$ the (unique) map that makes the following diagramn commutative

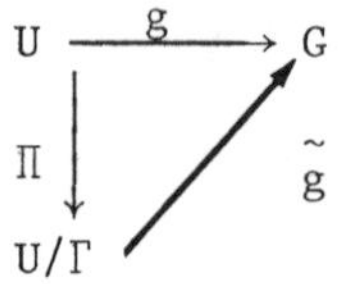

i.e., $\tilde{g}([x]) = g(\gamma x)$, $\gamma \in \Gamma$. Notice that $\tilde{g}$ can be written as $\tilde{g} = g \circ \Pi^{-1}$ which is single-valued since g is constant on every orbit (by the invariance of g).

Definition 3.2. A map $g: U \to G$ is called zero-epi invariant on $S \cap U$ if it is invariant and the equation $g(x)=h(x)$ has a solution in $S \cap U$ for any compact map $h:E \to G$ which is invariant with bounded support contained in U.

The following lemma gives a relation between g and $\tilde{g}$.

Lemma 3.1. An invariant map $g:U \to G$ is zero-epi on $S \cap U$ if and only if $\tilde{g}: U/\Gamma \to G$ is zero-epi on $(S/\Gamma)\cap(U/\Gamma)$.

Proof. (Only if). Let $\tilde{h}:E/\Gamma \to G$ be compact with bounded support contained in U/Γ. Let $h:E \to G$ be defined by $h(x) = \tilde{h}(\Pi(x))$. Clearly, h is compact invariant with bounded support contained in U (recall that the action is an isometry). Then the equation $g(x)=h(x)$ has a solution $\bar{x} \in S \cap U$. Thus $\tilde{x} = \Pi(\bar{x})$ is a solution of $\tilde{g}(x)=\tilde{h}(x)$.

(If). Let $h:E \to G$ be compact invariant with bounded support contained in U. Now, the map $\tilde{h}:E/\Gamma \to G$ defined by $\tilde{h}=h\cdot\Pi^{-1}$ is compact with bounded support contained in U/Γ . Therefore, the equation $\tilde{g}(x)=\tilde{h}(x)$ has a solution $\tilde{x} \in (S/\Gamma)\cap(U/\Gamma)$. Let $\bar{x} \in \tilde{x}$ be arbitrary. Clearly, $g(\bar{x})= h(\bar{x})$ since g and h are invariant.

Q.E.D.

Remark 3.1. On the basis of Lemma 3.1 we may extend all of the results contained in [I.M.P.V,1] and [I.M.P.V,2] to the context of Γ-epi maps. This can be done simply by rewriting them in terms of $\tilde{g}$, S/Γ and U/Γ (Cfr. Remark 1.1).

Of particular interest for our purposes is the following result which may be regarded as a completion of Theorem 1.1

Theorem 3.1. Let $g:U \to G$ and Σ be as in Theorem 1.1. Assume further that $G = G^\Gamma \oplus G_2$, where G_2 is such that $G_2^\Gamma = \{0\}$. Then, setting $g = (g^\Gamma, g_2)$, there exists an invariant minimal subset $\tilde{\Sigma}$ contained in $g_2^{-1}(o)\cap\Sigma$ such that

(i) The map g^Γ is zero-epi invariant on $\tilde{\Sigma} \cap U$. This implies in particular that $\tilde{\Sigma}$ is either unbounded, or $\tilde{\Sigma} \cap \partial U \neq \emptyset$ (provided $G^\Gamma \neq \{0\}$).

(ii) If $\tilde{\Sigma}/\Gamma = \Sigma_1 \cup \Sigma_2$ with Σ_1, Σ_2 closed and proper subsets of $\tilde{\Sigma}/\Gamma$, then $\dim (\Sigma_1 \cap \Sigma_2) \geq \dim G^\Gamma - 1$. This implies in particular that $\tilde{\Sigma}/\Gamma$ is connected and has dimension at each point at least $\dim G^\Gamma$.

(iii) The set $\tilde{\Sigma}/\Gamma$ is minimal for any map g_1 (invariantly) homotopic to g^Γ.

Proof. Since $g=(g^\Gamma, g_2)$ is Γ-epi on $\Sigma \cap U$, then by Property 1.5 the map g^Γ is Γ-epi on $(g_2^{-1}(0)\cap\Sigma)\cap U$. Now, g^Γ is invariant and, since the action on G^Γ is trivial, then g^Γ is zero-epi invariant. By Theorem 1.1 there exits a minimal closed subset $\tilde{\Sigma}$ with the properties listed in that theorem. This proves part (i).

By Lemma 3.1 the map g^Γ is such that $\tilde{g}^\Gamma$ is zero-epi on

$(\tilde{\Sigma}/\Gamma)\cap(U/\Gamma)$. Moreover, being $\tilde{\Sigma}$ minimal, so is $\tilde{\Sigma}/\Gamma$. In this situation Theorem 3.1 of [I.M.P.V,1] applies proving part (ii) and (iii) of our theorem.

Q.E.D.

Remark 3.2. One may not expect a result better than the above. In fact, if $G^\Gamma = \{0\}$ and $S = \{0\}\subset U$, then any equivariant map $g:U \to G$ is Γ-epi on $S\cap U$. But, $S/\Gamma = \{0\}$ (cfr. Remark 1.4).

Remark 3.3. The map g^Γ may be Γ-epi on $g_2^{-1}(0)\cap S\cap U$ and not Γ-epi on $S\cap U$. To see this consider the following example. Let $E = \mathbb{C}\times\mathbb{C}$ and $G = \mathbb{C}\times\mathbb{R}$ with actions $(z,\lambda) \longrightarrow (e^{i\varphi}z,\lambda)$ and $(z',x) \longrightarrow (e^{i\varphi}z',x)$ respectively. Let $U = \{(z,\lambda)\in\mathbb{C}\times\mathbb{C}: \varepsilon<|z|\}$ and let $S = \mathbb{C}\times\mathbb{C}$. Let $g:U \to G$ be defined by $g(z,\lambda) = (|\lambda|z, |z|^2-4\varepsilon^2)$. The map g is not Γ-epi on U since the equations

$$|\lambda|z + \varphi(|z| + |\lambda|)\, z = 0$$
$$|z|^2 - 4\varepsilon^2 = 0$$

where φ is as Example 1.1 of Part 1, has no solutions. On the other hand $G^\Gamma = \{0\}\times\mathbb{R}$ and, setting $g = (g_2,g^\Gamma)$, we have that $g^\Gamma(z,\lambda) = |z|^2 - 4\varepsilon^2$. Therefore, $g_2^{-1}(0)\cap U = \{(z,\lambda)\in\mathbb{C}\times\mathbb{C} : \lambda=0\}$. Now, let $h:E \to G^\Gamma$ be an invariant compact map with bounded support contained in U. To solve the equation $g^\Gamma(z,\lambda) = h(z,\lambda)$ in $g_2^{-1}(0)\cap U$ amounts to solve the following equation $r^2 - 4\varepsilon^2 = h(r,0)$ with $r>\varepsilon$. Since $h(r,0)$ has bounded support contained in $\{r>\varepsilon\}$, then the above equation is solvable for some $r>\varepsilon$.

4. GLOBAL CONTINUATION AND BIFURCATION PROBLEMS FOR Γ-EPI MAPS.

4.1 Global Continuation. Let E,F be Banach spaces and let Γ be a compact Lie group. Assume that E^Γ and F^Γ admit closed invariant complements E_1 and F_1 respectively (which is always the case in the Hilbert space setting).

Let Λ be a finite dimensional space (the parameter space) and let $f:E\times\Lambda \to F$ be a continuous map which can be written as $f=(f_1,f_0)$ where $f_1:E_1\times E^\Gamma\times\Lambda \to F_1$ and $f_0:E_1\times E^\Gamma\times\Lambda \to F^\Gamma$ are defined by $f_1(x_1,x_0,\lambda) = L_1x_1 - k_1(x_1,x_0,\lambda)$ and $f_0(x_1,x_0,\lambda) = L_0x_0 - k_0(x_1,x_0,\lambda)$ with $L_1\gamma x_1 = \gamma L_1x_1$, $k_1(\gamma x_1, x_0,\lambda) = \gamma\, k_1(x_1,x_0,\lambda)$ and $k_0(\gamma x_1,x_0,\lambda) = k_0(x_1,x_0,\lambda)$, $\gamma\in\Gamma$. (If L_1 is invertible, then E_1 and F_1 are two equivalent representations of Γ, thus $\rho(\gamma) = \tilde{\rho}(\gamma)$).

We shall make the following assumptions:

(1) L_i, is a bounded linear operator which is proper on $\bar{B}_i$, where $B_i = \{x_i\in E_i: \|x_i\| < r_i\}$, $i=0,1$ (here we identify E_0 with E^Γ). This is the case if L_i is a Fredholm operator of non negative index.

(2) k_i is compact, $i=0,1$.

(3) $f(x_1,x_0,0)\neq 0$ on $\partial(B_1\times B_0)$

(4) $f^{-1}(0)\cap(((B_1\times B_0)\setminus\{x\in E : x_1=0\})\times\{0\})$ is a closed set.

Under the above assumptions we have the following

Theorem 4.1. Let $(f(x_1,x_0,0),\lambda)$ *be* Γ*-epi on* $(B_1 \times B_2)\setminus A$, *where* $B_2 = \{(x_0,\lambda)\in E_0\times\Lambda : \|x_0\| < r_0,\ \|\lambda\| < r_2\}$ *and* $A = \{(x_1,x_0,\lambda)\in E_1\times E_0\times\Lambda : x_1 = 0,$ *or* $\|x_1\| \geqslant r_1,\ \lambda=0,$ *or* $\|x_0\| \geqslant r_0,\ \lambda=0\}$. *Put* $S=f^{-1}(0)$. *Then the set* $S\setminus A$ *has a minimal closed invariant set* Σ *such that the map* λ *(the projection onto the parameter space* Λ*) is* Γ*-epi on* $\Sigma\cap((E\times\Lambda)\setminus A)$ *with the following properties:*

(a) Σ *intersects the fiber* $\{\lambda=0\}$ *for* $\|x_0\| < r_0,\ 0<\|x_1\| <r_1$ *and, if* $f(x_1,x_0,\lambda)\neq 0$ *on* $\partial B_1 \times \bar{B}_2$ *and* $f^{-1}(0)\cap((B_1\times B_2)\setminus A)$ *is closed, then* Σ *intersects the fiber* $\{\lambda=\lambda_0\}$ *for* $\|x_0\| <r_0,\ 0<\|x_1\| <r_1$, *for any* λ_0 *with* $\|\lambda_0\| <r_2$.

(b) Σ *is either unbounded or* $\bar{\Sigma}\cap A \neq \emptyset$.

(c) *If* $\Sigma/\Gamma = \Sigma_1 \cup \Sigma_2$, *with* Σ_1,Σ_2 *proper subsets of* Σ/Γ, *then* $\dim(\bar{\Sigma}_1\cap\bar{\Sigma}_2) \geqslant \dim\Lambda-1$. *In particular,* Σ/Γ *is connected and it has dimension at each point at least* $\dim\Lambda$.

(d) *If* $\bar{\Sigma}\cap A = \emptyset$ *and* $\bar{\Sigma}\cap(E\times[0,\lambda_0])$ *is bounded for any* $\lambda_0\in\Lambda$, *then* $\bar{\Sigma}$ *covers* Λ.

Proof. Note first that the pair (L_1,L_0) is proper on closed and bounded subsets of E. Hence, from the assumptions (1) and (2) above, the map f is proper on closed and bounded subsets of $E\times\Lambda$. In particular the intersection of S with any closed bounded subset of $E\times\Lambda$ is a compact set. This fact combined with the above assumptions (3) and (4) above yield the existence of $\varepsilon>0$ such that the following inclusion holds $(f(x_1,x_0,\lambda),\lambda)^{-1}(0) \subset (B_1^{\varepsilon}\times B_2^{\varepsilon})\cup A$. Consider now the following homotopy $h:((B_1\times B_2)\setminus A)\times[0,1] \longrightarrow F\times\Lambda$ defined by $h(x_1,x_0,\lambda;t) = (f(x_1,x_0,t\lambda),\lambda)$. Clearly, this homotopy is admissible and thus the map (f,λ) is Γ-epi on $(B_1\times B_2)\setminus A$. From the scaling property (Property 1.7) we obtain that (f,λ) is Γ-epi on $E\setminus A$. Therefore, from Property 1.5, we obtain that the projection λ onto the parameter space Λ is zero-epi (invariant) on $S\cap(E\setminus A)$. We are now in a position of applying Theorem 1.1 and Theorem 3.1 with $G=G^{\Gamma}=\Lambda$ and the map $g=g^{\Gamma}=\lambda$, to obtain (a), (b) and (c) of our assertion except the second part of (a) and property (d). Let now $\varphi:\mathbb{R}^+ \longrightarrow \mathbb{R}^+$ be a nonincreasing function with $\varphi(r) = 1$, if $0\leqslant r\leqslant r_1-\varepsilon$ and $\varphi(r) = 0$, if $r\geqslant r_1$, where $\varepsilon>0$ is such that $f(x_1,x_0,\lambda)\neq 0$ on $(B_1\setminus B_1^{\varepsilon})\times\bar{B}_2$. Let λ_0 be such that $|\lambda_0|<r_2$ and consider the following homotopy $h(x,\lambda;t) =\lambda - t\varphi(\|x_1\|)\lambda_0$. Clearly, this homotopy is admissible so that the map $\lambda-\varphi(\|x_1\|)\lambda_0$ is zero-epi (invariant) on Σ and thus, from Property 1.1, we obtain that the set Σ must intersect the set of zeros of the map $\lambda-\varphi(\|x_1\|)\lambda_0$ restricted to $(E\times\Lambda)\setminus(A\cap\Sigma)$, hence in B_1^{ε}, where $\varphi(\|x_1\|) = 1$. This proves the second part of (a). A similar argument yields property (d) of our assertion.

Q.E.D.

4.2 *Some examples and sufficient conditions for* S^1 *- maps.* In this section we shall give a sample of the possibilities of the use of the

S^1 - degree. Let $E_1 = F_1$, $E_0 = F_0 \times R^{k-\ell}$, with $k-\ell$ odd; $x_0 = (\tilde{x}_0, \nu)$. ν in $\mathbb{R}^{k-\ell}$; $L_1 = I$, $L_0 x_0 = \tilde{x}_0$. k_i is compact, $i=1,0$. $f(x_1, x_0, 0) \neq 0$ on $\partial(B_1 \times B_0)$. $f_0(0, x_0, 0) \neq 0$, for $\| x_0 \| < r_0$.

Proposition 4.1. Assume $\deg^n_{S^1}(f(x_1,x_0,0)) \neq 0$, *then* $(f(x_1,x_0,0),\lambda)$ *is* S^1 - *epi on* $B_1 \times B_0 \setminus A$ *and one has all the properties of Theorem 4.1.*

Proof: From part 2, one has that $(F, F_0)(x_1,x_0,0)$ is S^1 - epi on $S^\infty \times \{r/0<r<r_1\} \times B_0$, hence $f(x_1,x_0,0)$ is S^1 - epi on $B_1 \times B_0 \setminus A$. Now, from Remark 2.3, λ acts as a suspension and does not alter the degree.

Q.E.D.

Proposition 4.2. Under the above hypothesis if f is a gradient mapping with non-zero Dancer's degree, or if f corresponds to a vector field with a non-zero Fuller's degree, then one has global continuation in the sense of Theorem 4.1.

Proof: It is enough to apply the equivalences given in sections 2.5 and 2.6. In case of the gradient map, one has to replace equation (2), with $F(Z,X)$ of the form $Z + k_1(z,x,\lambda)$, by $Z + (\nu A + I)^{-1} k_1(Z,X,\lambda)$, in order to satisfy conditions (1) and (2) of this section.

Q.E.D.

Remark 4.1. It is clear that one may relax the conditions of the above propositions, and that one may give global implicit function results in the spirit of [I.M.P.V.1].

4.3 Global bifurcation. Assume now that $\Lambda = \tilde{\Lambda} \times \tilde{\tilde{\Lambda}}$, and let $f: E \times \Lambda \to F$ satisfies the properties (1) and (2) listed in the introduction to this part of the paper.

Assume further that (3') $k_0(0,0,\lambda) = 0$ for all $\lambda \in \Lambda$

(4') If $f(x_1,x_0,\tilde{\lambda},0) = 0$ for $\|x_1\| \leqslant r_1$ and

$(\|x_0\| \leqslant r_0 + \varepsilon,\ r_2 - \varepsilon \leqslant \|\lambda\| \leqslant r_2 + \varepsilon)$, or

$(\|\lambda\| \leqslant r_2 + \varepsilon,\ r_0 - \varepsilon \leqslant \|x_0\| \leqslant r_0 + \varepsilon)$, then $x_1 = 0$.

We will use the following notation in Theorem 4.2 below

$$B_0 = \{x_0 \in E_0 = E^\Gamma : \|x_0\| < r_0\}$$

$$B_1 = \{x_1 \in E_1 : \|x\| < r_1\},\ B_2 = \{\lambda \in \Lambda : \|\lambda\| < r_2\}$$

Under the above assumptions we have the following

Theorem 4.2. Let $(f(x_1,x_0,\tilde{\lambda},0),\tilde{\tilde{\lambda}}, \|x_2\| - \frac{r_1}{2})$ *be* Γ - *epi on* $(B_1 \times B_0 \times B_2) \setminus \{(x_1,x_0,\lambda) : x_1 = 0\}$. *Then, if* $S = \{(x_1,x_0,\lambda) \in E \times \Lambda : f(x_1,x_0,\lambda) = 0,\ x_1 \neq 0\}$, *we have that S has a minimal closed (in S) invariant subset* Σ *such that the map* $(\tilde{\tilde{\lambda}}, \|x_1\| - \varphi(\|x_0\|, \|\lambda\|))$ *is* Γ - *epi (invariant) on* Σ, *where* $\varphi : \mathbb{R}^+ \times \mathbb{R}^+ \to [0, \frac{r_1}{2}]$ *is a continuous function such that* $\varphi(s,t) = \frac{r_1}{2}$, if $0 \leq s \leq r_0$ and $0 \leq t \leq r_2$, $\varphi(s,t) = 0$, if $s \geq r_0 + \varepsilon$, or if $t \geq r_2 + \varepsilon$. Moreover, the set Σ has the following properties:

(a) $(\Sigma \cap \{(x_1,x_0,\tilde{\lambda},0)\}/\Gamma$ intersects $\{0\} \times B_0^\varepsilon \times B_2^\varepsilon$ and contains a closed connected subset, having local dimension at least one, which is either unbounded, or intersects $\{0\} \times E_0 \times \Lambda$ outside $\{0\} \times B_0 \times B_2$.

(b) If $\Sigma/\Gamma = \Sigma_1 \cup \Sigma_2$, with Σ_1, Σ_2 proper closed subsets of Σ/Γ, then $\dim(\Sigma_1 \cap \Sigma_2) \geq \dim \tilde{\tilde{\Lambda}}$. In particular, Σ/Γ is connected and it has dimension at each point at least $\dim \tilde{\tilde{\Lambda}} + 1$.

Proof. Set $A = \{0\} \times E_0 \times \Lambda$. As in the proof of Theorem 4.1 the intersection of S with bounded and closed subsets of $E \times \Lambda$ is compact. Moreover, the zeros of the map $(f(x_1,x_0,\lambda),\tilde{\tilde{\lambda}},\|x_1\| - \varphi(\|x_0\|,\|\lambda\|))$ are the zeros of the map $f(x_1,x_0,\tilde{\lambda},0)$ with $\|x_1\| = \frac{r_1}{2}$, if $(x_0,\lambda) \in B_0^\varepsilon \times B_2^\varepsilon$ and the zeros of $f(0,x_0,\tilde{\lambda},0)$, if $\|x_0\| \geq r_0 + \varepsilon$, or $\|\lambda\| \geq r_2 + \varepsilon$ (which is contained in A). Thus, as in the proof of Theorem 4.1, the above map is Γ-epi on $(B_1 \times B_0 \times B_2)\backslash A$ and, from the scaling property, we have that this map is Γ-epi on $(E \times \Lambda)\backslash A$. Finally, from Property 1.5, it follows that the map $(\tilde{\tilde{\lambda}}, \|x_1\| - \varphi(\|x_0\|,\|\lambda\|))$ is Γ-epi on the set S. Property (b) of our assertion now follows from (ii) of Theorem 3.1.

Finally, by Property 1.5, the map $(\|x_1\| - \varphi(\|x_0\|,\|\lambda\|)$ is zero-epi on $(\Sigma/\Gamma) \cap \{(x,\tilde{\lambda},\tilde{\tilde{\lambda}}):\tilde{\tilde{\lambda}} = 0\}$. Therefore, by Theorem 3.1, one gets a minimal subset Σ_1 which is closed, connected with local dimension at least one and such that the map $\|x_1\| - \varphi(\|x_0\|,\|\lambda\|)$ is zero-epi on Σ_1. Moreover, Σ_1 intersects $\{0\} \times B_0^\varepsilon \times B_2^\varepsilon$ and if Σ_1 is bounded, then it intersects $\{0\} \times E_0 \times \Lambda$ outside $\{0\} \times B_0 \times B_2$. Indeed, assume the contrary, i.e.,

$$\text{let } d = \mathrm{dist}\,(\bar{\Sigma}_1, (\{0\} \times E_0 \times \Lambda)\backslash(\{0\} \times B_0 \times B_2)) > 0.$$

Let V be the intersection of a bounded open neighborhood of $\bar{\Sigma}_1$ with the set $\{(x_1,x_0,\lambda) : \|x_2\| > \frac{1}{2}\min(d,\frac{r_1}{2})\}$. By Property 1.2 the map $\|x_1\| - \varphi(\|x_0\|,\|\lambda\|)$ is zero-epi on $\Sigma_1 \cap V$. Then, by Corollary 3.1 of [I.M.P.V.1] the above map must change sign on $\Sigma_1 \cap \partial V$. But on this set $\|x_1\| - \varphi(\|x_0\|,\|\lambda\|)$ is negative. A contradiction.

Q.E.D.

We shall now give several conditions under which the map f satisfies the hypothesis of Theorem 4.2, all derived from the following result.

Lemma 4.1. Assume f satisfies properties (1), (2) and also

(a) $f_0\ (0,x_0,\tilde{\lambda},0) \neq 0$ on $\partial B_0 \times \bar{B}_2$,

(b) $f_1\ (x_1,x_0,\tilde{\lambda},0) \neq 0$ on $(B_1\backslash\{0\}) \times B_0 \times \{\lambda/r_2 - \varepsilon_2 \leq \|\lambda\| \leq r_2 + \varepsilon_2\}$

Then, for each ε_1, $0 < 2\varepsilon_1 \leq r_1$, small enough, there is an ε such that hypothesis (4') is satisfied, with r_1 replaced by $2\varepsilon_1$. The map $(f(x_1,x_0,\lambda),\tilde{\tilde{\lambda}},\|x_1\| - \varepsilon_1)$ is admissible on $B_1 \times B_0 \times B_2$ and on

$(B_1\setminus\{0\}) \times B_0 \times B_2$. Furthermore this map is Γ- deformable on the above sets to $(f_1(x_1,0,\tilde{\lambda},0),\ f_0(0,x_0,\lambda_0,0),\ \tilde{\tilde{\lambda}},\ \|x_1\| - \varepsilon_1)$, for any given λ_0 in $\bar{B}_2$.

Proof: Since $f^{-1}(0)$ intersects closed and bounded sets as compact sets then, from (a), $f^{-1}(0)$ is at a positive distance of $\partial B_0 \times \bar{B}_2$. Hence $f(x_1,x_0,\tilde{\lambda},0) \neq 0$ for $\{(x_1,x_0,\tilde{\lambda})/\|x_1\| \leq 2\varepsilon_1, r_0 - \varepsilon_0 \quad \|x_0\| \leq r_0 + \varepsilon_0, \|\lambda\| \leq r_2 + \varepsilon_0\}$, for some $\varepsilon_0, \varepsilon_1$ small enough. Taking $\varepsilon = \min(\varepsilon_0, \varepsilon_2)$, one gets hypothesis (4') for $\|x_1\| \leq 2\varepsilon_1$. Admissibility is then clear. Perform the compact and equivariant deformation: $(f_1(x_1,\tau x_0,\tilde{\lambda},\tau\tilde{\tilde{\lambda}}),\ f_0(\tau x_1,x_0,\tilde{\lambda},\tau\tilde{\tilde{\lambda}}),\ \tilde{\tilde{\lambda}},\ \|x_1\| - \varepsilon_1)$. Do next the homotopy $f_0(0,x_0,\tau\tilde{\lambda} + (1-\tau)\lambda_0,\ 0)$.

Q.E.D.

Corollary 4.1. If $f(x_1,x_0,\tilde{\lambda},0)$ satisfies (1), (2) and

(a') $f_0(0,x_0,\tilde{\lambda},0) \neq 0$ for $0 < \|x_0\| \leq r_0'$, $\|\lambda\| \leq r_2$ or

(a") $f_0(0,x_0,\tilde{\lambda},0) \neq 0$ for a sequence $\varepsilon_n \to 0$, $\|x_0\| = \varepsilon_n$, $\|\lambda\| \leq r_2$ (i.e. there is no bifurcation of stationary solutions)

(b') $f_1(x_1,x_0,\tilde{\lambda},0) = A_1(\tilde{\lambda})x_1 + k_1(x_1,x_0,\lambda,0)$, with $A_1(\tilde{\lambda})$ invertible, for $\|\lambda\| = r_2$, and continuous in the operator norm. $\|k_1(x_1,x_0,\tilde{\lambda},0)\| \leq c\|x_1\|(\|\lambda_1\|^{\alpha} + \|x_0\|^{\alpha})$, for some $\alpha > 0$, x_1, x_0 small.

Then f satisfies (a) and (b) for some set of constants r_0, r_1.

Proof: $A_1(\tilde{\lambda})$ is invertible for $r_2 - \varepsilon_2 \leq \|\tilde{\lambda}\| \leq r_2 + \varepsilon_2$. Property (b) follows from (b') if one chooses r_0, r_1 small enough.

Q.E.D.

Remark 4.2. If A is a linear Fredholm operator from E into F, then A has a block form (A_1, A_0), $\ker A = \ker A \cap E_1 \oplus \ker A \cap E_0$, Range A has a similar decomposition, and one may choose equivariant projections and have equivariant pseudo-inverses, see [V]. Furthermore if $A(\lambda) = A - T(\lambda)$ is a family of such operators, with $T(0) = 0$ and $T(\lambda)$ compact, then $A_i(\lambda) = A_i - T_i(\lambda)$ has spectral properties which are completely characterized by the spectral properties of a $d_i \times d_i$ matrix $B_i(\lambda)$, with $B_i(0) = 0$. $d_i = \dim\ker A_i$, $d_i^* = \operatorname{codim}$ Range A_i (in F_i), $i = 0, 1$. See [I.M.P.V.1, lemma 4.2.].

Proposition 4.3. Assume f satisfies (1), (2), (a') or (a"), (b') and that $A_1(\lambda) = A_1 - T_1(\lambda)$, A_1 a Fredholm operator of index 0, T_1 as in the above remark. Then, on $B_1 \times B_0 \times B_2$ or on $(B_1\setminus\{0\}) \times B_0 \times B_2$, the map $(f(x_1,x_0,\lambda),\tilde{\tilde{\lambda}},\|x_1\| - \varepsilon_1)$ is Γ-deformable to $(B_1(\tilde{\lambda},0)y_1, A_1 z_1,\ f_0(o,x_0,\lambda_0,o),\tilde{\lambda},\|y_1\| - \varepsilon_1)$, where $B_1(\lambda)$ is as in Remark 4.2, $x_1 = y_1 \oplus z_1$, y_1 in ker A.

Proof: After picking r_0, r_1 as in Corollary 4.1, ε_1 as in Lemma 4.1, one may deform $k_1(x_1,x_0,\lambda)$ to 0, $A(\lambda)$ to $\tilde{A}(\lambda,o)$ and to $(B_1(\tilde{\lambda},o)y_1, A_1 z_1)$ as in the proof of Proposition 4.5 in [I.M.P.V.1].

Q.E.D.

Proposition 4.4. If, in addition to the hypothesis of Proposition 4.3., $f_0(o,x_0,\lambda) = A_0(\lambda)x_0 - k_0(o,x_0,\lambda)$, with $\|k_0(o,x_0,\lambda)\| = o(\|x_0\|)$ and $A_0(\tilde{\lambda}_0)$ is invertible for some $\tilde{\lambda}_0$, then, for r_0 small enough, $(f(x_1,x_0,\lambda),\tilde{\tilde{\lambda}},\|x_1\| - \varepsilon_1)$ is Γ- epi on $(B_1\setminus\{0\}) \times B_0 \times B_2$ if and only if $(B_1(\tilde{\lambda},o)y_1,z_1,x_0,\tilde{\tilde{\lambda}}, \|y_1\| - \varepsilon_1)$ is Γ - epi.

Proof: For r_0 small enough, $f_0(o,x_0,\tilde{\lambda}_0,o)$ is deformable, for $\|x_0\| = r_0$, to $A_0(\tilde{\lambda}_0)x_0$. Since A_1, on the complement of ker A_1, and $A_0(\tilde{\lambda}_0)$ are isomorphisms onto Range A_1 and F_0 respectively, one gets the assertion.

Q.E.D.

Conditions (a') or (a") may not be easy to verify. Propositions 4.3 and 4.4 say in essence that if there is no local bifurcation of stationary solutions, then there is global bifurcation of non-stationary solutions. One may however guarantee the existence of global branches, even if one cannot be sure of the nature of the solutions, via the following result.

Theorem 4.3. Assume f satisfies (1), (2), (3') and (4"): if $f(x_1,x_0,\tilde{\lambda},o) = 0$ for $\|x_i\| \leq r_i$, $i=0,1,r_2 - \varepsilon_2 \leq \|\lambda\| \leq r_2 + \varepsilon_2$, then $(x_1,x_0) = (0,0)$.

Let $(f(x_1,x_0,\lambda),\tilde{\tilde{\lambda}},\|x_1\| + \|x_0\| - \varepsilon_1)$ be Γ - epi on $B_1 \times B_0 \times B_2 \setminus \{x_1 = x_0 = 0\}$. Take $S = \{x_1,x_0,\lambda) / f(x_1,x_0,\lambda) = 0, (x_1,x_0) \neq (0,0)\}$. Then S has a minimal closed (in S) invariant subset Σ such that the map $(\tilde{\tilde{\lambda}},\|x_0\|+\|x_1\| - \Phi(\|\lambda\|)$ is Γ - epi (invariant) on Σ, where $\Phi:\mathbb{R}^+ \to \mathbb{R}^+$ is a non-increasing function, with $\Phi(r) = \varepsilon_1$ for $0 \leq r \leq r_0$, $\Phi(r) = 0$ for $r \geq r_0 + \varepsilon$. Moreover Σ has properties (a) and (b) of theorem 4.2.

Proof. The proof is similar to that of Theorem 4.2.

Q.E.D.

As before, we shall give conditions under which the hypotheses of Theorem 4.3 are met. Note that if $(f_0(o,x_0,\lambda), \tilde{\tilde{\lambda}},\|x_0\| - \varepsilon_1)$ is 0-epi on $(B_0\setminus\{0\}) \times B_2$, then, by property 1.3 (c), $(f,\tilde{\tilde{\lambda}},\|x_1\| + \|x_0\| - \varepsilon_1)$ is Γ - epi on $(B_1 \times B_0 \setminus \{0,0\}) \times B_2$ and no new information is obtained.

Lemma 4.2. Assume f satisfies (1),(2),(3),(b') of Corollary 4.1. Assume that $f_0(o,x_0,\tilde{\lambda},o) \neq 0$ on $B_0 \setminus \{0\}$ and for $r_2 - \varepsilon_2 \leq \|\lambda\| \leq r_2 + \varepsilon_2$. Suppose that f_0 satisfies (c): $f_0(o,x_0,\tilde{\lambda},o)$ is not 0-epi on $\{\|x_0\| = \varepsilon_1\}$ $(B_0\setminus\{0\}) \times B_2$, for all ε_1 with $0 < \varepsilon_1 \lesssim \bar{\varepsilon}_1 < \min(r_0, r_1)$, for some ε_1. Then $(f(x_1,x_0,\lambda),\lambda, \|x_1\| + \|x_0\| - \varepsilon_1)$ is admissible on $(B_1 \times B_0 \setminus \{0,0\}) \times B_2$ and is Γ - deformable to $(B_1(\tilde{\lambda},o)y_1,A_1z_1,f_0(o,x_0,\tilde{\lambda}_0), \lambda, \|x_1\| - \varepsilon_1)$ for any $\tilde{\lambda}_0$ with $\|\lambda_0\| = r_2$. In particular, if $f_0(o,x_0,\lambda,o)$ has the form given in

Proposition 4.4, then the first map is Γ - epi if and only if $(B_1(\lambda,0)y_1, z_1, x_0, \tilde{\lambda}, \|x_1\| - \varepsilon_1)$ is Γ-epi and, thus Theorem 4.3 may be applied.

Proof: For admissibility, the only thing to check is the localization of the zeros for $\|\lambda\| = r_2$: if $f(x_1,x_0,\tilde{\lambda}) = 0$, with $\|x_1\| + \|x_0\| = \varepsilon_1$, ε_1 small enough, from (b') one has that $x_1 = 0$, and from the hypothesis on $f_0, x_0 = 0$ which leads to a contradiction. Furthermore the deformations of x_0 to 0 in f_1, x_1 to 0 in f_0, λ to $\tilde{\lambda}$ in both, $f_1(x_1,0,\tilde{\lambda},0)$ to $(B_1(\tilde{\lambda},0)y_1, A_1z_1)$ are all admissible. Now, from (c), there is a compact map $h(x_0,\tilde{\lambda}):E_0 \times \tilde{\Lambda} \to F_0$, with support contained in $(B_0\setminus\{0\}) \times B_2$, and $f_0(0,x_0,\tilde{\lambda}) - h(x_0,\tilde{\lambda}) \neq 0$, if $\|x_0\| = \varepsilon_1$. Perform then the homotopy $f_0(0,x_0,\lambda) - \tau h(x_0,\tilde{\lambda})$ (again the only point to check is for $\|\lambda\| = r_2$, where h is 0). Then, since for $\|x_0\| = \varepsilon_1$ this last map is never 0, one may do the homotopy $f_0(0,x_0,\tau\tilde{\lambda} + (1-\tau)\tilde{\lambda}_0) - h(x_0, \tau\tilde{\lambda} + (1-\tau)\lambda_0)$. Noting that for $\tau = 1$, $h(x_0,\tilde{\lambda}_0) = 0$, and that $f_0(0,x_0,\tilde{\lambda}_0) = 0$ implies $x_0 = 0$, one may deform $\|x_0\|$ to 0 in the last member of the equation.

Q.E.D.

Remark 4.3. The only fact used for E_0 is that, since E_0 is the fixed point subspace of Γ, one has $f_1(0,x_0,\lambda) = 0$ (the estimate for k_1 follows if k_1 is small and smooth). Hence the same argument works if E_0 is replaced by any fixed point subspace of H, an isotropy subgroup of Γ. This will be used extensively in a subsequent paper in order to analyse more carefully when $(f(x,\lambda), \|x\| - \varepsilon)$ is Γ - epi, in particular in order to get an equivariant version of the J - homomorphism.

4.4 Global Bifurcation for S^1 - maps. In this section we shall give a brief illustration of our abstract results in the case of S^1 - maps. As in part 2, if A_0, A_1 are Fredholm operators of index 0, then the bifurcation equation, in the respective kernels, have a natural splitting of the form

$$B_j(\tilde{\lambda},0)y_j + g_j(\tilde{\lambda},y) \quad j=0,1,\ldots,m, \; y_j \in \mathbb{C}^{m_j} \tag{11}$$

The action on y_j is of the form $e^{in_j\phi}$. Taking the conjugates of the equations for negative n_j's, we order the n_j with $\underset{\sim}{0} = n_0 < n_1 < n_2 < \ldots < n_m$. Assume $B_j(\tilde{\lambda},0)$ is invertible for $\tilde{\lambda} \neq 0$, j=1, ..., m. (Thus hypothesis (b') is satisfied if the original map is smooth enough). Let $k = \dim \tilde{\Lambda}$. Then each $B_j(\tilde{\lambda},0)$, for $\|\tilde{\lambda}\| = r_2$, defines an element of $\Pi_{k-1}(GL(\mathbb{C}^{m_j}))$ and, stably, an integer $N_j(B_j(\tilde{\lambda},0))$ in $\pi_{k-1}(U)$ (see [I] for details).

Assume one of the N_j's is non-zero (true if $\Sigma N_j/n_j^{k/2} \neq 0$). Note that $N_j \neq 0$, then k is even and for $k = 2$, $N_j \neq 0$ if and only if $\det B_j(\tilde{\lambda},0): \{\tilde{\lambda}/\|\tilde{\lambda}\| = r_2\} = S^1 \to \mathbb{C} \setminus \{0\}$ has a non-zero winding number.

Proposition 4.5. Assume one of the N_j's is non-zero and that

$f_0(o,x_0,\tilde{\lambda},o)$ satisfies (a') or (a") of Corollary 4.1 (for example if $A_0(0)$ is invertible, thus there is no bifurcation of stationary solutions). Assume also that: (d) either $A_0(\tilde{\lambda}_0)$ is invertible, for some $\tilde{\lambda}_0$ in B_2, or $A_0(o)x_0 - k_0(o,x_0,o)$ has a non-zero Leray-Schauder degree on B_0 (well defined by (a') or (a")). Then one has global bifurcation of non-stationary solutions with the properties of Theorem 4.2. Thus if $N_j \neq 0$ for some j and f_0 satisfies (d), then one has either local bifurcation of stationary solutions (i.e. a solution $(o,x_0,\tilde{\lambda},o)$ with $\|x_0\| = \varepsilon$, for all ε small enough) or the above global result for nonstationary solutions. On the other hand, if $A_i(\tilde{\lambda},o)$ are invertible for $\|\tilde{\lambda}\| \neq 0$, $\tilde{\lambda}$ small and $k < m_0$, if $N_j^i \neq 0$ for some j or if $B_0(\tilde{\lambda},0)$ has a non-trivial stable class in $\Pi^j_{k-1}(GL(\mathbb{R}^{m_0}))$, then one has the global result of Theorem 4.3.

Proof. The first part of the proposition follows from § 2 and [I] after noting that in the obstruction (either using the isotropy subspaces as in [I, Remark II - 5-3] or using the trick of part 2) the terms z_0, $\tilde{\tilde{\lambda}}$, $\tilde{z}_1$ act as a suspension for the usual Brouwer degree. The term $B_0(\tilde{\lambda},o)y_0 - g_0(o,x_0,\tilde{\lambda})$ can be deformed to $B_0(\tilde{\lambda}_0)y_0$, with index ± 1, or $k_0 - g_0(o,x_0,o)$: if $B_0(\tilde{\lambda})$ is always singular, then take $\lambda_0 = 0$ in Proposition 4.3. The new obstruction is the old one multiplied by the degree of $B_0(\tilde{\lambda}_0)y_0$ or of $-g_0(o,x_0,o)$. The second part of the proposition follows from Lemma 4.2, after noting that if $k<m_0$, then one is in the stable range and the triviality of the class of $B_0(\tilde{\lambda},o)$ is a necessary and sufficient condition for f_0 to satisfy hypothesis (c). See [I.M.P.V.1].

Q.E.D.

Remark 4.4 There have been some attempts by several authors (not always rigorous) to study the interplay between stationary and non-stationary bifurcation. We invite the reader to check for himself the different articles and to see that in many cases the results obtained by analytic methods are particular cases of Proposition 4.5. In a further publication we shall study more fully different examples. As a token we shall look at gradient mappings and at periodic solutions of autonomous equations.

4.5 Gradient Maps. We shall first give a slight extension of Dancer's bifurcation result. We leave it to the reader to explore other cases. As in § 2.5 assume that $f(x,\lambda)$ is a real valued C^2 - map, invariant under the S^1 action, x in $\mathbb{R}^{2n+\ell}$, λ in $\mathbb{R}^k$. Denote by $g(x,\lambda)$ its (real) gradient with respect to x and assume that

$$g(x,\lambda) = K(\lambda)x + k(x,\lambda), \quad k(x,\lambda) = o(\|x\|). \qquad (12)$$

Suppose that 0 is an eigenvalue of the symmetric matrix $K(0)$ (a necessary condition for bifurcation). Put $\lambda = (\mu,\tilde{\tilde{\lambda}})$, μ in $\mathbb{R}$, and assume that $K(\mu,o)$ is invertible for small, non-zero, μ. Let $\lambda_j(\mu), j = 1, \ldots, d$, be the eigenvalues of $K(\mu,o)$, such that $\lambda_j(o) = 0$ ($\lambda_j(\mu) \neq 0$ for $\mu \neq 0$ and are real). Let $0<m_1<\ldots<m_d$, be the order of the isotropy subgroups of the corresponding eigenvectors. Let N_j be the number of eigenvalues crossing 0 from

left to right minus the number of eigenvalues crossing from right to left, as μ passes through 0, corresponding to the order m_j. N_0 the same difference for the eigenvalues corresponding to stationary solutions, if any.

Proposition 4.6. a) If N_0 is odd, then one has global bifurcation of a k - dimensional Cantor manifold of stationary solutions, in the sense of [I.M.P.V.1].

b) If N_0 is even and $N_j \neq 0$ for some $j \geq 1$, then either one has local bifurcation of stationary solutions, or global bifurcation of a k - dimensional Cantor manifold in the orbit space, in the sense of Theorem 4.2.

Proof. From the complex structure induced by the group action, (12) is equivalent, as (2) in § 2.5, to the equation

$$i\nu n_j x_j + L_j(\mu,\tilde{\tilde{\lambda}})x_j + h_j(x_0, \ldots, x_m,\mu,\tilde{\tilde{\lambda}}) = 0 \qquad (13)$$

where $n_0 = 0 < n_1 < \ldots < n_m$ are the different orders of isotropy subgroups of $\mathbb{R}^{2n} \times \mathbb{R}^{\ell}$, x_j in $\mathbb{C}^{m_j}$, x_0 in $\mathbb{R}^{\ell}$. It is easy to check that the matrix $L(\mu,\tilde{\tilde{\lambda}})$ induced by $K(\mu,\tilde{\tilde{\lambda}})$ has the above splitting in diagonal blocks, that it is self-adjoint and that it has the same spectral properties as $K(\lambda)$. (a) is clear from [I.M.P.V.1] and (b) will follow from Proposition 4.5, once one computes the degree of $\det(i\nu n_j I + L_j(\mu,o)) : \{\tilde{\lambda} = (\mu,\nu), |\tilde{\lambda}| = r\} \to \mathbb{C} - \{0\}$. Now the term $i\nu n_j + \lambda(\mu)$ is deformable to 1 (in $\mathbb{C} - \{0\}$) if $\lambda(\mu)$ stays on one side of the origin, to $\mu + i\nu$, with index 1, if $\lambda(\mu)$ crosses from left to right and to $-\mu + i\nu$, with index -1, if the crossing is in the other direction (see [I] for this sort of computation). Thus the degree of the determinant is N_j.

Q.E.D.

Remark 4.5. a) If $\ker K(0) \cap E^{s^1} = \{0\}$, then one has global bifurcation of non-stationary solutions.

b) Assume that the component bifurcating from (0,0) is bounded, in the direction $(\mu,0)$, and has the property that whenever $p = (0,x_0)$ is a stationary point on it, then the corresponding linearization $K_p(\mu,o)$ is not singular on the fixed point subspace E^{s^1}. Then defining a local "Fuller's change of index" as (Sign $\det K_p(\mu,o)|E^{s^1}$) $\Sigma N_j/n_j$, one may prove, as in [I], that the sum of all these indices over the stationary points in the component is zero (this is done by the global obstruction used in [I, § III]). Note that this change of index is precisely $In^n_{s^1}(\mu<0) - In^n_{s^1}(\mu>0)$ (Same proof as in [Io p. 79], [I.M.P.V.1, Prop. 4.6] or [Da. Lemma 2].

c) Dancer has studied two particular cases:

(α) $K(\mu) = I - (\bar{\lambda} + \mu) B$, where $\bar{\lambda}$ is a characteristic value of B, $K(o)$ non singular on E^{s^1}. Since the application $\mu \to 1 - (\bar{\lambda} + \mu)/\bar{\lambda}$ has a degree $-$ sign $\bar{\lambda}$, then $N_j = -$ sign $\bar{\lambda}$ $\dim_{\mathbb{C}} \ker L_j(o)$. Thus, as in § 2.5, our formulae agree up to sign.

(β) $K(\mu) = I - B_0 - \mu B_1 - \mu B_2(\mu)$, with $B_2(o) = 0$ and QB_1Q invertible, where Q is the orthogonal projection onto $\ker(I - B_0)$. Then λ_j behaves like $\pm\mu$ on the subspaces of $\ker(I - B_0)$ where QB_1^jQ is negative (positive) definite, with the same result.

4.6 Hopf bifurcation. In this final section, we shall complete the results published in [I] and [I.M.P.V.1] by improving the dimension count and by having a quick look at the case where the linearized equation has non-trivial stationary solutions (a deeper study of the possible modes of bifurcation will be given in a forthcoming paper).

Assume that equation (3) of § 2.6 depends on $(\mu,\tilde{\tilde{\lambda}})$, μ in $\mathbb{R}^{k-1}$, and a linearization at (0,0) of the form

$$dX/dt = L(\mu,\tilde{\tilde{\lambda}})\, X + g(\mu,\tilde{\tilde{\lambda}},X) \qquad (14)$$

we shall make the following hypothesis

H.1. $g(\mu,\tilde{\tilde{\lambda}},o) = 0$ and $g(\mu,\tilde{\tilde{\lambda}},X) = o(X)$ for X small.

H.2. There is a $\beta>0$ and integers $m_0 = 0 < m_1 = 1 < m_2 < \ldots < m_\ell$ such that $\pm im_j\beta$ belong to the spectrum of $L(o)$ and no other eigenvalue of $L(o)$ has that form. Note that, contrary to the hypothesis of [I], we allow the possibility of $m_0 = 0$.

H.3. The eigenvalues of $L(\mu,o)$ corresponding to $\pm im_j\beta$, $j = 1, \ldots, \ell$, are off the imaginary axis for small and non-zero μ.

As in [I] and § 2.6, after a change of time scale, passage to Fourier series and reduction to the bifurcating modes $m_0, \ldots, m_\ell$, (14) is locally equivalent to

$$(i\nu m_jI - L(\mu,\tilde{\tilde{\lambda}}))X_j - g_j(\mu,\tilde{\tilde{\lambda}},X_0, \ldots,X_\ell) = 0 \quad (15)$$

with X_0 in $\mathbb{R}^m$, m_j in $\mathbb{C}^M$, $j = 1, \ldots, \ell$. Here $\tilde{\lambda}$ will stand for $(\mu, \nu - 2\pi/\beta)$ in $\mathbb{R}^k$. The previous hypothesis imply that $i\nu m_jI - L(\mu,o)$ is invertible for $\tilde{\lambda} \neq 0$, $j = 1, \ldots, \ell$, thus defining an element of $\Pi_{k-1}(U)$, characterized by the integer N_j. For $k=2$, N_j is just the crossing number at $im_j\beta$ (counted with algebraic multiplicity).

H.4. $L(\mu,o)$ is invertible for small and non-zero μ.

In this case $L(\mu,o)$ defines and element of $\Pi_{k-2}(O)$, and, through the J-homomorphism, $(L(\mu,o)X, \|X\| - \varepsilon)$, a stable class N_0 in $\Pi_{k-2}(S)$. If $k = 2$, N_0 is just the change of sign of det $L(\mu,o)$, as μ crosses 0, that is N_0 is non-zero if and only if there is an odd net change of crossings of eigenvalues at 0.

H.4'. There is a sequence $\varepsilon_n \to 0$, such that, for $\|\mu\| < r_2$, $\|x\| = \varepsilon_n$, $L(\mu,o)X + g(\mu,o,X) \neq 0$. Furthermore, there is a μ_0, small, such that $L(\mu_0,o)$ is invertible or $L(o)X + g(\mu,o,X)$ has a non-zero Brouwer degree for $\|x\| < \varepsilon_n$.

Proposition 4.7. (a) Assume H.1., H.2., H.4. and N_0 is non-trivial, then there is a $(\dim \tilde{\Lambda} + 1)$ - Cantor manifold of stationary solutions bifurcating globally from $(0,0)$, in the sense of [I.M.P.V.1]
(b) Assume H.1., H.2., H.3., H.4'. and $N_j \neq 0$ for some $j \geq 1$, then there is a $(\dim \tilde{\tilde{\Lambda}} + 1)$ - Cantor manifold in the space of orbits of truly periodic solutions bifurcating globally from $(0,0,\nu_0 = 2\pi/\beta)$, in the sense of Theorem 4.2. Thus if H.1 - H.4 hold one has either local bifurcation of stationary solutions or the global bifurcation of periodic solutions.

Proof. This is a straight application of [I.M.P.V.1], [I] and Proposition 4.5.

Q.E.D.

Example 4.1. Consider the three - dimensional system

$$\begin{pmatrix} x_1 \\ x_2 \\ x_3 \end{pmatrix}' = \begin{pmatrix} \mu & 1 & 0 \\ -1 & \mu & 0 \\ 0 & 0 & \mu^2 \end{pmatrix} \begin{pmatrix} x_1 \\ x_2 \\ x_3 \end{pmatrix} + \begin{pmatrix} P & (x,\mu) \\ Q & (x,\mu) \\ R & (x,\mu) \end{pmatrix}$$

where P,Q,R are C^2 functions with vanishing first derivatives at $x = 0$. Here $m_1 = 1$, $N_1 = 1$, $N_0 = 0$. Suppose $R(x,\mu) = a(\mu)x_1^2 + b(\mu)\,x_2^2 + c(\mu)x_1x_2 + x_3^3 + \text{H.O.T.}$, where H.O.T. means terms of order 3 (different from x_3^3) and higher. Clearly H.1. - H.4 are satisfied. For the stationary equation, one may solve the two first equations for x_1 and x_2 in terms of x_3, with $x_i = 0(x_3^2)$, $i = 1,2$. The last equation will be of the form $x_3(\mu^2 + x_3^2 + 0(x_3^3))$, which, for x_3 small, has the only solution $x_3 = 0$. This fixes r_0, r_2 and verifies H.4'. Thus one obtains a global branch of truly periodic solutions.

Example 4.2. Take the system above, but replace the third equation by $x_3' = R(x,\mu)$. Then H.4. is not satisfied but, as above, the only stationary solution, for μ small, is $x = 0$ and its index is 1 (from the term x_3^3). Thus one has the same behavior as in the preceding example.

Example 4.3. Take the system of Example 4.1., with $P = Q = 0$, $R(x,\mu) = x_1^2 + x_2^2$. Then any periodic solution gives either $\mu = 0$ or $x_1 = x_2 = 0$. But $x_3' = x_1^2 + x_2^2 \geq 0$, cannot have a periodic solution, unless $x_1 = x_2 = 0$. $\mu = 0$, x_3 in $\mathbb{R}$ is a global branch of stationary solutions.

Example 4.4. If H.4. holds and N_0 is trivial, one may have just local bifurcation of stationary solutions (i.e. H.4'. doesn't hold). With the same linear part as before, but with P=Q=0, $R = x_3(2x_3 + x_3^2) + x_1^2 + x_2^2$. If $x_1^2 + x_2^2 = r^2$ (and thus $\mu = 0$), then $R \geq x_1^2 + x_2^2 + x_3^2$, if $x_3 > -1$. Thus there is no small periodic solution, and the stationary solutions are $x_3 = 0$ and the circle $(x_3 + 1)^2 + \mu^2 = 1$, with no global bifurcation. Note that if the $\mu^2 x_3$ term is not present then, with the same P,Q,R, x = 0 is an isolated solution (here the index of the zero solution is 0).

Remark 4.6. For gradient maps and the Hopf bifurcation problem, the parameter ν has the property that the stationary part of the equation does not depend on ν, if $Z = \underset{\sim}{0}$: $L_0(\mu,\tilde{\tilde{\lambda}})\, x_0 + g_0(\mu,\tilde{\tilde{\lambda}},x_0,o)$. Thus the condition in Lemma 4.2: $f_0(o,x_0,\tilde{\lambda},o) \neq 0$ on $B_0 \setminus \{0\}$ and $\|\lambda\| \simeq r_2$ is not met (for $\mu = 0$, $|\nu| = r_2$) unless $L_0(0)$ is invertible (case already included in the above results). Thus in these examples one cannot use Theorem 4.3 and Lemma 4.2.

5. APPENDIX

Since the characterization of the generalized kernel is not readily available in standard text books, we shall give an outline of the proof.

If $X(t)$ is such that $(d/dt - (A/\nu_0 - \mu))^{\alpha} X = 0$, then

$$X(t) = e^{-\mu t}\, \Phi(t) \sum_{0}^{\alpha-1} W_k\, t^k/k!\,.$$

One needs that $e^{-\mu t}\, \Phi(t)\, Y^{(k)}(t)$ belong to E (i.e. 2π - periodic) for $k = 0, \ldots, \alpha - 1$. If K is the matrix $e^{-2\pi\mu}\, \Phi(2\pi)$, this requirement amounts to solving the system

$$(I-K)W_{\alpha-k} = \sum_{\ell=1}^{k-1} (2\pi)^{\ell}\, KW_{\alpha-k+\ell} \,/\, \ell!$$

$k = 1, \ldots, \alpha$. From this, it follows that

$$(I-K)^{k-1}\, W_{\alpha-k} = (2\pi)^{k-1}\, W_{\alpha-1} = (2\pi)^{k-\alpha}\, (I-K)^{\alpha-1}\, W_0 .$$

Thus $W_{\alpha-k}$ belongs to $\ker(I-K)^k$. Furthermore, by rewritting the system in matrix form, one has:

$$\begin{bmatrix} 2\pi I & (2\pi)^2\, I/2! & \cdot & \cdot & \cdot & (2\pi)^{\alpha-1}/(\alpha-1)! \\ I - K^{-1} & 2\pi\, I & \cdot & \cdot & \cdot & (2\pi)^{\alpha-2}/(\alpha-2)! \\ 0 & 0 & & & & \cdot \\ \cdot & & & & & \cdot \\ \cdot & & & & & \cdot \\ 0 & 0 & & & I-K^{-1} & 2\pi I \end{bmatrix} \begin{bmatrix} W_1 \\ W_2 \\ \cdot \\ \cdot \\ \cdot \\ W_{\alpha-1} \end{bmatrix} = \begin{bmatrix} (K^{-1} - I)W_0 \\ 0 \\ \cdot \\ \cdot \\ \cdot \\ 0 \end{bmatrix}$$

Then, if $(I-K)W_0 = 0$ and $\alpha > 1$, one will get $W_{\alpha-1} = 0$, $W_{\alpha-k}$ in $\ker\,(I-K)^{k-1}$, that is the same system with α replaced by $\alpha-1$. Thus:

$$(I-K)^{k-1} W_{\alpha-1-k} = (2\pi)^{k-1} W_{\alpha-2} = (2\pi)^{k-\alpha+1} (I-K)^{\alpha-2} W_0, \; k=1, \ldots, \alpha-1.$$

Again, if $\alpha > 2$, $W_{\alpha-2} = 0$ and so on ... Thus, if $(I-K)\, W_0 = 0$, then $W_1, \ldots, W_{\alpha-1} = 0.$ This implies that the matrix is invertible and $W_1, \ldots, W_{\alpha-1}$ are uniquely defined by $(I-K)W_0$. If W_0 is in $\ker (I-K)$ then, $W_1 = \ldots = W_{\alpha-1} = 0$ is a solution of the system. Conversely, it is clear that any solution of the system will generate an $X(t)$ with the periodicity conditions. Hence the multiplicities of both operators agree.

One may show that

$$(I-K)^{\alpha-k-2} W_k = (2\pi K)^{-k} (I-K)^{\alpha-2} (2-k+kK)\, W_0/2, \; k=0, \ldots, \alpha-2,$$

giving $W_{\alpha-2}$. The inversion of the whole matrix is much more tedious.

If $A(t)$ is a constant matrix, it is easy to check (for example from the dimensions) that, after complexification, the above generalized kernel is generated by $x(t) = e^{-int} W_0$, for W_0 in $\ker (A/\nu_0 - (\mu - in))^{\alpha}$ and for all such integers n's. This is a well known result.

REFERENCES

[B] G.E. Bredon. Introduction to compact transformation groups. Academic Press, New-York 1972.

[C.M.Y.] S.N Chow, J. Mallet-Paret and J.A. Yorke, 'Global Hopf bifurcation from a multiple eigenvalue', Nonlinear Anal. 2 (1978), 755-763.

[Da] E.N. Dancer, 'A new degree for S^1 - invariant gradient mappings and applications'. Preprint (1984).

[D] J. Dugundji, Topology, Allyn and Bacon, 1966.

[F] F.B. Fuller, 'An index of fixed point type for periodic orbits', Amer. J. Math. 89 (1967), 133-148.

[I_0] J. Ize, Bifurcation theory for Fredholm operators, Memoirs Amer. Math. Soc. 7, 174, (1976).

[I] J. Ize, 'Obstruction theory and multiparameter Hopf bifurcation', Trans, Amer. Math. Soc. 289 (1985), 757-792.

[I.M.P.V.1]. J. Ize, I. Massabo', J. Pejsachowicz and A. Vignoli, 'Structure and dimension of global branches of solutions to multiparameter nonlinear equations', to appear in Trans, Amer. Math. Soc..

[I.M.P.V.2]. J. Ize, I. Massabo', J. Pejsachowicz and A. Vignoli.

'Nonlinear multiparameter equations : structure and topological dimension of global branches of solutions'. To appear in Proc. Symp. in Pure. Math. Proc. Berkeley Summer Institute on nonlinear functional Analysis and Applications. F. Browder Ed.

[N] L. Nirenberg, 'Comments on nonlinear problems', Le Matematiche 36 (1981), 109-119.

[P] A.R. Pears, Dimension theory of general spaces, Cambridge Univ. Press, 1975.

[V] A. Vanderbauwhede, Local bifurcation and symmetry, Research notes in Math. Pitman, 1982.

EXISTENCE AND MULTIPLICITY FOR SEMI-LINEAR EQUATIONS BY THE DUALITY METHOD

J. Mawhin and M. Willem
Institut Mathématique
Université Catholique de Louvain
2, chemin du Cyclotron
B-1348 Louvain-la-Neuve
Belgium

ABSTRACT. The duality method is used to obtain necessary and sufficient conditions for the solvability of semi-linear equations and multiplicity results. Applications are given to a semi-linear wave equation and to asymptotically linear hamiltonian systems.

Introduction. Let $\Omega \subset \mathbb{R}^m$ be a bounded domain, $V \subset L^2(\Omega, \mathbb{R}^N)$ a closed vector subspace, $L : D(L) \subset V \to V$ a self-adjoint operator and $F : \Omega \times \mathbb{R}^N \to \mathbb{R}$ a sufficiently regular function such that $F(x,.)$ is convex for a.e. $x \in \Omega$. If $\nabla F(.,u(.)) \in V$ for all $u \in D(L)$, where $\nabla F(x,u) = (D_{u_1} F(x,u),\dots,D_{u_N} F(x,u))$, the semilinear equation

$$Lu(x) = \nabla F(x,u(x)) \tag{1}$$

can be considered as the Euler equation formally associated to the functional

$$\psi : u \mapsto \int_\Omega [(1/2)(-Lu(x)|u(x)) + F(x,u(x))]dx.$$

When L is indefinite, the same is true in general for ψ and the obtention of critical points of ψ may be difficult. We shall see that in the above situation, the obtention of solutions of (1) can be associated to the existence of minima of another functional, the so-called dual action introduced by Clarke and Ekeland [5], whose form can be motivated by the following oversimplified situation. If we assume that L is invertible as well as $\nabla F(x,.)$ for a.e. $x \in \Omega$, then, letting

$$Lu = v, \text{ i.e. } u = L^{-1}v,$$

we see that (1) is equivalent to the equation

$$L^{-1}v(x) = G(x,v(x)) \tag{2}$$

where

S. P. Singh (ed.), Nonlinear Functional Analysis and Its Applications, 113–129.

$$G(x,.) = [\nabla F(x,.)]^{-1}.$$

Now, it is well known that the Legendre transform (with respect to u) $F^*(x,.)$ of $F(x,.)$ defined by

$$F^*(x,v) = (u|v) - F(x,u)$$

with u given implicitely by the relation

$$v = \nabla F(x,u)$$

is such that, formally

$$\nabla F^*(x,v) = u.$$

Thus, (2) can be written

$$L^{-1}v(x) = \nabla F^*(x,v(x))$$

and is the Euler equation for the function

$$\varphi : v \mapsto \int_\Omega [(1/2)(-L^{-1}v(x)|v(x)) + F^*(x,v(x))]dx.$$

Under the above assumptions, L^{-1} is bounded and the negative part of the corresponding quadratic form can be compensated by the second term in χ to make the function bounded from below. Since the work of Clarke, Ekeland, Brézis, Coron, Nirenberg and others (see e.g. [2] for surveys and references), we know that the study of the solvability of (1) through the obtention of critical points of some dual action can be successfully achieved in situations where L is not invertible and $F(x,.)$ only convex, so that $\nabla F(x,.)$ is no more invertible. This paper surveys some other recent applications of this approach due to the authors, D. Costa and J.R. Ward. To minimize technicalities, we shall not state and prove the results in their greatest generality, for which we refer to the original papers [8, 9, 11, 12, 13].

1. THE CASE OF A STRICTLY CONVEX POTENTIAL

Let $\Omega \subset \mathbb{R}^m$ be a bounded domain, V a closed vector subspace of $L^2(\Omega, \mathbb{R}^N)$ with the usual inner product

$$(u,v) = \int_\Omega (u(x)|v(x))dx$$

(with $(.|.)$ the usual inner product in $\mathbb{R}^N$) and the corresponding norm $\|u\| = (u,u)^{1/2}$. Let $L : D(L) \subset V \to V$ be a linear self-adjoint operator with closed range, so that $V = \ker L \oplus R(L)$ (orthogonal direct sum).

Let $F : \Omega \times \mathbb{R}^N \to \mathbb{R}$, $(x,u) \mapsto F(x,u)$ be such that $F(x,.)$ is continuous and convex for a.e. $x \in \Omega$ and $F(.,u)$ is measurable for each

$u \in \mathbb{R}^N$. Assume moreover that there exists $\beta \in L^2(\Omega; \mathbb{R}_+)$ and $l \in L^2(\Omega; \mathbb{R}^N)$ such that

$$F(x,u) \geqslant (l(x)|u) - \beta(x)$$

for a.e. $x \in \Omega$ and all $u \in \mathbb{R}^N$. Finally, let us assume that $\nabla F(x,.)$ exists for a.e. $x \in \Omega$ and is such that $\nabla F(.,u(.)) \in V$ whenever $u \in D(L)$.

To motivate the conditions introduced later in our existence theorem, let us state and prove equivalent statements concerning the solvability of the equation

$$Lu = \nabla F(x,u) \tag{3}$$

when some supplementary conditions are made upon L and F.

Theorem 1. Assume that $\dim \ker L < \infty$ and that $F(x,.)$ is strictly convex for a.e. $x \in \Omega$. Then the three following conditions are equivalent and are necessary for the solvability of (3) :

(a) there exists $u \in D(L)$ such that

$$\int_\Omega (\nabla F(x,u(x))|v(x))dx = 0$$

for all $v \in \ker L$;

(b) there exists $w \in \ker L$ such that

$$\int_\Omega (\nabla F(x,w(x))|v(x))dx = 0$$

for all $v \in \ker L$;

(c)

$$\int_\Omega F(x,w(x))dx \to +\infty \text{ if } \|w\| \to \infty \text{ in } \ker L.$$

Proof. Let $u \in D(L)$ be a solution of (3); then $\nabla F(.,u(.)) = Lu$ is orthogonal to ker L and (a) holds. Trivially, (b) $\Rightarrow$ (a), and we show now that (a) $\Rightarrow$ (c). Let u satisfying (a) and write $u = \bar{u} + \tilde{u}$ where $\bar{u} \in \ker L$ and $\tilde{u} \perp \ker L$. Define on ker L the strictly convex functions G and $\tilde{G}$ respectively by

$$G(w) = \int_\Omega F(x,w(x))dx,$$

$$\tilde{G}(w) = \int_\Omega F(x,w(x) + \tilde{u}(x))dx.$$

By (a), $\nabla\tilde{G}(\bar{u}) = 0$ and hence, as $\dim \ker L < \infty$ and $\tilde{G}$ strictly convex, this implies that

$$\tilde{G}(w) \to +\infty \text{ if } \|w\| \to \infty.$$

But, by convexity,

$$\tilde{G}(w) \leqslant (1/2) \int_\Omega F(x,2w(x))dx + \frac{1}{2}\int_\Omega F(x,2\tilde{u}(x))dx = (1/2)G(2w) + C,$$

so that,

$$G(w) \to +\infty \text{ if } \|w\| \to \infty \text{ in } \ker L.$$

Now the strict convexity of G and the fact that dim ker $L < \infty$ imply that (b) $\Longleftrightarrow$ (c), which completes the proof.

Now conditions (a), (b) or (c) are not sufficient in general for the existence of a solution for (3). Indeed, if L has an isolated positive eigenvalue λ_1 and if we take

$$F(x,u) = (\lambda_1/2)|u|^2 - h(x)u$$

for some $h \in \ker(L-\lambda_1 I)$, then $F(x,.)$ is strictly convex and condition (c) holds trivially, but the problem

$$Lu - \lambda_1 u = h(x)$$

has no solution, as $h \notin R(L-\lambda_1 I)$.

Thus, supplementary conditions have to be added to L and F to eliminate such situations. On the other hand, if one wants to deal with situations where dim ker $L = +\infty$ or $F(x,.)$ is simply convex, one only has the implications (c) $\Rightarrow$ (b) $\Rightarrow$ (a) and it is then reasonable to try to prove the existence of a solution for (3) when the strongest condition (c) holds.

2. A PERTURBED PROBLEM AND ITS SOLVABILITY

In addition to the general assumptions made in the two first paragraphs of the previous section, let us make the following assumptions upon the spectrum $\sigma(L)$ of L :

(S_1) $0 \in \sigma(L)$

(S_2) $\sigma(L) \cap]0,+\infty[\neq \phi$ and consists of isolated eigenvalues having finite multiplicity.

Let us denote by λ_1 the smallest positive eigenvalue of L and by $K : R(L) \to R(L)$ the (bounded) right inverse of $(-L)$ given by $K = (-L|_{D(L)\cap R(L)})^{-1}$. If $\{P_\lambda : \lambda \in \mathbb{R}\}$ is the spectral resolution of $(-L)$, $P^- = \int_{-\infty}^{-\lambda_1/2} dP_\lambda$, $P^+ = \int_{-\lambda_1/2}^{+\infty} dP_\lambda$, $H^- = P^-(R(L))$, $H^+ = P^+(R(L))$, we see that $K\bar{H}^+ \subset \bar{H}^+$, $R(L) = H^- \oplus H^+$, KP^+ is semi-positive definite on $R(L)$ and, by (S_2), KP^- is compact on $R(L)$ and, moreover,

$$(Kv,v) \geqslant -\lambda_1^{-1} \|v\|^2 \tag{4}$$

for all $v \in R(L)$.

For $\varepsilon > 0$, define F_ε by

$$F_\varepsilon(x,u) = (\varepsilon/2)|u|^2 + F(x,u)$$

and the Legendre-Fenchel transform $F^*_\varepsilon(x,.)$ of $F_\varepsilon(x,.)$ by

$$F^*_\varepsilon(x,v) = \sup_{u\in\mathbb{R}^N} [(u|v) - F_\varepsilon(x,u)].$$

Then

$$F^*_\varepsilon(x,v) \leqslant \beta(x) + (2\varepsilon)^{-1}|v-1(x)|^2$$

and $\nabla F^*_\varepsilon(x,.)$ exists and is continuous for a.e. $x \in \Omega$. We define on $R(L)$ the functional φ_ε by

$$\varphi_\varepsilon(v) = \int_\Omega [(1/2)(Kv(x)|v(x)) + F^*_\varepsilon(x,v(x))]dx,$$

and prove the following useful result.

Lemma 1. Assume that there exists

$$0 < \alpha < \lambda_1$$

such that for each $\varepsilon > 0$ there exists $\beta_\varepsilon \in L^2(\Omega, \mathbb{R}_+)$ such that

$$F(x,u) \leqslant (\alpha+\varepsilon)(|u|^2/2) + \beta_\varepsilon(x)$$

for a.e. $x \in \Omega$ and all $u \in \mathbb{R}^N$. Then there exists $\varepsilon_0 > 0$ such that, for each $\varepsilon \in \,]0,\varepsilon_0]$, the equation

$$Lu = \nabla F_\varepsilon(x,u)$$

has a solution u_ε such that $v_\varepsilon = Lu_\varepsilon$ minimizes φ_ε on $R(L)$.

Proof. Let us take $\varepsilon_0 > 0$ such that

$$(\alpha+\varepsilon_0)^{-1} > \lambda_1^{-1}$$

and let $\beta_{\varepsilon_0} \in L^2(\Omega; \mathbb{R}_+)$ such that

$$F_\varepsilon(x,u) \leqslant (\alpha+\varepsilon_0)(|u|^2/2) + \beta_{\varepsilon_0}(x)$$

whenever $\varepsilon \in \,]0,\varepsilon_0]$, $x \in \Omega$ and $u \in \mathbb{R}^N$. Then,

$$F^*_\varepsilon(x,v) \geqslant \frac{|v|^2}{2(\alpha+\varepsilon_0)} - \beta_{\varepsilon_0}(x)$$

and

$$|\nabla F^*_\varepsilon(x,v)| \leqslant 2\varepsilon^{-1}[|v| + |1(x)| + \beta_{\varepsilon_0}(x) + \beta(x)]$$

for a.e. $x \in \Omega$, all $v \in \mathbb{R}^N$ and all $\varepsilon \in \,]0,\varepsilon_0]$.

Thus, $\varphi_\varepsilon : R(L) \to \mathbb{R}$ is well defined and of class C^1 for each $\varepsilon \in]0,\varepsilon_o]$. Moreover,

$$\begin{aligned}\varphi_\varepsilon(v) &\geqslant \left(\frac{1}{\alpha+\varepsilon_o} - \frac{1}{\lambda_1}\right)\frac{\|v\|^2}{2} - \int_\Omega \beta_{\varepsilon_o}(x)dx \\ &= \delta\,\frac{\|v\|^2}{2} - \int_\Omega \beta_{\varepsilon_o}(x)dx\end{aligned} \tag{5}$$

with $\delta > 0$, for all $v \in R(L)$. Now,

$$\begin{aligned}\varphi_\varepsilon(v) &= (1/2)(KP^-v,v) + [(1/2)(KP^+v,v) + \int_\Omega F^*_\varepsilon(x,v(x))dx] \\ &= \varphi^1(v) + \varphi^2_\varepsilon(v)\end{aligned}$$

with φ^1 sequentially weakly continuous (as KP^- is compact) and φ^2_ε weakly lower semi-continuous (as continuous and convex). Thus φ_ε is w.l.s.c. and coercive. It has therefore a minimum v_ε on $R(L)$ for each $\varepsilon \in]0,\varepsilon_o]$, which satisfies

$$Kv_\varepsilon + \nabla F^*_\varepsilon(.,v_\varepsilon(.)) = \bar{u}_\varepsilon \in \ker L.$$

Letting $u_\varepsilon = \bar{u}_\varepsilon - Kv_\varepsilon \in D(L)$ we deduce, by duality, that

$$Lu_\varepsilon = v_\varepsilon = \nabla F^*_\varepsilon(x,u_\varepsilon)$$

and the proof is complete.

Remark 1. When F satisfies the assumption of Lemma 1, we shall say shortly that

$$\limsup_{|u|\to\infty} |u|^{-2} F(x,u) \leqslant \alpha$$

uniformly a.e. in Ω.

3. THE BASIC EXISTENCE THEOREM

The idea is now to extract a solution u of (3) from the family of approximate solutions u_ε obtained in Lemma 1. This requires the obtention of a posteriori estimates on the found solutions u_ε.

Theorem 2. Let $\Omega \subset \mathbb{R}^m$ be a bounded domain, $V \subset L^2(\Omega; \mathbb{R}^N)$ a closed vector subspace, $L : D(L) \subset V \to V$ a linear self-adjoint operator with closed range and $F : \Omega \times \mathbb{R}^N \to \mathbb{R}$ a function such that $F(x,.)$ is convex and differentiable for a.e. $x \in \Omega$, satisfies the regularity assumptions listed at the beginning of Section 2 and is such that $\nabla F(.,u(.)) \in V$ whenever $u \in D(L)$. Assume moreover that the following conditions are satisfied :

(S_1) $0 \in \sigma(L)$

(S_2) $\sigma(L) \cap]0,\infty[\neq \phi$ and consists in isolated eigenvalues $\lambda_1 < \lambda_2 < \ldots$ with finite multiplicity.

(S_3) $\limsup_{|u| \to \infty} \frac{F(x,u)}{|u|^2} \leq \alpha$ uniformly a.e. in Ω for some $0 < \alpha < \lambda_1$.

(S_4) $\int_\Omega F(x,w(x))dx \to +\infty$ if $\|w\| \to \infty$ in ker L.

Then the problem

$$Lu = \nabla F(x,u)$$

has at least one solution.

Proof. 1. Estimate on v_ε. By condition (S_4), there exists $w \in \ker L$ such that

$$\nabla F(.,w(.)) \in R(L)$$

(see the remarks following Theorem 1). Therefore, if we set

$$\tilde{v} = \nabla F(.,w(.)),$$

we obtain, by duality (Fenchel relation)

$$F^*(x,\tilde{v}(x)) = (\tilde{v}(x)|w(x)) - F(x,w(x))$$

for a.e. $x \in \Omega$. Moreover, as

$$F(x,u) \leq F_\varepsilon(x,u),$$

we have

$$F^*(x,\tilde{v}(x)) \geq F^*_\varepsilon(x,\tilde{v}(x))$$

a.e. on Ω. Therefore, as v_ε minimizes φ_ε on R(L), we get, using (5)

$$(\delta|2) \|v_\varepsilon\|^2 - \int_\Omega \beta_{\varepsilon_0}(x)dx \leq \varphi_\varepsilon(v_\varepsilon) \leq \varphi_\varepsilon(\tilde{v}) \leq$$

$$\leq \int_\Omega [(1/2)(K\tilde{v}(x)|\tilde{v}(x)) + F^*(x,\tilde{v}(x))]dx =$$

$$= \int_\Omega [(1/2)(K\tilde{v}(x)|\tilde{v}(x)) + (\tilde{v}(x)|w(x)) - F(x,w(x))]dx =$$

$$= C_0.$$

Hence, for all $\varepsilon \in]0,\varepsilon_0]$,

$$\|Lu_\varepsilon\|^2 = \|v_\varepsilon\|^2 \leq (2/\delta)(C_0 + \int_\Omega \beta_{\varepsilon_0}(x)dx) = C_1^2, \tag{6}$$

which also implies, if $u_\varepsilon = \bar{u}_\varepsilon + \tilde{u}_\varepsilon$ with $\bar{u}_\varepsilon \in \ker L$, $\tilde{u}_\varepsilon \in R(L)$,

$$\|\tilde{u}_\varepsilon\| = \|KLu_\varepsilon\| \leq \|K\|C_1 = C_2. \tag{7}$$

2. Estimate on $\overline{u}_\varepsilon$. By the convexity of $F(x,.)$, we have

$$F(x,\overline{u}_\varepsilon(x)/2) = F(x,u_\varepsilon(x)/2 - \widetilde{u}_\varepsilon(x)/2) \leqslant$$

$$\leqslant (1/2)F(x,u_\varepsilon(x)) + (1/2)F(x,-\widetilde{u}_\varepsilon(x)) \leqslant$$

$$\leqslant (1/2)[F(x,0) + (\nabla F(x,u_\varepsilon(x))|u_\varepsilon(x)) + (\alpha+1)|\widetilde{u}_\varepsilon(x)|^2 + \beta_1(x)]$$

$$= (1/2)[F(x,0) + (Lu_\varepsilon(x)|u_\varepsilon(x)) - \varepsilon|u_\varepsilon(x)|^2 + (\alpha+1)|\widetilde{u}_\varepsilon(x)|^2 + \beta_1(x)]$$

$$\leqslant (1/2)[F(x,0) + (Lu_\varepsilon(x)|u_\varepsilon(x)) + (\alpha+1)|\widetilde{u}_\varepsilon(x)|^2 + \beta_1(x)].$$

Therefore,

$$\int_\Omega F(x,\overline{u}_\varepsilon(x)/2)dx \leqslant$$

$$\frac{1}{2}[\int_\Omega F(x,0)dx + (Lu_\varepsilon,\widetilde{u}_\varepsilon) + (\alpha+1)\|\widetilde{u}_\varepsilon\|^2 + \int_\Omega \beta_1(x)]$$

$$\leqslant (1/2)[C_1C_2 + (\alpha+1)C_2^2 + C_3] = C_4,$$

which, together with assumption (S_4) implies that

$$\|\overline{u}_\varepsilon\| \leqslant C_5, \quad \varepsilon \in]0,\varepsilon_0]. \tag{8}$$

3. Existence of a solution. By (6), (7) and (8), there exists $u \in V$, $v \in V$ and a sequence (ε_k) in $]0,\varepsilon_0]$ converging to 0 such that

$$u_k \equiv u_{\varepsilon_k} \rightarrow u, \quad Lu_k \rightarrow v$$

as $k \rightarrow \infty$, and hence $u \in D(L)$ and $v = Lu$ by the weak closedness of the graph of L. Now, $\nabla F(x,.)$ is monotone, and hence

$$(\nabla F(.,u_k(.)) - \nabla F(.,w(.)),u_k-w) \geqslant 0$$

for all $w \in D(L)$, i.e.

$$(Lu_k-\varepsilon_k u_k-\nabla F(.,w(.)),u_k-w) \geqslant 0$$

or,

$$(LP^-u_k,P^-u_k) + (LP^+u_k,P^+u_k) - (Lu_k-\varepsilon_k u_k,w)$$

$$- \varepsilon_k\|u_k\|^2 - (\nabla F(.,w(.)),u_k-w) \geqslant 0 \tag{9}$$

for all $w \in D(L)$ and $k \in \mathbb{N}^*$.
By the compactness of KP^-,

$$P^-u_k = -KP^-Lu_k \rightarrow -KP^-Lu = P^-u$$

as $k \rightarrow \infty$, and hence

$$(LP^-u_k, P^-u_k) \to (LP^-u, P^-u)$$

if $k \to \infty$. By the convexity of $u_k \mapsto (-LP^+u_k, u_k)$ we have

$$\liminf_{k \to \infty}(-LP^+u_k, P^+u_k) \geq -(LP^+u, P^+u).$$

Hence, it follows from (9) that

$$(LP^-u, P^-u) - (Lu, w) - (\nabla F(.,w(.)), u-w) \geq -(LP^+u, P^+u),$$

i.e.

$$(Lu - \nabla F(.,w(.)), u-w) \geq 0$$

for all $w \in D(L)$. Taking $w = u-tz$ with $t > 0$ and $z \in D(L)$ and letting $t \to 0_+$, we get

$$(Lu - \nabla F(.,u(.)), z) \geq 0$$

for all $z \in D(L)$ and hence, as $D(L)$ is dense in V,

$$Lu - \nabla F(.,u(.)) = 0,$$

which completes the proof.

Remark 2. It can be shown in addition that $v = Lu$ minimizes

$$\varphi : R(L) \to]-\infty, +\infty],$$
$$v \mapsto \int_\Omega [(1/2)(Kv(x) \,|\, v(x)) + F^*(x, v(x))]dx$$

on $R(L)$.

Corollary 1. Let L and F satisfy the general assumptions of Theorem 2 as well as assumptions (S_1), (S_2) and (S_3). Assume moreover that $F(x,.)$ is strictly convex and that $\dim \ker L < \infty$. Then the problem

$$Lu = \nabla F(x,u)$$

has a solution if and only if there exists $w \in \ker L$ such that

$$\int_\Omega (\nabla F(x,w(x)) \,|\, v(x))dx = 0$$

for each $v \in \ker L$.

Proof. Follows directly from Theorem 1 and Theorem 2.

4. THE PERIODIC-DIRICHLET PROBLEM FOR SEMI-LINEAR DISPERSIVE WAVE EQUATIONS AT RESONANCE

Let us take $\Omega =]0,2\pi[\times]0,\pi[$, $V = L^2(\Omega, \mathbb{R}^N)$ and for L the abstract realization in $L^2(\Omega, \mathbb{R}^N)$ of the dispersive wave operator $u \mapsto \Box u - \mu_k u = u_{tt} - u_{xx} - \mu_k u$ with the periodic boundary conditions $u(0,x) - u(2\pi,x) = u_t(0,x) - u_t(2\pi,x) = 0$ in t and the Dirichlet boundary conditions $u(t,0) = u(t,\pi) = 0$ in x and with μ_k a positive eigenvalue of $\Box$ with the periodic-Dirichlet conditions on Ω. It is well known that, with those boundary conditions, $\sigma(\Box) = \{m^2-j^2 : m \in \mathbb{N}, j \in \mathbb{N}^*\}$, so that $\sigma(\Box)$ is unbounded from below and from above and can be written $\{\ldots < \mu_{-1} < \mu_0 = 0 < \mu_1 < \ldots\}$. Moreover, each μ_j with $j \neq 0$ has finite multiplicity. Consequently, $\sigma(L) = \{\ldots < -\mu_k < \mu_1-\mu_k < \ldots < \mu_{k-1}-\mu_k < 0 < \mu_{k+1}-\mu_k < \ldots\}$ and hence $\sigma(L) \cap]0,\infty[$ is made of isolated eigenvalues with finite multiplicity. Moreover, L is self-adjoint, has closed range and $\sigma(L) = \text{span}\{\cos jt \sin mx\, e_r, \sin jt \text{ sim } mx\, e_r : j \in \mathbb{N}, m \in \mathbb{N}^*, m^2-j^2 = \mu_k, 1 \leq r \leq N\}$ where the e_r are the unit vectors in $\mathbb{R}^N$. Similar considerations obviously hold if we take for L the abstract realization of $-\Box + \mu_{-k}I$ with the periodic-Dirichlet conditions on Ω with μ_{-k} a negative eigenvalue of $\Box$.

We take now $F : \Omega \times \mathbb{R}^N \to \mathbb{R}$ satisfying the convexity and regularity assumptions of Theorem 2. Condition (S_3) becomes here, if we write (t,x) the elements of Ω,

$$(S_3') \qquad \limsup_{|u| \to \infty} |u|^{-2} F(t,x,u) \leq \alpha$$

uniformly a.e. in Ω for some $0 < \alpha < \mu_{k+1} - \mu_k$. The condition (S_4) becomes here

$$(S_4') \quad \int_0^{2\pi}\int_0^{\pi} F\Big(t,x, \sum_{r=1}^{N} \sum_{\substack{j\in\mathbb{N}\\ m\in\mathbb{N}^*\\ m^2-j^2=\mu_k}} [a_{jmr}(\cos jt \sin mx) + b_{jmr}(\sin jt \sin mx)]e_r\Big)dt\,dx$$

$$\to +\infty \text{ if } \Sigma(|a_{jmr}|^2 + |b_{jmr}|^2) \to \infty.$$

In particular, using the fact that for a convex function $F(t,x,.)$,

$$F(t,x,u) \to +\infty \text{ if } |u| \to \infty$$

$$\Longleftrightarrow F(t,x,u) \geq \gamma|u| - \delta(t,x)$$

for some $\gamma > 0$, we easily see that (S_4') will be satisfied if the following condition holds :

(S_4'') $F(t,x,u) \to +\infty$ if $|u| \to \infty$, uniformly for a.e. $(t,x) \in \Omega$.

Consequently, under the assumptions (S_3')-(S_4') (and in particular (S_3')-(S_4''), the problem

$$\Box u - \mu_k u = \nabla F(t,x,u), \tag{10}$$

with the periodic-Dirichlet boundary conditions on $]0,2\pi[\times]0,\pi[$, has a (weak) solution for each positive eigenvalue μ_k of $\Box$ with the same boundary conditions. This result constitutes a partial generalization of a theorem of Rabinowitz [14] who deals with the scalar case, and under the more restrictive assumptions that F_u' is bounded (which of course implies assumption (S_3')). On the other hand, $F(t,x,.)$ needs not to be convex, but another assumption in [ ] namely

$$|\mu_k^{-1} F_u''(t,x,u)| < 1$$

for a.e. $(t,x) \in \Omega$ and all $u \in \mathbb{R}$, implies that $u \mapsto (1/2)\mu_k u^2 + F(t,x,u)$ has to be convex. Notice also that, in contrast to Rabinowitz result, all our assumptions on the nonlinear term imply only the potential F.

Remark 3. The reader will easily apply Theorem 2 to the problem

$$u''(t) + k^2 u(t) = \nabla F(t,u(t))$$

$$u(0) - u(2\pi) = u'(0) - u'(2\pi) = 0$$

($k \in \mathbb{N}$) and state and prove the corresponding existence theorem.

Remark 4. The reader can consult the references for further applications of the ideas and results stated here to Hamiltonian systems, Dirichlet and Neumann problems for elliptic equations and the periodic-Dirichlet problem for semi-linear beam equations.

Remark 5. When $F(t,x,.)$ is strictly convex, condition (S_4') is indeed necessary and sufficient for the solvability of (10) under condition (S_3').

5. AN ABSTRACT MULTIPLICITY THEOREM

When the functional associated to a variational problem is invariant under some group action, it is in general possible to obtain strong multiplicity results. In this section we recall a particular case of an abstract multiplicity result from [8].

Let $\{T(\theta) : \theta \in S^1\}$ be a representation of S^1 over a real Banach space X. A function $\varphi : X \to \mathbb{R}$ is invariant if $\varphi \circ T(\theta) = \varphi$ for every $\theta \in S^1$. A subset Y of X is invariant if $T(\theta)Y = Y$ for every $\theta \in S^1$. Let us define

$$\text{Fix}(S^1) = \{u \in X : T(\theta)u = u, \forall \theta \in S^1\}.$$

The S^1-orbit of a point u is the set

$$\theta(u) = \{T(\theta)u : \theta \in S^1\}.$$

A function $\varphi \in C^1(X, \mathbb{R})$ satisfies the Palais-Smale condition if every sequence (u_n) such that $(\varphi(u_n))$ is bounded and $\varphi'(u_n) \to 0$ contains a convergent subsequence.

Theorem 3. Let $\varphi \in C^1(X, \mathbb{R})$ be an invariant functional satisfying the Palais-Smale condition. Let Y and Z be closed invariant subspaces of X with codim Y and dim Z finite and codim Y < dim Z. Assume that the following conditions are satisfied :

$$\mathrm{Fix}(S^1) = \{0\}, \ \varphi(0) = 0 \tag{11}$$

$$\inf_Y \varphi > -\infty \tag{12}$$

there exists $r > 0$ such that $\varphi(u) < 0$ whenever $u \in Z$ and $\|u\| = r$. (13)

Then there exists at least $\frac{1}{2}$ (dim Z − codim Y) distinct S^1-orbits of non-zero critical points of φ.

Remarks. 1. Theorem 2 is the S^1 version of a multiplicity theorem proved by D.C. Clark for even functionals ([3]). The finite dimensional version of theorem 2 is due to Amann and Zehnder ([1]).

2. Various applications to nonlinear elliptic and hyperbolic problems are given in [7] and [8].

6. PERIODIC SOLUTIONS OF ASYMPTOTICALLY LINEAR HAMILTONIAN SYSTEMS

This section is concerned with multiplicity results for periodic solutions of asymptotically linear hamiltonian systems. In [4] D.C. Clark obtains lower bounds for the number of T-periodic solutions of the system

$$\frac{d}{dt}(A(t)\dot{x}(t)) + D_x V(t,x(t)) = 0$$

where V is T-periodic in t and even in x. It is assumed that A(t) is positive definite and T-periodic, that

$$D_x V(t,x) = B_0(t) + o(|x|) \text{ as } |x| \to 0$$

and that

$$D_x V(t,x) = B_\infty(t) + o(|x|) \text{ as } |x| \to \infty.$$

These results were extended by Amann and Zehnder to general hamiltonian systems

$$J\dot{u}(t) + D_u H(t,u(t)) = 0$$

where H is T-periodic in t and even in u. It is assumed that

$$D_u H(t,u) = A_o u + O(|u|) \text{ as } |u| \to 0$$

$$D_u H(t,u) = A_\infty u + P(|u|) \text{ as } |u| \to \infty$$

for two symmetric matrices A_o, A_∞. The autonomous system

$$J\dot{u} + \nabla H(u) = 0$$

is also treated in [1] by using S^1 symmetry instead of $\mathbb{Z}^2$ symmetry.

The lower bound for the number of T-periodic solutions is always obtained by comparing the Morse index of quadratic functionals related to the linearization at zero and at infinity. We shall first define the index of a linear positive definite hamiltonian system. Let A be a continuous mapping from $\mathbb{R}$ into the space of symmetric positive definite matrices of order 2N and consider the periodic boundary value problem

$$J\dot{u}(t) + A(t)\, u(t) = 0 \qquad (14)$$
$$u(0) = u(T)$$

where $T > 0$ is fixed. The corresponding hamiltonian is given by

$$H(t,u) = \frac{1}{2}(A(t)u,u).$$

It is easy to verify that the Legendre transform (with respect to u) $H^*(t,v)$ is of the form

$$H^*(t,v) = \frac{1}{2}(B(t)v,v)$$

where $B(t) = A(t)^{-1}$. Let L be the abstract realization in $L^2(0,T;\mathbb{R}^{2N})$ of the operator $-J\dot{u}$ with periodic boundary conditions $u(0)= u(T)$. Then L is self-adjoint, has closed range and

$\sigma(L) = \frac{2\pi}{T}\mathbb{Z}$. Each eigenvalue is of multiplicity 2N. In particular

L satisfies (S_1) and (S_2). The dual action corresponding to (14) is defined on

$$R(L) = \{u \in L^2(0,T;\mathbb{R}^{2N}) : \int_0^T u(t)dt = 0\}$$

by

$$\varphi_T(v) = \int_0^T \frac{1}{2}[(Kv(t)|v(t)) + (B(t)v(t)|v(t)]dt.$$

The following definition is due to I. Ekeland.

Definition [10]. The index $i(A,T)$ is the Morse index of φ_T.

It is easy to verify that the index $i(A,T)$ is finite and that $R(L)$ is the orthogonal sum of V^0, V^+ and V^- with φ_T degenerate on H^0, positive definite of H^+ and negative definite on H^-. Thus there is $\delta > 0$ such that

$$\varphi_T(v) \geq \frac{\delta}{2} \|v\|^2, \ \forall v \in H^+ \tag{15}$$

$$\varphi_T(v) \leq -\frac{\delta}{2} \|v\|^2, \ \forall v \in H^-. \tag{16}$$

Clearly $i(A,T) = \dim H^-$. It is proved in [10] that the dimension of H^0 is equal to the number of linearly independent solutions of (14). In particular, if (14) has no nontrivial solution, there $i(A,T) = \operatorname{codim} H^-$.
Let us assume that A is a constant matrix such that

$$\sigma(JA) \cap \frac{2i\pi}{T} \mathbb{N} = \emptyset.$$

In this case the Conley-Zehnder index $j(A,T)$ is also defined ([6]). It is proved in [9] that

$$j(A,T) = i(A,T) + 2N.$$

We now consider the autonomous hamiltonian system

$$J\dot{u} + \nabla H(u(t)) = 0 \tag{17}$$

where $H \in C^1(\mathbb{R}^{2N}, \mathbb{R})$ is strictly convex and satisfies the conditions

$$\nabla H(u) = A_0 u + 0(|u|) \text{ as } |u| \to 0 \tag{18}$$

$$\nabla H(u) = A_\infty u + o(|u|) \text{ as } |u| \to \infty \tag{19}$$

with symmetric positive definite matrices A_0 and A_∞.

Theorem 4. Assume that $T > 0$ is such that

A1. $$\sigma(JA_\infty) \cap \frac{2i\pi}{T} \mathbb{N} = \emptyset$$

A2. $$i(A_0,T) > i(A_\infty,T).$$

Then system (17) has at least

$$\frac{1}{2} [i(A_0,T) - i(A_\infty,T)]$$

non-zero T-periodic orbits.

Remarks. 1. Theorem 4 is due to Costa and Willem ([9]). This result was first proved in [1] under the assumption that $H \in C^2(\mathbb{R}^{2N}, \mathbb{R})$ and

$$\sup_{\mathbb{R}^{2N}} \|H''(u)\| < \infty.$$

The method of proof of [1] is not applicable without this assumption.

2. It follows from (A1) that the linear system

$$J\dot{u}(t) + A_\infty u(t) = 0$$

has no non-trivial T-periodic solution.
Thus (A1) is a non-resonance condition "at infinity".

3. Assumption (A2), which requires a distinct behaviour of ∇H at the origin and at infinity is similar to the twist condition is the Poincaré-Birkhoff-theorem.

4. Since H is strictly convex and $\nabla H(0) = 0$ by (18), 0 is the unique equilibrium point of (17).

Sketch of proof of the theorem.

1) Without loss of generality, we can assume that $H(0) = 0$. Since $\nabla H(0) = 0$, this implies that $H^*(0) = 0$.
Since H is strictly convex and, because of (19), such that

$$H(u)/|u| \to +\infty \text{ as } |u| \to \infty,$$

$H^* \in C^1(\mathbb{R}^{2N}, \mathbb{R})$. Moreover (19) implies also that

$$\nabla H^*(v) = B_\infty v + O(|v|) \text{ as } |v| \to \infty, \qquad (20)$$

where $B_\infty = A_\infty^{-1}$. The dual action defined by

$$\varphi(v) = \int_0^1 [\tfrac{1}{2}(Kv(t),v(t)) + H^*(v(t))]dt$$

is there continuously differentiable on R(L). Since φ is invariant for the representation of $S^1 \simeq \mathbb{R}/T\mathbb{Z}$ defined over R(L) by the translations in time

$$(T(\theta)v(t) = v(t+\theta)$$

we are in position to apply theorem 3. (The functions of R(L) are extended to $\mathbb{R}$ by T-periodicity).

2) Let us define the functional

$$\varphi_T^\infty(v) = \int_0^T \tfrac{1}{2}[(Kv(t)|Kv(t)) + (B_\infty v(t)|v(t))]dt$$

on R(L). By assumption (A1), φ_T^∞ is non-degenerate. Formula (15) implies the existence of a closed invariant subspace, $Y = N^+$ of R(L) with codimension $i(A_\infty,T)$ and of $\delta > 0$, such that, for every $v \in Y$, one has

$$\varphi_T^\infty(v) \geqslant \frac{\delta}{2}\|v\|^2.$$

It follows then easily from (20) that φ is bounded from below on Y.

3) Assumption (18) imply that

$$\nabla H^*(v) = B_o v + O(|v|) \text{ as } |v| \to 0 \tag{21}$$

where $B_o = A_o^{-1}$. By (16) there exists an invariant subspace $Z = V^-$ of R(L) with dimension $i(A_o,T)$ and some $\delta > 0$ such that

$$\varphi_T^o(v) \equiv \int_0^T \frac{1}{2}[(Kv(t),v(t)) + (B_o v(t),v(t))]dt$$

$$\leqslant -\frac{\delta}{2}\|v\|^2$$

whenever $v \in Z$. It follows then easily from (21) that there exists $r > 0$ such that $\varphi(u) < 0$ whenever $u \in Z$ and $\|u\| = r$.

4) The verification of the Palais-Smale condition is simple and left to the reader. We now apply theorem 3 to the invariant functional φ. The spaces Y and Z satisfy the assumption

$$\operatorname{codim} Y = i(A_\infty,T) < i(A_o,T) = \dim Z$$

and conditions (12), (13). Moreover $\operatorname{Fix}(S^1) = \{0\}$ and $\varphi(0) = 0$ so that (11) is satisfied. Thus theorem 3 implies the existence of at least

$$\frac{1}{2}(i(A_o,T) - i(A_\infty,T))$$

distinct S^1-orbits of non-zero critical points of φ. It is then easy to prove by duality the existence on n non-zero T-periodic orbits of (17). □

References

[1] H. Amann - E. Zehnder, *Periodic solutions of asymptotically linear hamiltonian systems*, Manuscripta Math. 32 (1980) 149-189.

[2] H. Brézis, *Periodic solutions of nonlinear vibrating strings and duality principles*, Bull. Amer. Math. Soc. (NS) 8 (1983) 409-426.

[3] D.C. Clark, *A variant of the Lusternik-Schnirelman theory*, Indiana Univ. Math. J. 22 (1972) 63-74.

[4] D.C. Clark, *Periodic solutions of variational systems of ordinary differential equations*, J. Diff. Equ. 28 (1978) 354-368.

[5] F. Clarke - I. Ekeland, *Hamiltonian trajectories with prescribed minimal period*, Comm. Pure Appl. Math. 33 (1980) 103-116.

[6] C. Conley - E. Zehnder, *Morse-type index theory for flows and periodic solutions of hamiltonian systems*, Comm. Pure Appl. Math. 37 (1984) 207-253.

[7] D.G. Costa - M. Willem, *Points critiques multiples de fonctionnelles invariantes*, C. R. Acad. Sc. Paris 298 (1984) 381-384.

[8] D.G. Costa - M. Willem, *Multiple critical points of invariant functionals and applications*, Nonlinear Analysis TMA, to appear.

[9] D.G. Costa - M. Willem, *Lusternik-Schnirelman theory and asympto-*

tically linear hamiltonian systems, to appear in "Differential Equations : Qualitative theory", North-Holland.

[10] I. Ekeland, *Une théorie de Morse pour les systèmes hamiltoniens*, Ann. Inst. H. Poincaré - Analyse non linéaire 1 (1984) 19-78.

[11] J. Mawhin - J. Ward - M. Willem, *Necessary and sufficient conditions for the solvability of a nonlinear two-point boundary value problem*, Proc. of the Amer. Math. Soc. 93 (1985) 667-674.

[12] J. Mawhin - J. Ward - M. Willem, *Variational methods and semi-linear elliptic equations*, Arch. Rat. Mech. Anal., to appear.

[13] J. Mawhin - M. Willem, *Critical of convex perturbations of some indefinite quadratic forms and semi-linear boundary value problems at resonance*, to appear.

[14] P. Rabinowitz, *Some minimax theorems and applications to nonlinear partial differential equations*, in "Nonlinear Analysis", Cesari, Kannan and Weinberger ed., Academic Press, 1978, 161-177.

SPECIAL PROBLEMS INVOLVING UNIQUENESS AND MULTIPLICITY IN HYPERELASTICITY

C.A. Stuart
Département de Mathématiques
Ecole Polytechnique Fédérale de Lausanne
1015 Lausanne
Switzerland

ABSTRACT. In these lectures I shall describe a few recent results concerning some problems related to nonlinear elasticity. They deal with questions of uniqueness and multiplicity of solutions of the equilibrium equations and focus upon the regularity of the solutions. This material is organised as follows.

Lecture 1: An informal presentation of the basic notions from elasticity is followed by a more precise statement of the mathematical problems.

Lecture 2: We discuss the uniqueness of classical solutions of the equilibrium equations for some special boundary conditions. This covers joint work with R.J. Knops [9].

Lecture 3: For deformations of a ball we describe all weak radial solutions of the equilibrium equations. For a prescribed displacement of the boundary there is exactly one classical solution but an infinite number of weak solutions. These results are based on [10] and lead to an alternative approach to the problem of cavitation discussed by J.M. Ball [3].

Lecture 1. FORMULATION OF THE PROBLEMS

Let $\Omega \subset \mathbb{R}^N$ be a bounded domain which is taken to be the reference configuration for a piece of elastic material. (The physical remarks concern the cases $N = 2,3$.) A deformation of this body is described by a function $u: \Omega \to \mathbb{R}^N$ which should be one-to-one with $\det \nabla u(x) > o$ for $x \in \Omega$. The internal forces due to such a deformation are described by the Piola-Kirchhoff stress tensor $S(x)$ for $x \in \Omega$ or by the Cauchy stress tensor $\tilde{S}(y)$ for $y \in u(\Omega)$. They are related by the formula

$$\tilde{S}(u(x)) = \frac{S(x)\nabla u(x)^t}{\det\nabla u(x)} \quad \text{for} \quad x \in \Omega$$

where t denotes the transpose of a matrix.

S. P. Singh (ed.), Nonlinear Functional Analysis and Its Applications, 131–145.

If there are no applied forces acting on the interior of the body the conditions for equilibrium can be expressed as follows

$$\text{(EE)} \begin{cases} \sum_{j=1}^{N} \frac{\partial}{\partial x_j} \{S_{ij}(x)\} = o & \text{for } x \in \Omega, \quad 1 \leqslant i \leqslant N \\ \tilde{S}(y) = \tilde{S}(y)^t & \text{for } y \in u(\Omega). \end{cases}$$

The first equation represents the condition that the sum of all forces on the boundary of every sub-body of $u(\Omega)$ is zero and the second equation then ensures that the sum of the moments of these forces is also zero. This second condition is equivalent to

$$S(x)\nabla u(x)^t = \nabla u(x)S(x)^t \quad \text{for } x \in \Omega. \tag{1.1}$$

To proceed we must specify how S or $\tilde{S}$ depends upon the deformation u from which it arises. For this we need a little notation.
M denotes the space of all $(N\text{x}N)$ real matrices with scalar product
$\ll A,B \gg = \text{trace } AB^t = \sum_{i,j=1}^{N} A_{ij}B_{ij}$.
$M_+ = \{A \in M: \det A > o\}$ is an open subset of M.
For $W \in C^1(M_+,\mathbb{R})$, the gradient of W at $F \in M_+$ is the matrix defined by $DW(F)G = \ll \text{grad } W(F),G \gg \ \forall G \in M$ where $DW(F)$ is the derivative of W at F.

A material is said to be homogeneous and hyperelastic if there exists $W \in C^1(M_+,\mathbb{R})$ such that

$$S(x) = T(\nabla u(x)) \text{ for all deformations where } T = \text{grad } W. \tag{1.2}$$

In this case, W is called the *stored energy* function and T is called the *response function* for the material. The relationship (1.2) is called the constitutive equation for the material. In order that it be independent of the choice of co-ordinates, W should satisfy the condition

$$W(QF) = W(F) \quad \forall F \in M_+ \text{ and } Q \in SO(N) \tag{1.3}$$

which is referred to as frame-indifference. It then follows that for a homogeneous hyperelastic material $T(F)F^t = FT(F)^t \ \forall F \in M_+$ and so (1.1) is satisfied for all deformations. Thus for a homogeneous hyperelastic material the equilibrium conditions reduce to

$$\sum_{j=1}^{N} \frac{\partial}{\partial x_j} \{T_{ij}(\nabla u(x))\} = o \quad \text{for } x \in \Omega, \quad 1 \leqslant i \leqslant N. \tag{1.4}$$

We note that the variational character of the constitutive equation (1.2) is reflected in the equilibrium equations (1.4) in the following way. Let

$E(u) = \int_\Omega W(\nabla u(x))dx$. Then (formally)

$$DE(u)v = \int_\Omega \ll T(\nabla u),\nabla v \gg dx$$

and if integration by parts can be justified

$$= - \int_\Omega \sum_{i,j=1}^{N} \frac{\partial}{\partial x_j} \{T_{ij}(\nabla u)\} v_i \, dx$$

provided $v(x) = o$ on $\partial\Omega$.
Thus at least formally, solutions of the equilibrium equations (1.4) correspond to stationary points of the functional E, which is called the *total energy*.

For a more complete and careful discussion of these aspects of mechanics one can consult [1,2].

We turn now to a more precise statement of some mathematical problems motivated by the above considerations.

Let $W \in C^2(M_+, \mathbb{R})$ and set $T = \text{grad } W$.

Definitions. A function $u\colon \Omega \subset \mathbb{R}^N \to \mathbb{R}^N$ is a *classical solution* of the equilibrium equations (EE)

$$\Longleftrightarrow \begin{cases} u \in C^2(\Omega) \cap C^1(\bar\Omega) \\ \nabla u(x) \in M_+ \quad \text{for} \quad x \in \bar\Omega \\ \sum_{j=1}^{N} \frac{\partial}{\partial x_j} \{T_{ij}(\nabla u(x))\} = o \quad \text{for} \quad x \in \Omega, \quad 1 \leq i \leq N . \end{cases}$$

It is a *weak solution* of the (EE)

$$\Longleftrightarrow \begin{cases} u \in W^{1,1}(\Omega) \\ \nabla u(x) \in M_+ \quad \text{a.e. } \Omega \\ T(\nabla u) \in L^1(\Omega) \\ \int_\Omega \langle\langle T(\nabla u), \nabla v \rangle\rangle \, dx = o \qquad \forall v \in C_0^\infty(\Omega) \end{cases}$$

It is also convenient to define two special types of deformation.

A function $u\colon \bar\Omega \to \mathbb{R}^N$ of the form

$$u(x) = Fx + a \quad \text{for} \quad x \in \bar\Omega$$

where $F \in M_+$ and $a \in \mathbb{R}^N$ is called a *homogeneous deformation*. We note that every homogeneous deformation is a classical solution of the (EE).

A function $u\colon \bar\Omega \to \mathbb{R}^N$ is called a *radial deformation* if it has the form, $\Omega = \{x \in \mathbb{R}^N : \|x\| < 1\}$ and

$$u(x) = U(r) \frac{x}{r} \quad \text{for} \quad o < \|x\| = r < 1 \tag{1.5}$$

where $U(r) > o$ and $U'(r) > o$ for $r > o$. As we shall see a radial deformation is a solution of the (EE) if and only if U satisfies a second order ordinary differential equation. We note that for a radial deformation (sufficiently smooth) $\nabla u(x)$ is a symmetric matrix with eigenvalues

$U'(r)$ (multiplicity 1) and $\frac{U(r)}{r}$ (multiplicity N-1).

Clearly a homogeneous radial deformation has the form

$$u(x) = \lambda x \quad \text{for} \quad \lambda > o \quad \text{and} \quad o \leqslant \|x\| < 1 .$$

In terms of these definitions we can now state the problems which will be discussed.

Problem 1. (Uniqueness of classical solutions)

Give conditions on Ω and W which imply that the homogeneous deformation $u(x) = Fx + a$ for $x \in \bar{\Omega}$ is the only classical solution of the (EE) satisfying the displacement boundary condition (DBC)

$$u(x) = Fx + a \quad \text{for} \quad x \in \partial\Omega \quad \text{where}$$

$$F \in M_+ \quad \text{and} \quad a \in \mathbb{R}^N .$$

As a special case of Problem 1, we have the following question concerning radial deformations.

Let $\Omega = \{x \in \mathbb{R}^N : \|x\| < 1\}$. Give conditions on W which imply that the homogeneous radial deformation $u(x) = \lambda x$ for $x \in \bar{\Omega}$ is the only classical solution of the (EE) satisfying the DBC

$$u(x) = \lambda x \quad \text{for} \quad \|x\| = 1 \quad \text{where } \lambda > o.$$

It is natural to inquire to what extent the regularity of the solutions is important for an answer to Problem 1. We shall investigate this aspect of the problem in the context of radial displacements.

Problem 2. Discuss all weak radial solutions of the (EE).

When dealing with radial deformations it is reasonable to restrict attention to materials which have no preferred directions. Such isotropic hyperelastic materials are characterised by the property

$$W(FQ) = W(F) \quad \forall F \in M_+ \quad \text{and} \quad Q \in SO(N) \tag{1.6}$$

of the stored energy. For a hyperelastic material the properties of frame-indifference and isotropy of W imply that W can be expressed in the following form

$$W(F) = \Phi(v_1, v_2, \ldots, v_N) \tag{1.7}$$

where $\Phi: (o,\infty)^N \to \mathbb{R}$ is a symmetric function of $v_1, \ldots, v_N$ which are the eigenvalues of the matrix $(FF^t)^{\frac{1}{2}}$.

For a radial deformation we have

$$W(\nabla u(x)) = \Phi(U'(r), \frac{U(r)}{r}, \ldots, \frac{U(r)}{r})$$

and the classical form of the (EE) reduces to

$$r \frac{d}{dr} \Phi_1 = (N-1) [\Phi_2 - \Phi_1] \quad o < r < 1$$

where Φ_i denotes the i-th partial derivative of Φ evaluated at

$$(U'(r), \frac{U(r)}{r}, \ldots, \frac{U(r)}{r}).$$

Furthermore, under mild assumptions on Φ, Ball [3] has established that a radial deformation $u(x) = U(r)\frac{x}{r}$ is a weak solution of the (EE) $\Longleftrightarrow$

$$\text{(REE)} \begin{cases} U \in C^2((o,1]) & \\ U(r) > o \quad \text{and} \quad U'(r) > o & \text{for } o < r \leqslant 1 \\ r^{N-1}\,\Phi_i \in L^1(o,1) & \text{for } i = 1,2 \\ r\frac{d}{dr}\Phi_1 = (N-1)\,[\Phi_2 - \Phi_1] & \text{for } o < r < 1 \end{cases}$$

Thus for an isotropic hyperelastic material, Problem 2 amounts to discussing all solutions of (REE). We note that a radial deformation can be a classical solution of (EE) only if $U(o) = o$.

Lecture 2. UNIQUENESS OF CLASSICAL SOLUTIONS

The results in this lecture are taken from joint work with R.J. Knops [9]. We begin by observing that, as is well-known from examples of buckling, uniqueness of classical solutions to problems in elastostatics is not to be expected in general. Usually it will occur only under conditions which limit the size of the deformations or stresses imposed. There are, however, certain types of problem for which there is uniqueness without any kind of restriction of smallness; namely, when the boundary of a star-shaped body is subjected to an affine deformation. To formulate this result more precisely, we recall some standard terminology.

A matrix $F \in M$ is said to be of rank-one if and only if $\dim \operatorname{Im} F = 1$. In this case, there exist $a,b \in \mathbb{R}^N\setminus\{o\}$ such that $F_{ij} = a_i b_j$ for $1 \leqslant i,j \leqslant N$ and we shall write $F = a \otimes b$ to indicate this.

A function $W \in C(M_+,\mathbb{R})$ is *rank-one convex*

$$\Longleftrightarrow \quad W(tF + (1-t)G) \leqslant t\,W(F) + (1-t)W(G) \quad \forall t \in [o,1]$$

whenever $F,G \in M_+$ and $F-G$ is rank-one.

Under appropriate smoothness conditions this condition can be expressed in terms of derivatives of W.

For $W \in C^1(M_+,\mathbb{R})$, W is rank-one convex

$$\Longleftrightarrow \quad W(G) \geqslant W(F) + \langle\!\langle T(F), G-F \rangle\!\rangle \tag{2.1}$$

whenever $F,G \in M_+$ and $F-G$ is rank-one with $T = \operatorname{grad} W$.

For $W \in C^2(M_+,\mathbb{R})$, W is rank-one convex

$$\Longleftrightarrow \quad D^2W(F)(G,G) \geqslant o \tag{2.2}$$

whenever $F \in M_+$ and $G \in M$ is rank-one.

The inequality (2.2) is known as the Legendre-Hadamard condition. If there is strict inequality for $G \neq o$ (2.2) is called *strong ellipticity* for the system (EE).

A function $W \in C(M_+, \mathbb{R})$ is *quasi-convex*

$$\Longleftrightarrow \quad \forall F \in M_+ , \quad \int_D W(F + \nabla w(x))dx \geq W(F)\int_D dx$$

for all open bounded $D \subset \mathbb{R}^N$, $D \neq \phi$ and all $w \in C^1(\bar{\Omega})$ such that $w \equiv o$ on $\partial\Omega$ and $F + \nabla w(x) \in M_+$ $\forall x \in \bar{\Omega}$.

Strict quasi-convexity means that there is equality only for $w \equiv o$. This definition is due to Morrey [5] and plays an essential role in the study of the weak continuity of integrals such as the energy $E(u)$,[5,4]. As it stands the definition is not easy to check for a given function W. In elasticity an important step in this direction was taken by Ball [4] with the introduction of the following notion. For $N = 3$, a function $W \in C(M_+, \mathbb{R})$ is *polyconvex*

$$\Longleftrightarrow \quad \exists \text{ a convex function } g \in C(M \times M \times (o,\infty), \mathbb{R}) \text{ such that}$$

$$W(F) = g(F, \operatorname{adj} F, \det F) \quad \forall F \in M_+ .$$

A similar notion of polyconvexity can be introduced for any $N \in \mathbb{N}$ and it is known that

$$\text{polyconvex} \Rightarrow \text{quasi-convex} \Rightarrow \text{rank-one convexe.}$$

We shall further discuss these concepts at the end of this lecture. For the moment, we turn to the reference configuration, Ω, which will be supposed henceforth to have a boundary $\partial\Omega$ which is piecewise C^1. Then for almost all $x \in \partial\Omega$, $\nu(x)$ will denote the outward unit normal to $\partial\Omega$ at a regular point x of $\partial\Omega$.

Theorem 2.1. (A.E. Green, 1973). Let $\partial\Omega$ be piecewise C^1. Suppose that $W \in C^2(M_+, \mathbb{R})$ and that u is a classical solution of (EE). Then

$$E(u) = \frac{1}{N} \int_{\partial\Omega} \xi W(\nabla u) + \langle\langle T(\nabla u), (u - r\frac{\partial u}{\partial r}) \otimes \nu \rangle\rangle \, ds$$

where E is the total energy,

$$\xi(x) = \langle \nu(x), x \rangle \quad \text{a.e. on } \partial\Omega \text{ and}$$

$$r\frac{\partial}{\partial r} \equiv x \cdot \nabla .$$

Remarks 1. This result shows that the volume integral E can be replaced by a surface integral over $\partial\Omega$ when u is a classical solution of (EE).

2. For the system (EE) it plays the role of what is known as the Pohazaev identity in the theory of semilinear elliptic equations.

3. It can be established as follows. Multiply (EE) by u and

integrate by parts. Then multiply (EE) by $r\frac{\partial u}{\partial r}$ and integrate by parts. The result then follows by eliminating common terms.

The usefullness of rank-one matrices for Problem 1 is suggested by the following lemma from calculus.

Lemma 2.2. Let $\partial\Omega$ be piecewise C^1 and let

$$u,v \in C^1(\bar{\Omega}) \text{ with } u(x) = v(x) \quad \forall x \in \partial\Omega.$$

Then (i) $\nabla u - \nabla v = \frac{\partial}{\partial\nu}(u-v)\otimes\nu$ a.e. on $\partial\Omega$

(ii) $r\frac{\partial}{\partial r}(u-v) = \xi\frac{\partial}{\partial\nu}(u-v)$ a.e. on $\partial\Omega$

where u,v take values in $\mathbb{R}^N$ and

$$\xi(x) = \langle\nu(x),x\rangle \text{ at regular points of } \partial\Omega.$$

Remark. Part (i) shows that $\nabla u - \nabla v$ is rank-one on $\partial\Omega$. Combining these results in an obvious way we obtain the following inequality concerning the energies of classical solutions.

Lemma 2.3. Let Ω be star-shaped with boundary $\partial\Omega$ which is piecewise C^1. Suppose that $W \in C^2(M_+,\mathbb{R})$ is rank-one convex and that, u,v are classical solutions of (EE) with $u \equiv v$ on $\partial\Omega$. Then,

$$E(u) - E(v) \leq \frac{1}{N}\int_{\partial\Omega} \langle\langle T(\nabla u) - T(\nabla v), (v - r\frac{\partial v}{\partial r})\otimes\nu\rangle\rangle\, ds .$$

Remarks. The proof can be reduced to the case where Ω is star-shaped with respect to $o \in \Omega$. Then $\xi(x) \geq o$ for all regular $x \in \partial\Omega$ and the inequality can be obtained by expressing $E(u) - E(v)$ as an integral over $\partial\Omega$ and then using the rank-one convexity in the form (2.1).

Theorem 2.5. Let Ω be star-shaped with boundary $\partial\Omega$ which is piecewise C^1. Let v be the homogeneous deformation defined by $v(x) = Fx + a$ for $x \in \bar{\Omega}$ where $F \in M_+$ and $a \in \mathbb{R}^N$. Let u be any classical solution of (EE) satisfying the boundary condition $u(x) = Fx + a$ for $x \in \partial\Omega$.

(a) If $W \in C^2(M_+,\mathbb{R})$ is rank-one convex, $E(u) \leq E(v)$.

(b) If $W \in C^2(M_+,\mathbb{R})$ is strictly quasi-convex, $u \equiv v$ on $\bar{\Omega}$ and so v is the unique classical solution of the boundary value problem.

Remarks 1. The proof can be reduced to the case where $a = o$. Then $v - r\frac{\partial v}{\partial r} \equiv o$ and part (a) follows immediately from the preceeding result.

Using $w \equiv u - v$ and $\nabla v = F$ in the definition of strict quasi-convexity we find that $E(v) < E(u)$ unless $u \equiv v$ on $\bar{\Omega}$.

2. In the case where $\Omega = \{x \in \mathbb{R}^N : \|x\| < 1\}$ and $W \in C^2(M_+, \mathbb{R})$ is strictly quasi-convex, it follows that the homogeneous radial deformation $u(x) = \lambda x$ for $x \in \bar{\Omega}$ is the unique classical solution of the (EE) satisfying the boundary condition $u(x) = \lambda x$ for $x \in \partial\Omega$ where $\lambda > o$ is fixed. In the next section we shall show that this boundary value problem can have an infinite number of weak (non-classical) solutions.

As noted in Lecture 1, when dealing with radial deformations it is natural to consider isotropic materials. Then the stored energy can be written in the form (1.7) and it would be convenient to expres the hypotheses concerning W in terms of Φ. Except for $N = 2$, necessary and sufficient conditions for the rank-one convexity or quasi-convexity of W in terms of Φ do not seem to be known. We note, however, that if W is rank-one convex then

(i) $\Phi_{ii}(v) \geqslant o \quad \forall v \in (o,\infty)^N$ and $1 \leqslant i \leqslant N$

and (ii) $\dfrac{v_i\Phi_i(v) - v_j\Phi_j(v)}{v_i - v_j} \geqslant o \quad \forall v \in (o,\infty)^N$ with $v_i \neq v_j$ and $1 \leqslant i \neq j \leqslant N$.

Thus (i) and (ii) constitute necessary conditions on Φ for the rank-one convexity of W. A useful set of sufficient conditions are provided by the so called *Ogden materials*. For $N = 3$, this class is defined as follows:

$$W(F) = \Phi(v_1,v_2,v_3) = \sum_{i=1}^{3} \phi(v_i) + \sum_{j=1}^{3} \psi(v_i v_{i+1}) + h(v_1v_2v_3)$$

where $\phi,\psi,h \in C^2((o,\infty),\mathbb{R})$ and $v_4 = v_1$.

Then if (a) $\phi''(t) \geqslant o$, $\psi''(t) \geqslant o$, $h''(t) \geqslant o \quad \forall t > o$

and (b) $\phi'(t) \geqslant o$, $\psi'(t) \geqslant o \quad \forall t > o$

it follows that W is polyconvex.

Lecture 3. WEAK RADIAL SOLUTIONS, MULTIPLICITY AND CAVITATION

Throughout this lecture $\Omega = \{x \in \mathbb{R}^N : \|x\| < 1\}$ and we consider radial deformations of a homogeneous isotropic hyperelastic material. The stored energy function W is written in the form (1.7) and Φ is supposed to satisfy the following hypotheses.

(H1) $\Phi \in C^3((o,\infty)^N, \mathbb{R})$ is symmetric.

(H2) $\Phi_{11}(q,t,\ldots,t) > o \quad \forall q,t > o$

(H3) $\Phi \geqslant o$ and $\dfrac{\Phi_2 - \Phi_1}{q - t} - \Phi_{12} < o$ for $q,t > o$ with $q \neq t$

where the partial derivatives Φ_1, Φ_2 and Φ_{12} are evaluated at $(q,t,\ldots,t)$.

(H4) $\forall b > o, \lim_{(q,t)\to(o,b)} \Phi_1(q,t,\ldots,t) = -\infty$

and $\lim_{(q,t)\to(\infty,b)} \Phi_1(q,t,\ldots,t) = +\infty$.

Setting $R(q,t) = \dfrac{q\Phi_1(q,t,\ldots,t) - t\Phi_2(q,t,\ldots,t)}{q-t}$

for $q \neq t$, we note that (H1) implies that R has a C^1 extension to $(o,\infty)^2$.

(H5) $\exists$ positive constants A,B and β such that
$o < R(q,t) \leqslant A + Bt^{\beta}$ for $o < q \leqslant t$.

In the case of N = 3 and an Ogden material, we find that (H1) to (H5) are satisfied provided that:

(i) $\phi, \psi, h \in C^3((o,\infty), \mathbb{R}_+)$

(ii) ϕ'', ψ'' and $h'' > o$ on (o,∞)

(iii) $\lim_{s\to o} h'(s) = -\infty$ and $\lim_{s\to\infty} h'(s) = +\infty$

(iv) $\exists$ positive constants A,B and β such that

$o < \{s\phi'(s)\}' \leqslant A + Bt^{\beta}$ for $o < s \leqslant t$

$o < t\{s\psi'(s)\}' \leqslant A + Bt^{\beta}$ for $o < s \leqslant t^2$ with $t > o$

An interesting discussion of constitutive assumptions is given in [8].

Lemma 3.1. Let (H1) to (H5) be satisfied and consider a radial deformation $u(x) = U(r)\frac{x}{r}$ for $o < \|x\| = r < 1$. Then u is a weak solution of (EE) $\Longleftrightarrow$ U satisfies (REE).

Proof. See Theorem 4.2 and Proposition 6.1 of [3].

Ball has studied the existence of solutions of (REE) using the direct method of the calculus of variations. For $\lambda > o$, he sets

$$A(\lambda) = \{U \in W^{1,1}(o,1) : U(r) > o,\ U'(r) > o \text{ on } (o,1] \text{ and } U(1) = \lambda\}$$

and considers the problem of minimising the total energy

$$J(U) = E(u) = \omega_N \int_o^1 r^{N-1} \Phi\left(U'(r), \frac{U(r)}{r}, \ldots, \frac{U(r)}{r}\right) dr$$

on $A(\lambda)$, where ω_N is the surface area of the unit sphere in $\mathbb{R}^N$. The main results can be summarised as follows.

(i) J attains its minimum on $A(\lambda)$ at a function $U_\lambda \in A(\lambda)$

(ii) U_λ satisfies (REE) and so corresponds to a weak solution of (EE)

(iii) $\exists\ \lambda^* > o$ such that

for $\lambda \leqslant \lambda^*$, U_λ is a classical solution of (EE),
for $\lambda > \lambda^*$, $U_\lambda(o) > o$ and so U_λ is not a classical solution of (EE).
For $\lambda > \lambda^*$, $\lim_{r\to o} [\frac{U(r)}{r}]^{1-N}\ \Phi_1(U'(r),\frac{U(r)}{r},\ldots,\frac{U(r)}{r}) = o$ where $U = U_\lambda$.

The results in (iii) mean that, for $\lambda > \lambda^*$, the deformed body corresponding to the minimiser U_λ has a spherical cavity of radius $U_\lambda(o)$ in the middle. Furthermore the Cauchy stress is zero on the boundary of the cavity.

Ball establishes these results for a restricted class of Ogden materials. His approach has recently been extended by J. Sivaloganathan [7].

We describe an alternative method [10] which has the advantage of yielding all weak radial solutions of (EE) rather than only the minimisers of the energy. We use the shooting method to discuss all solutions of (REE).

For $\lambda > o$ and $\alpha > o$, we consider the initial value problem:

$$\text{(IVP)} \begin{cases} U(r) > o,\ U'(r) > o \text{ for } r \leqslant 1 \\ r\dfrac{d\Phi_1}{dr} = (N-1)[\Phi_2 - \Phi_1] \text{ for } r < 1 \\ U(1) = \lambda \text{ and } U'(1) = \alpha \end{cases}$$

where here and henceforth Φ_i denotes the i-th partial derivative of Φ evaluated at

$$(U'(r),\frac{U(r)}{r},\ldots,\frac{U(r)}{r}).$$

Let $U(\lambda,\alpha): J(\lambda,\alpha) \subset (o,1] \to \mathbb{R}$ denote the unique maximal solution of (IVP) and set

$$\tilde{T}(\lambda,\alpha)(r) = [\frac{U(r)}{r}]^{1-N}\ \Phi_1 \quad \text{for} \quad U = U(\lambda,\alpha).$$

Then $\tilde{T}(\lambda,\alpha)(r)$ gives the radial Cauchy stress at r for the deformation corresponding to $U(\lambda,\alpha)$.

Lemma 3.2. Let (H1) to (H5) be satisfied and let $U = U(\lambda,\alpha)$.

(a) For $o < \alpha < \lambda$ we have that $J(\lambda,\alpha) = (o,1]$ and $(\frac{U(r)}{r})' < o$, $U''(r) > o$ and $\tilde{T}'(r) > o$ for $o < r \leqslant 1$.

In particular, $R(\lambda,\alpha) \equiv \lim_{r\to o} U(r) \geq \lambda - \alpha > o$.

(b) For $o < \alpha = \lambda$, we have $U(\lambda,\lambda)(r) = \lambda r$ on $(o,1]$, and $\tilde{T}(r) = \Phi_1(\lambda,\lambda,\ldots,\lambda)$.

(c) For $o < \lambda < \alpha$, we have that $\inf J(\lambda,\alpha) \geq \frac{\alpha-\lambda}{\alpha} > o$ and $(\frac{U(r)}{r})' > o$, $U''(r) < o$ on $J(\lambda,\alpha)$.

Proof. See [10].

Remarks. It follows that for $o < \lambda < \alpha$, $U(\lambda,\alpha)$ cannot generate a weak solution of (EE). For $\alpha = \lambda$, we obtain a homogeneous radial deformation which is a classical solution of (EE). For $o < \alpha < \lambda$ we now show that we obtain a weak radial solution of (EE) which cannot be a classical solution since $R(\lambda,\alpha) > o$.

Lemma 3.3. Let (H1) to (H5) be satisfied with $\beta < N$. For $o < \alpha < \lambda$, let $U = U(\lambda,\alpha)$ and $u(x) = U(x)\frac{x}{r}$.

Then (i) $u \in W^{1,p}(\Omega)$ for $1 \leq p < N$.

(ii) u is a weak solution of (EE).

(iii) u is not a classical solution of (EE) since $R(\lambda,\alpha) \equiv \lim_{r\to o} U(r) \geq \lambda - \alpha$.

Proof: (i) By Lemma 4.1 of [3], $u \in W^{1,p}(\Omega)$

$$\iff \int_0^1 r^{N-1}\{|U'(r)|^p + \left|\frac{U(r)}{r}\right|^p\}dr < \infty .$$

By Lemma 3.2, $o < U'(r) \leq \alpha$ and $o < \lambda - \alpha \leq U(r) \leq \lambda$. Hence $u \in W^{1,p}(\Omega)$ if and only if $1 \leq p < N$.

(ii) By Lemma 3.1, we need only show that $r^{N-1}\Phi_i \in L^1(o,1)$ for $i = 1,2$.
From (IVP) we obtain,

$$\frac{d}{dr}\tilde{T}(r) = -(N-1)[\frac{U(r)}{r}]^{-N} R(U'(r),\frac{U(r)}{r})[\frac{U(r)}{r}]'$$

for $o < r < 1$ and so

$$\tilde{T}(r) = \tilde{T}(1) + (N-1)\int_r^1 [\frac{U(s)}{s}]^{-N} R(U'(s),\frac{U(s)}{s})[\frac{U(s)}{s}]' ds .$$

Using (H5) and Lemma 3.2, this yields

$$|\tilde{T}(r)| \leq |\tilde{T}(1)| + (N-1)\int_r^1 Cs^N s^{-\beta}[\frac{1}{s}+\frac{1}{s^2}]ds$$

$$= |\tilde{T}(1)| + C(N-1)\int_r^1 \{s^{N-\beta-1} + s^{N-\beta-2}\}ds .$$

It follows that there exist positive constants A and B such that $|\tilde{T}(r)| \leq A + B\, r^{N-\beta-1}$ for $o < r < 1$. Since $r^{N-1}\Phi_1 = U^{N-1}\tilde{T}$, we have that $r^{N-1}\Phi_1 \in L^1(o,1)$ provided that $\beta < N$.

However, IVP can be written as

$$\frac{d}{dr}\{r^{N-1}\Phi_1\} = (N-1)r^{N-2}\Phi_2 \quad \text{for} \quad o < r < 1$$

and so

$$(N-1)\int_\varepsilon^1 |r^{N-1}\Phi_2|dr = \int_\varepsilon^1 r\left|\frac{d}{dr}\{r^{N-1}\Phi_1\}\right|dr$$

$$\leq |r^N\Phi_1|\Big|_{r=1} + |r^N\Phi_1|\Big|_{r=\varepsilon} + \int_\varepsilon^1 |r^{N-1}\Phi_1|dr .$$

We have already shown that

$$|r^N\Phi_1| \leq \lambda^{N-1}\{Ar + Br^{N-\beta}\} \quad \text{for} \quad o < r < 1$$

and so $r^{N-1}\Phi_2 \in L^1(o,1)$.

Lemma 3.4. Let (H1) to (H5) be satisfied with $\beta < N-1$. For $o < \alpha < \lambda$, let U and u be as in Lemma 3.3 and set

$$\tau(\lambda,\alpha) = \lim_{r\to o} \tilde{T}(r),$$

$$E(\lambda,\alpha) = \int_\Omega W(\nabla u)dx = \omega_N \int_0^1 r^{N-1}\Phi\, dr$$

where ω_N is the surface area of the unit sphere in $\mathbb{R}^N$.

Then (i) $-\infty < \tau(\lambda,\alpha) < \lambda^{1-N}\Phi_1(\alpha,\lambda,\ldots,\lambda)$

(ii) $$E(\lambda,\alpha) = \frac{\omega_N}{N}\{\Phi(\alpha,\lambda,\ldots,\lambda) + (\lambda-\alpha)\Phi_1(\alpha,\lambda,\ldots,\lambda) - R(\lambda,\alpha)^N\tau(\lambda,\alpha)\}.$$

Proof. (i) is proved in [10].
Furthermore, as in Lemma 9 of [10], we have that,

$$N\int_\varepsilon^1 r^{N-1}\Phi dr = \{r^N\Phi - r^N U'(r)\Phi_1 + r^{N-1}U(r)\Phi_1\}\Big|_{r=\varepsilon}^1$$

From the proof of Lemma 3.3, we recall that

$$|r^{N-1}\Phi_1| \leq U(r)^{N-1}\{A + Br^{N-\beta-1}\} \quad \text{for} \quad o < r < 1$$

and $$r^{N-1}U(r)\Phi_1 = U(r)^N\,\tilde{T}(r).$$

Hence $$\lim_{\varepsilon\to o}\{N\int_\varepsilon^1 r^{N-1}\Phi dr + \varepsilon^N\Phi\Big|_{r=\varepsilon}\} = \Phi(\alpha,\lambda,\ldots,\lambda) + (\lambda-\alpha)\Phi_1(\alpha,\lambda,\ldots,\lambda) - R(\lambda,\alpha)^N\tau(\lambda,\alpha).$$

By (H3), we have that $\Phi \geqslant o$ and then it follows that

$$\lim_{\varepsilon \to o} \int_{\varepsilon}^{1} r^{N-1}\Phi dr \quad \text{and} \quad \lim_{\varepsilon \to o} \varepsilon^{N}\Phi\Big|_{r=\varepsilon} \quad \text{exist.}$$

But this in turn implies that $\lim_{\varepsilon \to o} \varepsilon^{N}\Phi\Big|_{r=\varepsilon} = o$ and the lemma is proved.

These results lead to the following conclusions concerning Problem 2 of Lecture 1.

Theorem 3.5. Let $\Omega = \{x \in \mathbb{R}^N : \|x\| < 1\}$ and consider the boundary value problem posed by (EE) and the displacement boundary condition $u(x) = \lambda x$ for $x \in \partial\Omega$. Let Φ satisfy (H1) to (H5) where W is related to Φ by (1.7).

(i) There is exactly one classical radial solution. It is the homogeneous radial deformation $u(x) = \lambda x$ for all $x \in \bar{\Omega}$.

(ii) Suppose that $\beta < N$ in (H5). The weak radial solutions form a continuum given by

$$u(\lambda,\alpha)(x) = U(r)\frac{x}{r} \quad \text{for} \quad o < \|x\| \leqslant 1$$

where $o < \alpha \leqslant \lambda$ and $U = U(\lambda,\alpha)$. For $o < \alpha < \lambda$, these weak solutions are not classical solutions. If $\beta < N-1$ in (H5), then all these weak solutions have finite energy.

In view of this multiplicity of weak solutions, it seems natural to inquire whether some additional mechanical restrictions should be imposed. We note that the deformed configuration corresponding to a weak non-classical solution $u(\lambda,\alpha)$, has the form of a ball of radius λ with a spherical cavity of radius $R(\lambda,\alpha)$ in the middle. It is reasonable to prescrible the pressure on the boundary of this cavity and this leads to the following problem.

Problem 3. (cavitation). Given $\lambda > o$ find all solutions of (REE) such that either $U(o) = o$ or

$$U(o) > o \text{ and } \lim_{r \to o} \tilde{T}(r) = o \qquad (3.1)$$

The condition (3.1) means that if there is a cavity then the Cauchy stress on its boundary vanishes. In terms of the results already stated, Problem 3 reduces to the following:

Find $\alpha \in (o,\lambda)$ such that $\tau(\lambda,\alpha) = o$.

It is shown in [10] that τ is strictly increasing in λ and in α on

$$\Delta = \{(\lambda,\alpha) \in \mathbb{R}^2 : o < \alpha < \lambda\}$$

and that

$$\lim_{\alpha \to o} \tau(\lambda,\alpha) = -\infty, \quad \lim_{\alpha \to \lambda} \tau(\lambda,\alpha) = g(\lambda)$$

where g is a continuous strictly increasing function with

$$\lim_{\lambda\to 0} g(\lambda) = -\infty, \quad \lim_{\lambda\to\infty} g(\lambda) = +\infty$$

This analysis of τ requires some extra hypotheses on Φ.

(H6) $\exists\ C > o$ and $t_0 > o$ such that

$$\Phi_{11}(q,t,\dots,t) \geqslant Ct^{2(N-1)} \quad \text{for } o<q<t \text{ and } t \geqslant t_0 .$$

(H7) $\forall b > o, \quad \lim_{(q,t)\to(b,\infty)} t^{1-N}\Phi_1(q,t,\dots,t) = +\infty .$

(H8) $\lim_{t\to 0} t^{1-N}\Phi_1(t,t,\dots,t) = -\infty$

$\lim_{t\to\infty} t^{1-N}\Phi_1(t,t,\dots,t) = +\infty$

(H9) $\exists$ positive constants ε, t_0, K and γ with

$\gamma < 2(N-1)$ such that $\left|\frac{\partial R}{\partial q}(q,t)\right| \leqslant Kt^{\gamma}$

whenever $o < q < \varepsilon$ and $t \geqslant t_0$.

The main results concerning Problem 3 can now be summarised.

Theorem 3.6. Let (H1) to (H9) be satisfied with $\beta < N-1$ in (H5). There exist $\lambda^* > o$ and a strictly decreasing function $w \in C^1((\lambda^*,\infty),\mathbb{R})$ such that

(i) $o < w(\lambda) < \lambda$ for $\lambda > \lambda^*$

(ii) $\lim_{\lambda\to\lambda^*} w(\lambda) = \lambda^*$ and $\lim_{\lambda\to\infty} w(\lambda) = o$,

(iii) $\{(\lambda,\alpha) \in \Delta: \tau(\lambda,\alpha) = o\} = \{(\lambda,w(\lambda)): \lambda > \lambda^*\}$.

This means that for $\lambda \leqslant \lambda^*$, Problem 3 has exactly one solution. It is the homogeneous radial deformation $u(\lambda,\lambda)$. For $\lambda > \lambda^*$, Problem 3 has exactly two solutions, namely, $u(\lambda,\lambda)$ and $u(\lambda,w(\lambda))$. The second solution has a cavity of radius $R(\lambda,w(\lambda))$ and a considerable amount of additional information is obtained in [10]. We simply note that the energy $E(\lambda,\lambda)$ of the solution $u(\lambda,\lambda)$ is greater that the energy $E(\lambda,w(\lambda))$ of $u(\lambda,w(\lambda))$ for all $\lambda > \lambda^*$. The critical value λ^* is the unique solution of the equation $g(\lambda) = o$ and the family $u(\lambda,w(\lambda))$ bifurcates from the family $u(\lambda,\lambda)$ at this value.

References

[1] Gurtin, M.E.: *An Introduction to Continuum Mechanics*, Academic Press, New York, 1981.

[2] Ciarlet, P.G.: *Three-Dimensional Elasticity*, Tata Institute Lecture Notes, Springer, Berlin, 1983.

[3] Ball, J.M.: 'Discontinuous equilibrium solutions and cavitation in nonlinear elasticity', *Phil. Trans. R. Soc. London*, A306 (1982), 557-611.

[4] Ball, J.M.: 'Constitutive inequalities and existence theorems in nonlinear elastostatics', in *Nonlinear Analysis and Mechanics, Heriot-Watt Symposium* Vol.1, Ed. R.K. Knops, Pitman, London, 1977.

[5] Morrey, C.B.: *Multiple Integrals in the Calculus of Variations*, Springer, Berlin, 1966.

[6] Green, A.E.: 'On some general formulae in finite elastostatics', *Arch. Rational Mech. Anal.*, 50 (1973), 73-80.

[7] Sivaloganathan, J.: 'Uniqueness of regular and singular equilibria in radial elasticity', preprint Heriot-Watt University, 1985.

[8] Antman, S.S.: 'Regular and singular problems for large elastic deformations of tubes, wedges and cylinders', *Arch. Rational Mech. Anal.*, 83 (1983), 1-52.

[9] Knops, R.J. and Stuart C.A.: 'Quasiconvexity and uniqueness of equilibrium solutions in nonlinear elasticity, *Arch. Rational Mech. Anal.*, 86 (1984), 233-249.

[10] Stuart, C.A.:'Radially symmetric cavitation for hyperelastic materials', *Ann. Inst. H. Poincaré, Anal. Non linéaire*, 2 (1985), 33-66.

AN INDEX FOR HAMILTONIAN SYSTEMS WITH A NATURAL ORDER STRUCTURE

J. F. Toland
School of Mathematics
University of Bath
Claverton Down
Bath BA2 7AY England

ABSTRACT. A brief account is given of degree theoretical considerations which arise when studying Hamiltonian systems whose 'kinetic energy' is indefinite with a negative cone of ellipsoidal cross-section in $\mathbb{R}^n$. Such Hamiltonian systems enjoy a powerful monotonicity property and the purpose of these notes is to explain how this is so, and how it leads to very natural theorems on the existence of homoclinic and periodic orbits.

1. INTRODUCTION

Recently [1] Hofer and I studied a class of Hamiltonian systems in n-dimensional real Euclidian space which are of the form

$$\dot{q}(t) = Sp(t), \qquad \dot{p}(t) = -V'(q(t)),$$

where S is a real symmetric operator with (n-1) positive and one negative eigenvalue, and V is a potential. The corresponding Hamiltonian $H = \frac{1}{2} \langle Sp,p \rangle + V(q)$ is conserved on trajectories. As a consequence if $V > 0$ and $H = 0$, then $\langle S^{-1}\dot{q},\dot{q} \rangle = \langle p,Sp \rangle < 0$, and this is a constraint that $\dot{q}$ must lie in the set $\{q : \langle S^{-1}q,q \rangle < 0\}$. This is a cone of ellipsoidal cross-section, since S has the special form cited above. Hence if $H = 0 < V$ trajectories are monotonic in q-space with respect to a natural ordering induced by S.

The present lecture is to give a brief account of theory for a certain class of such Hamiltonians which unifies the homoclinic and periodic theories of [1], and the main conclusion is that under quite general and natural hypotheses either a homoclinic or a periodic must occur. (In [1] we gave two theorems one covering each possibility.) It is then often easy to say precisely which of the alternatives occurs in practice. More significantly the proofs here are explicitly a consequence of degree theoretical calculations, and hence the existence theory is, in a topological sense, stable. It leads in the usual way to the existence of continua of solutions in parameter dependent problems and to multiplicity results when the local index can be calculated.

S. P. Singh (ed.), Nonlinear Functional Analysis and Its Applications, 147–160.

Problems of this type arise in problems from applied mathematics. For example, Toland [2], Klassen and Troy [3] have recently analysed parameter-dependent problems which fit into the general framework. Though classical Hamiltonian particle mechanics falls outside this general framework, similar methods do work in certain limited circumstances there also [5].

2. PRELIMINARIES

2.1 *Notation and terminology*

Let $(E,<,>)$ denote an n-dimensional real inner produce space, and let $\mathcal{B}(E)$ denote the space of all bounded linear operators on E. If $V: E \to \mathbb{R}$ is differentiable then its derivative at x, $dV[x] \in E^*$, and $V'(x) \in E$ is then defined by the formula

$$dV[x](y) = \langle V'(x), y\rangle , \quad x,y \in E.$$

By a Lorenz operator on E we will mean a symmetric operator with one negative and (n-1) positive eigenvalues, and the set of all such operators we denote by $L(E)$. If $S \in L(E)$, the spectrum $\sigma(S)$ of S is $\lambda_1 \geq \lambda_2 \geq \ldots \geq \lambda_{n-1} > 0 > \lambda_n$, $\{e_i : i = 1,\ldots n\}$ is a corresponding set of eigenvectors, and

$$K(S) = \{q \in E : \langle S^{-1}q,q\rangle < 0\} = K^+(S) \cup K^-(S)$$

where

$$K^{\pm}(S) = \{q \in K(S) : \pm\langle q,e_n\rangle \geq 0\}, \text{ whence } K^+(S) \cap K^-(S) = \emptyset.$$

(Note that $K^+(S)$ and $K^-(S)$ depend on e_n or $-e_n$ being chosen as the eigenvector of S corresponding to $\lambda_n < 0$. It will be assumed throughout that once the set $\{e_i\}$ is chosen, it is fixed, thus avoiding ambiguity.)

Now $K^{\pm}(S)$ are open, and their closure $\overline{K^{\pm}(S)}$ are closed cones in E. For the sake of having a convenient notation we denote them by $P^{\pm}(S)$ respectively. Now a partial ordering $\leq_S$ can be defined on E by putting

$$q_1 \leq_S q_2 \text{ if and only if } q_2 - q_1 \in P^+(S),$$

and we will write

$$q_1 <_S q_2 \text{ if and only if } q_2 - q_1 \in K^+(S).$$

Now if $U \subset E$, then $\hat{q} \in U$ is said to be S-maximal (S-minimal) in U if it is a maximal (minimal) element of U with respect to $\leq_S$. We say that two elements are S-commensurate if they are ordered relative to one another by $\leq_S$. The set of S-maximal (S-minimal) elements of U will be referred to as the S-maximal (S-minimal) set in U. An element $\hat{q} \in U$ will be called weakly S-maximal (weakly S-minimal) in U if the set $\{q \in U : \hat{q} <_S q\}(\{q \in U : q <_S \hat{q}\})$ is empty.

DEFINITION. A functional $H : E \times E \to \mathbb{R}$ is called an admissible Hamiltonian if

$$H(q,p) = \tfrac{1}{2}\langle Sp,p\rangle + V(q), \quad (q,p) \in E \times E,$$

where $S \in L(E)$ and there exists a bounded open convex set $C \subset E$ such that

(i) V is thrice continuously differentiable on $\overline{C}$ (the closure of C);

(ii) $V > 0$ on C and $V = 0$ on ∂C (the boundary of C);

(iii) every weakly S-maximal (weakly S-minimal) element of $\overline{C}$ is S-maximal (S-minimal), and if q_0 lies on the boundary of either the S-maximal or the S-minimal set and $\langle S\eta,\eta\rangle = 0$ for some $\eta \in N(q_0)$, then $V'(q_0) \neq 0$ and $\langle V''(q_o)SV'(q_o),SV'(q_o)\rangle < 0$. (Here $N(q_0) = \{\eta \in E : \|\eta\| = 1, \langle q-q_o,\eta\rangle \geq 0, q \in \overline{C}\}, q_0 \in \partial C$.)

Remarks 1. If $q_0 \in \partial\overline{C}$ is S-maximal, then $(q_0 + P^+(S)) \cap \overline{C} = \{q_0\}$, while if q_0 is weakly S-maximal then $(q_0 + K^+(S)) \cap \overline{C} = \emptyset$.

2. From now on the prefix and subscript S will be omitted from the notation. A given admissible Hamiltonian supplies $S \in L(E)$ and the ordering is automatically determined up to sign.

There now follows a few results which help the determination of maximal and minimal sets when H is admissible.

PROPOSITION 1.(a) *The maximal and the minimal sets are closed, non-empty subsets of* ∂C.

(b) *If* $\langle SV'(q_o),V'(q_o)\rangle < 0$, $q_0 \in \partial C$ *then* q_0 *is maximal or minimal; if* $\langle SV'(q_o),V'(q_o)\rangle > 0$, *then* q_0 *is neither maximal nor minimal; if* $V'(q_o) \neq 0$, $\langle SV'(q_o),V'(q_o)\rangle = 0$, *then* q_0 *is on the boundary of the maximal or minimal sets.*

Proof (a) Let $\{q_n\}$ be a sequence of maximal points such that $q_n \to q_0$, and let $q \in \overline{C}$. If $q \in q_0 + K^+(S)$, then $q \in q_n + K^+(S)$ for all n sufficiently large since $K^+(S)$ is open. This contradicts the maximality of q_n. Hence q_0 is weakly maximal, and so is maximal by the admissible of H. Hence the maximal set is closed; it is non-empty because any "north pole" of ∂C lies in it, and it is clearly a subset of ∂C. Similarly for the minimal set.

(b) Suppose $\langle SV'(q_o),V'(q_o)\rangle < 0$, $q_0 \in \partial C$. Now $V'(q_0)$ is an inward normal to the set $\overline{C}$ at a point $q_0 \in \partial C$; i.e. $\langle q-q_o,V'(q_o)\rangle \geq 0$ for all $q \in \overline{C}$. However if $\langle S^{-1} q,q\rangle \leq 0$, then

$$\begin{aligned}\langle q,V'(q_o)\rangle &= \sum_1^n \langle q,e_i\rangle\langle V'(q_o),e_i\rangle \\ &= \sum_1^{n-1} \lambda_i^{-\frac{1}{2}}\langle q,e_i\rangle\lambda_i^{\frac{1}{2}}\langle V'(q_o),e_i\rangle + \langle q,e_n\rangle\langle V'(q_o),e_n\rangle \\ &\geq \langle q,e_n\rangle\langle V'(q_o),e_n\rangle - \Big\{\Big(\sum_1^{n-1} \lambda_i^{-1}\langle q,e_i\rangle^2\Big)\Big(\sum_1^{n-1} \lambda_i\langle V'(q_o),e_i\rangle^2\Big)\Big\}^{\frac{1}{2}},\end{aligned}$$

whose sign changes with that of $\langle q,e_n\rangle$ since

$$\lambda_n < 0 < \lambda_{n-1} \le \ldots \le \lambda_1,$$

$$0 > \langle SV'(q_o), V'(q_o)\rangle = \lambda_n \langle V'(q_o), e_n\rangle^2 + \sum_1^{n-1} \lambda_i \langle V'(q_o), e_i\rangle^2,$$

and $$0 \ge \langle S^{-1}q, q\rangle = \lambda_n^{-1}\langle q, e_n\rangle^2 + \sum_1^{n-1} \lambda_i^{-1}\langle q, e_i\rangle^2.$$

Therefore $P^+(S) + q_0$ and $P^-(S) + q_0$ are separated by the supporting hyperplane normal to ∂C at q_0. As a consequence q_0 is either maximal or minimal.

A similar argument shows that if $\langle SV'(q_o), V'(q_o)\rangle > 0$, then q_0 is neither maximal nor minimal.

If $V'(q_0) \neq 0$, and $\langle SV'(q_o), V'(q_o)\rangle = 0$ then the argument just given ensures that q_0 is maximal or minimal. Without loss of generality suppose the former. If q_0 is an interior point of the maximal set then $\langle SV'(q), V'(q)\rangle \le 0$ for all q in a neighbourhood of q_0 in ∂C. Let H^* denote the plane spanned by $V'(q_0)$ and $SV'(q_0)$ which passes through q_0, and let $\mathcal{C}$ denote the convex curve where it intersects ∂C. Then the line $\{q_0 + tSV'(q_0) : t \in \mathbb{R}\}$ is tangent to $\mathcal{C}$ at q_o, and $\langle S^{-1}SV'(q_o), SV'(q_o)\rangle = 0$ (i.e. $SV'(q_0)$ lies on the boundary of $P(S)$). Moreover $(q + P^+(S)) \cap \mathcal{C} = q$ for all q in a neighbourhood of q_0 in $\mathcal{C}$ since $\langle SV'(q), V'(q)\rangle \le 0$ and each weakly maximal point is maximal. However, this contradicts the convexity of $\mathcal{C}$. q.e.d.

Recall that if $q_0 \in \partial C$, $N(q_0)$ denotes the set of inward normals to hyperplanes supporting C at q_0; i.e.

$$N(q_0) = \{\eta : \|\eta\| = 1,\ \langle q - q_o, \eta\rangle \ge 0,\ q \in C\}.$$

PROPOSITION 2. *Suppose* $q_0 \in \partial C$. *If there exists* $\eta \in N(q_0)$ *such that* $\langle S\eta, \eta\rangle = 0$ *then* q_0 *is on the boundary of either the maximal or the minimal set. In particular,* $V'(q_0) \neq 0$ *and* $\langle V''(q_o)SV'(q_o), SV'(q_o)\rangle < 0$.

Proof Let $q_0 \in \partial C$ and $\eta \in N(q_0)$ be such that $\langle S\eta, \eta\rangle = 0$. Then the argument of Proposition 1 ensures that q_0 is maximal or minimal. Since $q_0 + tS\eta \in q_0 + P^+(S)$ for all $t \in \mathbb{R}$, and since $q_0 + tS\eta$ lies on H_0, the supporting hyperplane normal to η at q_0, it follows that $\{q_0 + tS\eta : t \in \mathbb{R}\} \cap \overline{C} = \{q_0\}$ since every weakly maximal or minimal point of $\overline{C}$ is maximal or minimal.

Now for any $\varepsilon > 0$ sufficiently small the parallel hyperplane $H_\varepsilon = \{q : \langle q - q_o, \eta\rangle = \varepsilon\}$ intersects C; say $q_\varepsilon \in H_\varepsilon \cap C$, and let t_ε^i be such that $q_\varepsilon + t_\varepsilon^i S\eta \in \partial C$, $i = 1,2$. Then $t_\varepsilon^i \to 0$ as $\varepsilon \to 0$, since otherwise $\{q_0 + tS\eta : t \in \mathbb{R}\} \cap \overline{C} \neq \{q_0\}$. Clearly either $q_\varepsilon + t_\varepsilon^1 S\eta \ge q_\varepsilon + t_\varepsilon^2 S\eta$, or *vice versa*. In any case q_0 cannot be an interior point of either the maximal or the minimal sets. q.e.d.

Now we give two technical results in preparation for the next section.

PROPOSITION 3. *Suppose* $q_0 \in \partial C$, *and* $\langle S\eta, \eta\rangle > 0$ *for all* $\eta \in N(q_0)$. *Then there exists* $\eta^* \in N(q_0)$ *such that* $q_0 + tS\eta^* \in C$ *for all* $t > 0$ *sufficiently small.*

Proof Let $K = \{q_o + tS\eta : \eta \in N(q_o),\ t \geq 0\}$. Then K is a convex set. Suppose that $K \cap C = \emptyset$. Then the Hahn-Banach separation theorem ensures the existence of $\eta \in N(q_o)$ such that $<q-q_o, \eta> \leq 0$ for all $q \in K$. But this means that $<S\eta,\eta> \leq 0$ which contradicts the hypotheses. Hence $K \cap C \neq \emptyset$ and the result follows. q.e.d.

PROPOSITION 4. *Let* $q_o \in \partial C, <S\eta,\eta> > 0$ *for all* $\eta \in N(q_o)$, *let* η^* *be given by Proposition 3, and let* $W = \{q-tS\eta^* : t \geq 0,\ q \in C\}$. *Then* $W \cap \partial C = U$ *is an open neighbourhood of* q_o *in* ∂C, *and* $W \cap H^*$ *is an open neighbourhood of* q_o *in* H^*, *the hyperplane through* q_o *normal to* η^*. *Let* $Q^* : W \setminus C \to \partial C$ *denote the affine projection onto* ∂C *parallel to* $S\eta^*$ *defined by*

$$Q^*(q) = q + tS\eta^* \in \partial C \text{ if } q + sS\eta^* \notin C, \quad s < t.$$

Then $<V'(Q^*(q)),S\eta^*> \geq 0$ *for all* $q \in W \setminus C$.

Proof That U is open in ∂C is clear from the construction; that $W \cap H^*$ is open in H^* follows because $<S\eta^*,\eta^*> \neq 0$.

Now for $q \in W \setminus C$, $V(Q^*(q)) = 0$, whence

$$(d/dt)V(Q^*(q) + tS\eta^*)\big|_{t=0} \geq 0.$$

Hence $\quad V'(Q^*(q),S\eta^*) \geq 0.$ q.e.d.

2.2 *Initial-Value Problems*

Here we consider the initial value problem for admissible Hamiltonian systems

$$\dot{q}(t) = Sp(t), \quad \dot{p}(t) = -V'(q(t)), \quad t > 0,$$
$$q(0) = q_o, \qquad p(0) = p_o. \tag{IVP}$$

PROPOSITION 5. *Suppose* $q_o \in \partial C$ *and* $p_o = 0$.

(a) *If* $V'(q_o) = 0$, *then* $q(t) = q_o$, $p(t) = 0$, *for all* $t > 0$.

(b) *If* $V'(q_o) \neq 0$, *then*

(i) *there exists* $\varepsilon > 0$ *such that* $q(t) \in C$ *for all* $t \in (0,\varepsilon)$ *if and only if* $<V'(q_o),SV'(q_o)> < 0$;

(ii) *there exists* $\varepsilon > 0$ *such that* $q(t) \notin \overline{C}$ *for all* $t \in (0,\varepsilon)$ *if and only if* $<V'(q_o),SV'(q_o)> \geq 0$.

Proof Part (a) is obvious. Part (b) follows at once from [1; Lemma 4] in the light of Proposition 1. q.e.d.

The next result is quite straightforward.

PROPOSITION 6. *Suppose* $q_0 \in \partial C$ *and* $\langle Sp_0, p_0 \rangle = 0$.
(a) *If* $\langle V'(q_o), Sp_o \rangle \neq 0$, *then there exists* $\varepsilon > 0$ *such that* $q(t) \in C$, $t \in (0,\varepsilon)$ *if and only if* $\langle V'(q_0), Sp_0 \rangle > 0$.
(b) *If* $V'(q_0) = 0$ *and* $\langle \eta, Sp_0 \rangle < 0$ *for some* $\eta \in N(q_0)$, *then* $q(t) \notin \overline{C}$, $t \in (0,\varepsilon)$ *for some* $\varepsilon > 0$.

Proof Since $V(q(0)) = 0$ and

$$(d/dt)V(q)(t) = \langle V'(q(t)), \dot{q}(t) \rangle = \langle V'(q_0), Sp_0 \rangle \text{ at } t = 0,$$

part (a) is immediate. Moreover in part (b)

$$0 > \langle \eta, Sp_0 \rangle = (d/dt)\langle \eta, q(t) - q_0 \rangle$$

implies that $q(t) \notin \overline{C}$, $t \in (0,\varepsilon)$ since η is an inward normal to ∂C.

q.e.d.

The next result along the same lines is somewhat more subtle.

PROPOSITION 7. *Suppose that* $q_0 \in \partial C$, $p_0 \neq 0$, $\langle Sp_0, p_0 \rangle = 0$, *and* $\langle \eta, Sp_0 \rangle = 0$ *for all* $\eta \in N(q_0)$. *Then there exists no* $\varepsilon > 0$ *such that* $q(t) \in C$, $t \in (0,\varepsilon)$.

Proof First of all consider the case when $V'(q_0) \neq 0$. The hypothesis of the theorem then says that $\langle Sp_0, p_0 \rangle = \langle V'(q_0), Sp_0 \rangle = 0$. Suppose also that $\langle SV'(q_0), V'(q_0) \rangle < 0$. Then

$$-\lambda_n \langle V'(q_0), e_n \rangle \langle p_0, e_n \rangle = \sum_1^{n-1} \lambda_i \langle V'(q_0), e_i \rangle \langle p_0, e_i \rangle$$

$$\leq \left\{ \left(\sum_1^{n-1} \lambda_i \langle V'(q_0), e_i \rangle^2 \right) \left(\sum_1^{n-1} \lambda_i \langle p_0, e_i \rangle^2 \right) \right\}^{\frac{1}{2}}$$

(by Hölder's inequality)

$$= \left\{ \left(\sum_1^{n-1} \lambda_i \langle V'(q_0), e_i \rangle^2 \right) \left(|\lambda_n| \langle p_0, e_n \rangle^2 \right) \right\}^{\frac{1}{2}}$$

$$< \quad |\lambda_n \langle V'(q_0), e_n \rangle \langle p_0, e_n \rangle|,$$

and replacing $V'(q_0)$ by $-V'(q_0)$ in the above calculation gives the opposite (strict) inequality. This contradiction ensures that $\langle SV'(q_0), V'(q_0) \rangle \geq 0$. The result is now immediate from [1, Lemma 4].

So suppose that $V'(q_0) = 0$ and $\langle \eta, Sp_0 \rangle = 0$ for all $\eta \in N(q_0)$. The argument just given ensures that $\langle S\eta, \eta \rangle \geq 0$ for all $\eta \in N(q_0)$, and Proposition 2 ensures that $\langle S\eta, \eta \rangle > 0$ for all $\eta \in N(q_0)$. Therefore by Proposition 3 there exists $\eta^* \in N(q_0)$ such that $q_0 + tS\eta^* \in C$ for all $t > 0$ sufficiently small, and let Q^* be the corresponding affine projection.

Now define a vector field F on W by

$$F(q) = -V'(q), \qquad q \in C$$

$$F(q) = -V'(Q^*(q)), \quad q \in W\backslash C.$$

We note that F is Lipschitz continuous because V′ is, and that $\langle SF(q), \eta^* \rangle = \langle F(q), S\eta^* \rangle \leq 0$, $q \in W\backslash C$.

Now consider the initial-value problem

$$\dot{q}(t) = Sp(t), \quad \dot{p}(t) = F(q(t)), \quad t > 0,$$

$$q(0) = q_o \quad , \quad p(0) = p_o,$$

where (q_o, p_o) is given in the statement of the theorem. Then by standard theory there exists a unique solution which satisfies

$$(q(t),p(t)) = \int_0^t (Sp(w), F(q(w))dw + (q_o, p_o).$$

Moreover if $T > 0$ is chosen sufficiently small then the iteration

$$(q^o(t), p^o(t)) = (q_o, p_o)$$

and for $n \geq 1$,

$$(q^n(t), p^n(t)) = \int_0^t \left(Sp^{n-1}(w), F(q^{n-1}(w))\right) dw + (q_o, p_o)$$

converges uniformly on $[0,T]$ to the solution, and $q^n(t) \in W$ for $t \in [0,T]$. Note that

$$\langle q^o(t) - q_o, \eta^* \rangle = 0, \quad t \in [0,t] \quad \text{and so } q^o(t) \in W\backslash C,$$

and

$$\langle Sp^o(t), \eta^* \rangle = 0, \quad t \in [0,T].$$

Now suppose that

$$\langle q^{n-1}(t) - q_o, \eta^* \rangle \leq 0 \text{ and } \langle Sp^{n-1}(t), \eta^* \rangle \leq 0, \quad t \in [0,T].$$

Then

$$\langle q^n(t) - q_o, \eta^* \rangle = \int_0^t \langle Sp^{n-1}(w), \eta^* \rangle dw \leq 0$$

whence $q^n(t) \in W\backslash C$, and

$$\langle Sp^n(t), \eta^* \rangle = \int_0^t \langle F(q^{n-1}(w)), S\eta^* \rangle dw + \langle Sp_o, \eta^* \rangle$$

$$\leq 0 \text{ since } q^{(n-1)}(t) \in W\backslash C, \quad t \in [0,T].$$

Therefore in the limit

$$\langle q(t) - q_o, \eta^* \rangle \le 0, \quad \text{and so } q(t) \notin C, \quad t \in [0,T].$$

Hence, by the uniqueness theorem for initial-value problems, the solution of (IVP) cannot have $q(t) \in C, t \in (0,\varepsilon)$, for any $\varepsilon > 0$. q.e.d

Let Γ^+ and Γ^- denote respectively the maximal and minimal sets, let γ^+ and γ^- denote their boundaries in ∂C, and let $\Gamma = \Gamma^+ \cup \Gamma^-$, $\gamma = \gamma^+ \cup \gamma^-$.

With the convention that $\sup(\phi) = 0$ we define $\tau : \partial C \to \mathbb{R} \cup \{+\infty\}$ as follows:

$\tau(q_0) = \sup\{t > 0 : q(s) \in C, s \in (0,t)\}$ *where* $(q(t),p(t))$ *solves* (IVP) *with* $q_0 \in \partial C$ *and* $p_0 = 0$.

Remark It is an immediate consequence of Proposition 5 that $\tau(q_0) > 0$ if and only if $\langle V'(q_o), SV'(q_o)\rangle < 0$. We abbreviate $\tau(q_0)$ by τ where appropriate.

PROPOSITION 8. *There exists a constant* M *such that*

$$\int_0^{\tau(q_0)} \|\dot{q}(t)\|\, dt \le M \text{ and } \sup\{\|\dot{q}(t)\| : t \in (0,\tau(q_0))\} \le M$$

where M *is independent of* $q_0 \in \partial C$.

Proof The first *a priori* bound is proved in [1; Lemma 1]. Since $\{\|V'(q)\| : q \in \overline{C}\}$ is bounded, it follows that $\sup\{\|\ddot{q}(t)\| : t \in (0,\tau)\}$ is bounded independently of q_0. Hence $\|\dot{q}(t)\| \ge \frac{1}{2} \sup\{\|\dot{q}(t)\| : t \in (0,\tau)\}$ on an interval of length at least $\min\{\frac{1}{2}\tau, (2M)^{-1}\|\dot{q}\|_{L_\infty(0,\tau)}\}$.

Hence by the first inequality

$$\|\dot{q}\|_{L_\infty(0,\tau)} \min\{\tau, M^{-1}\|\dot{q}\|_{L_\infty(0,\tau)}\} \le \text{const.}$$

However, since $\dot{q}(0) = 0$, and $\|\ddot{q}\|$ is uniformly bounded, it follows that $\|\dot{q}\|_{L_\infty(0,\tau)} \le (\text{const.})\tau$, whence $\|\dot{q}\|_{L_\infty(0,\tau)} \le$ const. This completes the proof. q.e.d.

3. THE INDEX

3.1 *The definition of the degree*

Suppose that H is an admissible Hamiltonian and let $\{(q(t),p(t)) : t \ge 0\}$ denote the solution of (IVP) where $q_0 \in \partial C$ and $p_0 = 0$. It is an immediate consequence of Proposition 5 that $\tau(q_0) > 0$ if and only if $\langle V'(q_o), SV'(q_o)\rangle < 0$. Now we define a function θ on ∂C as follows:

$$\theta(q_o) = \begin{cases} \tau^{-1}Sp(\tau) & \text{if } \tau = \tau(q_o) > 0 \\ -SV'(q_o) & \text{if } \tau = \tau(q_o) = 0. \end{cases}$$

PROPOSITION 9. *The mapping* θ *is continuous from* ∂C *into* E, *and* $\langle S^{-1}\theta(q_o), \theta(q_o)\rangle = 0$ *for all* $q_o \in \partial C$. *In other words* $\theta : \partial C \to \partial P(S)$.

Proof Let q_o^k, $q_o \in \partial C$ with $q_o^k \to q_o$ as $k \to \infty$. First we consider the behaviour of $\tau_k = \tau(q_o^k)$ as $k \to \infty$ when $p(\tau(q_o)) \neq 0$.

If $0 < \tau = \tau(q_o) < \infty$, then $q(\tau) \in \partial C$, $p(\tau) \neq 0$, $\langle \eta, \dot{q}(\tau)\rangle = \langle \eta, Sp(\tau)\rangle \leq 0$ for all $\eta \in N(q(\tau))$, and $V(q(\tau)) = \langle Sp(\tau), p(\tau)\rangle = 0$. By Proposition 7 $\langle \eta, Sp(\tau)\rangle < 0$ for some $\eta \in N(q(\tau))$ (for otherwise $(\hat{q}(t), \hat{p}(t)) = (q(2\tau - t), -p(2\tau - t))$ solves the differential equations with $\hat{q}(0) = q(\tau) \in \partial C$, $\hat{p}(0) = -p(\tau)$, and $\hat{q}(t) \in C$, $t \in (0,\tau)$ which contradicts Proposition 7). Hence $\langle \eta, Sp(\tau)\rangle < 0$ for some $\eta \in N(q(\tau))$ and so by Proposition 6 there exists $\varepsilon > 0$ such that $q(t) \notin \overline{C}$, $t \in (\tau, \tau+\varepsilon)$. By Proposition 5 $\langle V'(q_o^k), SV'(q_o^k)\rangle < 0$, and it follows by the standard theory of continuous dependence for initial-value problems that $\tau_k \to \tau$ as $k \to \infty$ (see [1, Section III]).

If $\tau(q_o) = \infty$, then $\langle SV'(q_o), V'(q_o)\rangle > 0$ and the same method ensures that $\tau_k \to \infty$ as $k \to \infty$.

Therefore if $0 < \tau(q_o) \leq \infty$ it follows from standard theory and Proposition 8 that $\theta(q_o^k) \to \theta(q_o)$ as $k \to \infty$ when $p(\tau(q_o)) \neq 0$.

Now suppose $\tau(q_o) = 0$, and first of all consider the case when $V'(q_o) = 0$. Then $\theta(q_o) = 0$. If $\tau(q_o^k) \to \infty$ as $k \to \infty$, then $\theta(q_o^k) \to \theta(q_o) = 0$ as $k \to \infty$, by Proposition 8. If $\tau(q_o^k) = 0$ for all k, then $\theta(q_o^k) = -V'(q_o^k) \to -V'(q_o) = \theta(q_o)$ as $k \to 0$. If $0 < \tau_k \to 0$ as $k \to \infty$, then

$$\begin{aligned} \theta(q_o^k) &= \tau_k^{-1}\, Sp^k(\tau_k) \\ &= \tau_k^{-1}\int_0^{\tau_k} -SV'(q^k(w))\,dw \to -SV'(q_o) \\ &= 0 = \theta(q_o) \quad \text{as } k \to \infty. \end{aligned}$$

The possibility that $\tau_k \to \tau > 0$ can be excluded for $\tau_k > 0$ implies that $q_o \in \Gamma$ and $q^k(\tau_k) \notin \Gamma$, and from $V'(q_o) = 0$ it follows that $q^k(\tau_k) \to q_o$ as $k \to \infty$. Hence $q_o \in \gamma$ and so $V'(q_o) \neq 0$, by the admissibility of H, which contradicts the assumption.

Finally, consider the case $\tau(q_o) = 0$ and $V'(q_o) \neq 0$. Then, by Proposition 5, $q(t) \notin \overline{C}$, $t \in (0,\varepsilon)$ for $t \in (0,\varepsilon)$ whence it follows that $\tau_k \to 0$ as $k \to \infty$. Without loss of generality suppose that $\tau_k > 0$ for all k.

Then

$$\theta(q_o^k) = \tau_k^{-1} \ Sp^k(\tau_k)$$

$$= \tau_k^{-1} \int_0^{\tau_k} -SV'(q^k(w))\,dw \to -SV'(q_o)$$

$$= 0 = \theta(q_o) \quad \text{as} \quad k \to \infty.$$

Thus θ is continuous at q_o. When $p(\tau(q_0)) = 0$, and $\tau(q_0) \in (0,\infty)$ then $q(2\tau-t) = q(t)$, $p(2\tau-t) = -p(t)$. The continuity is then an easy consequence of standard existence theory for initial-value problems. Thus $\theta : \partial C \to E$ is continuous. Now

$$<S^{-1}\theta(q_o),\theta(q_o)> = \begin{cases} \tau^{-2}<Sp(\tau),p(\tau)>, & 0 < \tau \le \infty \\ <V'(q_o),\ SV'(q_o)>, & \tau = 0 \end{cases} \ge 0. \qquad \text{q.e.d.}$$

Let P denote the projection on E defined by

$$P(q) = q - <q,e_n> e_n,$$

and let R denote the stereographic projection of the hyperplane $F_n = \{q \in E : <q,e_n> = 0\}$ onto the unit sphere $S^{n-1} = \{q \in E : \|q\| = 1\}$. Then the mapping $\Phi = R \circ P \circ \theta : \partial C \to S^{n-1}$ is a continuous function between (n-1)-dimensional topological manifolds. In order to standardise orientations let $(a + rS^{n-1})$ denote an (n-1)-dimensional sphere, of radius r and centre a, lying in C, and let R^* denote the radial mapping of S^{n-1} onto ∂C given by the mapping $q \to a + rq$ followed by radial retraction onto ∂C. Then we unambiguously define the degree of our admissible Hamiltonian system as follows.

DEFINITION. *If* H *is an admissible Hamiltonian system then its degree is defined to be*

$$\deg_B(S^{n-1},\ R \circ P \circ \theta \circ R^*) = d(H)$$

where $\deg_B$ *denotes the Brouwer degree, and* $R \circ P \circ \theta \circ R^* : S^{n-1} \to S^{n-1}$.

Remark Since $P \circ \theta : \partial C \to F_n$ is bounded, it is immediate that $d(H) = 0$ for any admissible Hamiltonian system H. The significance of the definition is that it can be calculated locally. In particular we are interested in calculating the index of $R \circ P \circ \theta \circ R^*$ on subsets of S^{n-1}.

3.2 *Calculation of the index*

PROPOSITION 10. *If* H *is an admissible Hamiltonian and* $V'(q) \neq 0$, $q \in \Gamma^+$ (or Γ^-) *then* Γ^+ (or Γ^-) *is contractible in itself to a point.*

Proof Let $A = \{q \in \partial C : -e_n \in N(q)\}$. Then A is a convex subset of Γ^+. Hence there exists an open neighbourhood B of A in ∂C such that B is contractible to a point $q \in A$. Since $V'(q) \neq 0$, $q \in \Gamma^+$ it follows that B may be chosen as a subset of Γ. It suffices to define $\Phi : \Gamma^+ \times [0,T] \to \Gamma^+$ such that $\Phi(q,0) = q$, $\Phi(q,T) \in B$ for all $q \in \Gamma^+$, for some $T > 0$. To this end let $\Phi(q,t)$ be the solution $q(t)$ of the initial-value problem

$$\dot{q}(t) = SV'(q(t)) - (\langle SV'(q(t)), V'(q(t))\rangle / \|V'(q(t))\|^2) V'(q(t)), t > 0$$

$$q(0) = q \in \Gamma^+.$$

The proof will be complete once it is shown that $q(t) \in \Gamma^+$ for all $t > 0$ and that, for some $T > 0, q(T) \in B$, for any initial data $q \in \Gamma^+$.

We begin by noting that

$$(d/dt)V(q(t)) = \langle V'(q(t)), \dot{q}(t)\rangle = 0$$

whence $V(q(t)) = V(q(0)) = 0$, $t \geq 0$. Also, by Proposition 1, $\langle SV'(q(0)), V'(q(0))\rangle \leq 0$, and

$$(d/dt)\langle SV'(q(t)), V'(q(t))\rangle$$

$$= 2\left\{\langle V''(q(t))SV'(q(t), SV'(q(t))\rangle - \frac{\langle V''(q(t))V'(q(t)), SV'(q(t))\rangle\langle SV'(q(t), V'(q(t))\rangle}{\|V'(q(t))\|^2}\right\}.$$

Therefore

$$0 > (d/dt)\langle SV'(q(t)), V'(q(t))\rangle \quad \text{if } \langle SV'(q(t)), V'(q(t))\rangle = 0$$

by the admissibility of H. Therefore since

$$\langle SV'(q(0)), V'(q(0))\rangle \leq 0, \quad q(0) \in \Gamma^+,$$

then

$$\langle SV'(q(t)), V'(q(t))\rangle < 0, \quad t > 0.$$

Now
$$(d/dt)\langle q(t),e_n\rangle = \langle \dot{q}(t),e_n\rangle$$
$$= \lambda_n\langle V'(q(t)),e_n\rangle\left(1-\frac{\langle SV'(q(t)),V'(q(t))\rangle}{\lambda_n\|V'(q(t))\|^2}\right)$$
$$= \frac{\langle V'(q(t)),e_n\rangle}{\|V'(q(t))\|^2}\left(\sum_{k=1}^{n}(\lambda_n-\lambda_k)\langle V'(q(t)),e_k\rangle^2\right)$$
$$\geq 0 \quad \text{if } \langle V'(q(t)),e_n\rangle \geq 0$$

and the inequality is strict if $q(t)\in\Gamma^+$ unless $(-V'(q(t))/\|V'(q(t))\|) = e_n$. Hence Φ defined by the flow maps Γ^+ to Γ^+. It is now a simple matter to obtain estimates which ensure that the mapping Φ defined by this flow has all the required properties. In particular the existence of T with the required property is a consequence of the above estimates. q.e.d.

PROPOSITION 11. *If H is admissible and* $V'(q)\neq 0$, $q\in\Gamma^+$, *then* $\Gamma^+\backslash\gamma^+$ *is an* (n-1) *dimensional submanifold of* ∂C *and* γ^+, *the boundary of* γ^+, *is an* (n-2) *dimensional submanifold of* ∂C.

Proof If $q\in\Gamma^+$, then by Proposition 1, $\langle SV'(q),V'(q)\rangle\leq 0$, $\langle V'(q),e_n\rangle<0$, and if $q\in\gamma^+$, $\langle SV'(q),V'(q)\rangle = 0$, $\langle V''(q)SV'(q),SV'(q)\rangle<0$. Hence $\Gamma^+\backslash\gamma^+$ is an (n-1) dimensional submanifold of ∂C. Moreover, by the implicit function theorem γ^+ is an (n-2) dimensional submanifold of ∂C. q.e.d.

PROPOSITION 12. *If* $q\in\Gamma^+$, *then the vectors* $SV'(q)$ *and*

$$\nu(q) = V''(q)SV'(q) - \frac{\langle V''(q)SV'(q),V'(q)\rangle}{\|V'(q)\|^2}\,V'(q)$$

lie on the tangent plane to ∂C *at* q, *and* $\nu(q)$ *is normal to* γ *and outward from* Γ.

Proof This follows from the implicit function and elementary considerations of the level set $V(q)=0$, $\langle SV'(q),V'(q)\rangle = 0$, $q\in\Gamma^+$. Clearly $\langle SV'(q),V'(q)\rangle = 0$, $q\in\Gamma^+$. q.e.d.

PROPOSITION 13. *The projection* P *maps* Γ^+ *homeomorphically onto a closed subet of* F_n, $P(\Gamma^+\backslash\gamma^+)$ *is open, and* $\partial P(\Gamma^+\backslash\gamma^+) = P(\gamma^+)$. *Moreover if* $V'(q)\neq 0$, $q\in\Gamma^+$, *then*

$$\deg_B(P(\Gamma^+),\quad P\circ\theta\circ P^{-1},\ 0) = 1,$$

and hence there exists $q\in\Gamma^+$ *with* $\theta(q)=0$.

Proof First observe that P is injective on Γ^+, for otherwise $Pq_1 = Pq_2, q_i \in \Gamma^+$, $i = 1,2$, whence $q_2 - q_1, = \langle q_2 - q_1, e_n\rangle e_n$. Hence $q_1 \geq q_2$ or *vice-versa*, contradicting their maximality. Clearly P is continuous, and $P(\Gamma^+)$ is compact. Hence $P^{-1} : P(\Gamma^+) \to \partial C$ is continuous. If $q \in \Gamma^+$, then $\langle V'(q), e_n\rangle \neq 0$ and so $P(\Gamma^+ \backslash \gamma^+)$ is open in F_n, and $\partial P(\Gamma^+ \backslash \gamma^+) = P(\gamma^+)$. From Proposition 12 it follows that the normal to $P(\gamma^+)$ in F_n which is outward to $P(\Gamma^+)$ is $\mu(P(q)) =$

$$V''(q)SV'(q) - (\langle V''(q)SV'(q), e_n\rangle / \langle V'(q), e_n\rangle)V'(q),$$

whence

$$\langle \mu(Pq)), \quad P \circ \theta \circ P^{-1}(q)\rangle > 0, \quad q \in \gamma^+.$$

Since Γ^+ is contractible, so is $P(\Gamma^+)$ and hence the result follows.

q.e.d.

PROPOSITION 14. *If* $V'(q) \neq 0$, $q \in \Gamma^+$ *and* H *is an admissible Hamiltonian, then either*

(i) *there exists* $q_0 \in \Gamma^+$ *such that the solution of* (IVP) *with* $q(0) = q_0$, $p(0) = 0$ *has* $q(t) \in C$ *for all* $t > 0$ *and* $(q(t), p(t)) \to (q^*, 0) \in \partial C \times \{0\}$ *where* $V'(q^*) = 0$;

or

(ii) *there exists* $q_0 \in \Gamma^+$ *such that the solution of* (IVP) *with* $q(0) = q_0, p(0) = 0$ *has* $(q(T), p(T)) \in \partial C \times \{0\}$ *for some* $T > 0$, *and the corresponding solution is periodic with period* 2T.

Proof By Proposition 12, $\theta(q_0) = 0$ for some $q_0 \in \Gamma^+$. Hence either $\tau(q_0) = \infty$ which corresponds to alternative (i) or $\tau(q_0) = T < \infty$ and $\dot{p}(\tau(q_o)) = 0$ which corresponds to alternative (ii). (See [1] for further details.) q.e.d.

3.3 *Concluding remarks*

Thus for admissible systems a dichotomy has been established: either a periodic or a homoclinic orbit exists and which occurs in practice will depend on the circumstances. Some general observations may, however, be made; see Toland [4]. These methods may be developed so one obtains continuous dependence theory in parameter dependent problems, and that is a matter which will be pursued elsewhere. Finally, we remark that the definition of admissibility is not yet quite the best that can be done. A complete theory is a great deal more elaborate in technicalities, and so it is inappropriate for this brief survey of the lectures given.

REFERENCES

[1] HOFER, H. and TOLAND, J.F., Homoclinic, heteroclinic and periodic orbits for a class of Hamiltonian systems. *Math. Annal.*, **268**(1984), 387-403.

[2] TOLAND, J.F., On the bifurcation of waves in a Boussinesq system of equations. (To appear)

[3] KLASSEN, G.A. and TROY, W.C., Stationary wave solutions of a system of reaction-diffusion equations derived from the Fitzhugh-Nagumo Equations. *SIAM J. Appl. Math.*, **44**(1984), 96-110.

[4] TOLAND, J.F., Necessary and sufficient conditions for the existence of homoclinic orbits for certain pairs of Euler-Lagrange equations. (To appear)

[5] HOFER, H. and TOLAND, J.F., Free oscillations of prescribed energy at a saddle point of the potential in Hamiltonian dynamics. (To appear)

G. DARBO'S FIXED POINT PRINCIPLE AFTER 30 YEARS

Jürgen Appell
University of Augsburg
Department of Mathematics
Memminger Str. 6
D - 8900 Augsburg

1. Let A be a continuous nonlinear operator in a Banach space X which leaves some convex closed bounded set M in X (e.g. a closed ball) invariant. There are two classical conditions which guarantee the existence of a fixed point u for A in M:

- *Schauder*: AM compact,
- *Banach-Caccioppoli*: $\|A\| < 1$.

(Here and in what follows $\|A\|$ denotes the minimal Lipschitz constant for A; if there is not any, we set $\|A\| = \infty$).

A unified approach to these two classical fixed point theorems was given by M.A. Krasnosel'skij [1] in 1955 who proved a corresponding result for operators of the form $A = A_1 + A_2$, with $\|A_1\| < 1$ and A_2 compact. In the same year, G. Darbo [2] considered continuous operators A which satisfy an estimate

$$(1) \qquad \gamma(AM) \le k\gamma(M) ,$$

where $\gamma(M)$ denotes the *Kuratowski measure of noncompactness* of M, i.e. the smallest number $d > 0$ with the property that M admits a finite covering with sets of diameter $\le d$ (see e.g. [3]). Such operators are nowadays called *k-set-contractions*, and the smallest k in (1) could referred to as measure of noncompactness $\gamma(A)$ of the operator A, since it is

S. P. Singh (ed.), Nonlinear Functional Analysis and Its Applications, 161–167.

zero if and only if A is compact. Since obviously $\gamma(A) \leq \|A\|$, a theorem like

> *"Any continuous operator which leaves a convex closed bounded set M in a Banach space X invariant has fixed points in M, provided $\gamma(A) < 1$"*

would generalize both the Schauder principle and Banach-Caccioppoli theorem (as well as Krasnosel'skij's result) in a very natural way. This is in fact the statement of the famous *fixed-point theorem of G. Darbo* [2].

2. Both the theory and applications of k-set-contractions have been largely developed since 1955, mainly by R.D. Nussbaum [4] and B.N. Sadovskij [5]. For instance, there exists a topological degree theory for k-set-contractions [4,5,6,7], bifurcation results can be obtained [8], and various applications have been given, mostly to functional-differential equations [9].

Let us give a typical (though very simple) example, where k-set contractions occur. Consider the problem

$$Lu = Fu \ , \tag{2}$$

where L is some linear *self-adjoint differential operator* in L_2 (a typical example is $Lu(x) = -\Delta u(x) + q(x)u(x)$), and F is the nonlinear *Nemytskij operator*

$$Fu(x) = f(x,u(x)) \ . \tag{3}$$

If 0 does not belong to the spectrum $\sigma(L)$, the operator L has a bounded inverse $R = L^{-1}$ in L_2. Consequently, equation (2) can be written equivalently as fixed point problem $u = Au$ with $A = RF$.

In order to apply Darbo's theorem, it suffices to show that $\gamma(R)\gamma(F) < 1$. Concerning the linear part R, the number

$\gamma(R)$ can be estimated by means of the inverse of the *inner radius of the essential spectrum* of L, i.e. $\gamma(R) \leq 1/\sigma$, where $\sigma = \inf \sigma_e(L)$, as was pointed out by C.A. Stuart [8]: In fact, if $\{E_\lambda\}_\lambda$ is the spectral decomposition of L and if σ is positive (in the above example this holds if the function q is strictly positive), then for any $\varepsilon \in (0,\sigma)$

$$R = \int_{-\infty}^{+\infty} \lambda^{-1} dE_\lambda = \int_{|\lambda|>\sigma-\varepsilon} \lambda^{-1} dE_\lambda + \int_{|\lambda|\leq\sigma-\varepsilon} \lambda^{-1} dE_\lambda$$

$$= R_1 + R_2 ,$$

where $\|R_1\| \leq 1/(\sigma-\varepsilon)$ and R_2 is compact (since R_2 is of finite rank). Consequently, since $\varepsilon > 0$ is arbitrary, we have

$$\gamma(R) \leq \gamma(R_1) + \gamma(R_2) \leq \|R_1\| \leq 1/\sigma ,$$

as claimed.

Concerning the number $\gamma(R)$, we have the following

Proposition 1 [10]: *In the space* L_p $(1 \leq p < \infty)$,

$$(4) \qquad \gamma(F) = \|F\| = \mathrm{Lip}(f) ,$$

where

$$(5) \qquad \mathrm{Lip}(f) := \inf \{k : |f(x,u)-f(x,v)| \leq k|u-v|\} .$$

3. Proposition 1 is quite disappointing: In fact, it shows that, whenever it is possible to apply Darbo's theorem (since $\gamma(F) < 1$), it is already possible to apply the Banach-Caccioppoli principle (since $\|F\| < 1$). There is another flaw in the above example: the linear part R is of the form "contraction + compact", and thus Krasnosel'skij's theorem applies. In general, as mentioned above,

$$\gamma(R) \leq \|R\| , \quad \gamma(F) \leq \|F\| , \tag{6}$$

and whenever equality holds for both operators in (6), Darbo's theorem does not provide any new information in comparison with the Banach-Caccioppoli principle.

Let us point out that this is the case not only in L_p, but also in other function spaces; for example, the statement of Proposition 1 holds also for continuous functions:

Proposition 2 [10]: *In the space* C,

$$\gamma(F) = \|F\| = \mathrm{Lip}(f) . \tag{7}$$

We shall now provide an example where strict inequality holds in the first part of (6).

Consider the equation

$$u(x) = \frac{1}{\pi i} \int_{|y|=1} \frac{f(y,u(y))}{x-y} \, dy \tag{8}$$

in the space L_p on the complex unit circle S^1. Equation (8) can obviously be written in the form u = RFu, where R is the singular integration

$$Rv(x) = \frac{1}{\pi i} \int_{|y|=1} \frac{v(y)}{x-y} \, dy . \tag{9}$$

This operator acts in L_p for $1 < p < \infty$ (see e.g. [11]) and its norm $\|R\|_p$ in L_p satisfies

$$\|R\|_p = \begin{cases} \tan \frac{\pi}{2p} & 1 < p \leq 2 , \\ \cot \frac{\pi}{2p} & 2 \leq p < \infty . \end{cases}$$

Since $\lim_{p\to 1} \|R\|_p = \lim_{p\to\infty} \|R\|_p = \infty$, it becomes "more and more difficult" to apply Banach's theorem, if we must consider the equation $u = RFu$ in a small $(p \to \infty)$ or large $(p \to 1)$ L_p-space: In fact, by Proposition 1, the Banach-Caccioppoli principle applies only if $\mathrm{Lip}(f) < 1/\|R\|_p$, and the right-hand side tends to zero as $p \to 1$ or $p \to \infty$.

On the other hand, the operator R has the remarkable property that R^2 is the identity (in any space), and thus $\gamma(R) = 1$, independently of $p \in (1,\infty)$. This means that the Darbo principle applies, again by Proposition 1, if only $\mathrm{Lip}(f) < 1$, which is less restrictive.

4. One point in the preceding example is still somewhat unsatisfactory: The Lipschitz constant (= norm) of R, although tending to infinity, is always *finite*; this means that the Banach-Caccioppoli theorem gives the solvability of the equation $u = \lambda RFu$ at least for small $|\lambda|$ (more precisely, for $|\lambda| \; \|R\| \; \mathrm{Lip}(f) < 1$). It would therefore be interesting, in order to have a *real application* of Darbo's theorem, to find an example where in the inequalities (6) one obtains a finite number on one of the left-hand sides, but infinity on the corresponding right-hand side. Since both operators R and F must be continuous, we always have $\|R\| < \infty$, and thus any possible example in this spirit must be due to the nonlinearity F.

It is the aim of the final section to provide such an example. To this end, consider the equation

$$u(x) = \int_0^1 \frac{k(x,y)}{x-y} f(y,u(y))dy \tag{10}$$

which again can be written as operator equation $u = RFu$ with

$$Rv(x) = \int_0^1 \frac{k(x,y)}{x-y} v(y)dy \; . \tag{11}$$

As an appropriate function space we may choose here the little Hölder space H_α^O $(0 < \alpha \leq 1)$ of all continuous functions u with $|u(x)-u(y)| = o(|x-y|^\alpha)$ $(|x-y| \to 0)$. Under some natural conditions, the operator (11) maps H_α^O into itself and is bounded (see e.g. [12]). Similarly, one can give continuity and boundedness conditions for F in H_α^O (see e.g. [13]). On the other hand, the contrast between the conditions $\gamma(F) < \infty$ and $\|F\| < \infty$ is extremely sharp in this case:

Proposition 3: *If* F *is bounded in* H_α^O, F *is always a k-set-contraction. On the other hand,* F *satisfies a Lipschitz condition in* H_α^O *if, and only if, the generating function* f *is of the form*

$$f(x,u) = g(x)u + h(x) \qquad (g,h \in H_\alpha^O) \;,$$

i.e. linear in u.

This shows, surprisingly enough, that already such a "harmless" nonlinearity like $f(u) = |u|$ does not generate a Lipschitzian operator in H_α^O (nor a compact operator either, of course). This means, in particular, that *no classical fixed point theorem* applies to the singular equation

$$u(x) = \lambda \int_0^1 \frac{k(x,y)}{x-y} |u(y)| dy$$

in this space, but Darbo's fixed point principle does.

References

[1] M.A. Krasnosel'skij: 'Two remarks on the method of successive approximations'(Russian). *Uspehi Mat. Nauk* **10**, 1 (1955), 123-127

[2] G. Darbo: 'Punti uniti in trasformazioni a codominio non compatto'. *Rend. Sem. Mat. Univ. Padova* **24** (1955), 84-92

[3] C. Kuratowski: 'Sur les espaces complets'. *Fund. Math.* **15** (1930), 301-309

[4] R.D. Nussbaum: 'The fixed point index and fixed point theorems for k-set-contractions'. *Ph. D. Thesis, Univ. Chicago* (1969)

[5] B.N. Sadovkij: 'Limit-compact and condensing operators' (Russian). *Uspehi Mat. Nauk* **27**, 1 (1972), 81-146

[6] R.R. Ahmerov, M.I. Kamenskij, A.S. Potapov, B.N. Sadovskij: 'Condensing operators' (Russian). *Itogi Nauki Tehniki* **18** (1980), 185-250

[7] C.A. Stuart, J.F. Toland: 'The fixed-point index of a linear k-set-contraction'. *J. London Math. Soc.* **6** (1973), 317-320

[8] C.A. Stuart: 'Some bifurcation theory for k-set-contractions'. *Proc. London Math. Soc.* **27** (1973), 531-550

[9] R.R. Ahmerov, M.I. Kamenskij, A.S. Potapov, A.Je. Rodkina, B.N. Sadovskij: ' On the theory of equations of neutral type' (Russian). *Itogi Nauki Tehniki* **19** (1982), 55-126

[10] J. Appell: 'Implicit functions, nonlinear integral equations, and the measure of noncompactness of the superposition operator'. *J. Math. Anal. Appl.* **83**, 1 (1981), 251-263

[11] I.Ts. Gohberg, N.Ja. Krupnik: *Introduction to the theory of one-dimensional singular integral operators* (Russian). Stiinka, Kishinjov 1973

[12] A.I. Gusejnov, H.Sh. Muhtarov: *Introduction to the theory of nonlinear singular integral equations* (Russian). Nauka, Moskva 1980

[13] V.A. Bondarenko, P.P. Zabrejko: 'The superposition operator in Hölder function spaces' (Russian). *Dokl. Akad. Nauk SSSR* **222**, 6 (1975), 1265-1268.

BEST APPROXIMATION AND CONES IN BANACH SPACES

M. Baronti
University of Parma
Maratea, Italy

P. L. Papini
University of Bologna
Maratea, Italy

SUMMARY. Two results concerning best approximation in Banach spaces are proved. A characterization of best approximation from a closed cone in a Hilbert space is given. Then we generalize a theorem of Phelps concerning the conic differential of best approximation map from closed convex sets in some Banach spaces.

1. INTRODUCTION AND NOTATIONS

Let X be a normed space over the real field R. Given a nonempty subset G of X and a point $x \in X$, consider the (possibly empty) set of best approximations to x from G:

$$P_G(x) = \{ x_0 \in G;\ \| x-x_0 \| \leq \| x-g \| \text{ for every } g \in G \}.$$

The operator P_G (with domain $\{ x \in X;\ P_G(x) \neq \emptyset \}$) has been studied for many different kinds of sets G (see [7]). Many results have been given under the assumption of convexity for G; in particular, the case of G a convex cone has been considered e.g. in [2], [4], [11], [12], [13], [14]. On the contrary, the case of general (non convex) cones has been disregarded. In Section 2 here we discuss such problem, mainly for X a pre-hilbert space.

The best approximation map onto a closed convex set G cannot be differentiable at boundary points. Anyway, under some assumptions on X it has been proved the existence of a nonlinear "conical differential" of P_G at these points. In Section 3 we shall show that such results hold under slightly weaker assumptions on X.

The following notations will be used.
We shall denote be [y] the linear span of a vector $y \in X$.
We use orthogonality in the sense of Birkhoff and James, that is: $x \perp y$ will mean $\| x+\lambda y \| \geq \|x\|$ for every $\lambda \in R$.
Consider, for x,y in X, the right norm derivative:

S. P. Singh (ed.), Nonlinear Functional Analysis and Its Applications, 169–176.

$$\tau(x,y) = \lim_{t \to 0^+} (\|x+ty\| - \|x\|)/t.$$

Note that $x \perp y$ is equivalent to $-\tau(x,-y) \leq 0 \leq \tau(x,y)$, while in pre-hilbert spaces $\|x\|\ \tau(x,y)$ is equal to the inner product (x,y).

In what follows we shall call <u>cone</u> (with vertex at θ) a closed subset C of X satisfying the following condition

$$\text{if} \quad y \in C \quad \text{and} \quad \lambda \geq 0, \text{ then } \lambda y \in C. \tag{0}$$

2. PROJECTIONS ONTO CONES

In our opinion, the class of approximation problems having a cone as a natural setting is not void. The following examples (and mainly the last one) bring some support to this statement. We denote by $C_F[a,b]$ the space of all continuous functins on the real interval $[a,b]$.

<u>Example 1</u>. Let $G\{f \in C_F[0,1];\ f = \alpha x + \beta;\ \alpha,\ \beta \in R;\ \alpha \cdot \beta > 0\}$.

<u>Example 2</u>. Let $G \subset C_F[a,b]$ be the set of all polynomials of the types: $p(x) = \sum_{k=0}^{n} \alpha_k x^{2k}$ or $p(x) = \sum_{k=0}^{n} \beta_k x^{2k+1}$ $(n \in N;\ \alpha_k,\ \beta_k \in R)$.

<u>Example 2 bis</u>. If $a = -b$, let $G = \{f \in C_F[a,b];\ f \text{ is odd or even}\}$.

<u>Example 3</u>. Let $G = f \in C_F[a,b]$; f is a rational function corresponding to degrees n,m $(f(x) = (\sum_{k=0}^{n} \alpha_k x^k)/(\sum_{k=0}^{m} \beta_k x^k)$, with n and m fixed in N).

Unfortunately, the following example -communicated to us by G. Godini- show that we cannot expect to give existence results for best approximation from cones, also in the Hilbert space setting.

<u>Example 4</u>. Let $X = \ell^2$ and for $n \geq 2$ let $x_n = (1, 0, \ldots, 0, \frac{n+1}{n}, 0, \ldots)$; let $C = \bigcup_{n=2}^{\infty} \{\lambda x_n;\ \lambda \geq 0\}$. Then C is a closed cone. Let $x = (1, 0, 0, \ldots)$ and denote by $C_n = \{\lambda x_n;\ \lambda \geq 0\}$, $n \geq 2$. We have $\operatorname{dist}(x,C_n) = d(x, \frac{n^2 x_n}{2n^2 + 2n + 1}) = \frac{n+1}{\sqrt{2n^2 + 2n + 1}}$ and $\operatorname{dist}(x,C_n) < \operatorname{dist}(x,C_m)$ for $n > m$. Hence $\operatorname{dist}(x,C) = 1/\sqrt{2}$ and $P_C(x) = \emptyset$.

Now we want to consider a few conditions concerning this problem. It is easy to see that $x_0 \in P_C(x)$ implies (for C a cone)

$$x - x_0 \perp x_0. \tag{1}$$

In fact, (1) is trivial if $x_0 = \theta$. Otherwise, $x_0 \in P_C(x)$ implies that the convex function of $\lambda \in R$, $f(\lambda) = \| x - \lambda x_0 \|$ has a local, then also a global minimum, for $\lambda = 1$. This implies (1). Note that $x_0 \in P_C(x)$ always implies $\lambda x_0 \in P_C(\lambda x)$ for $\lambda \geq 0$, and $\| x - x_0 \| \leq \| x \|$. Also, when X is strictly convex, the last inequality is strict when $x_0 \neq \theta$, while $\theta \in P_C(x)$ implies $P_C(x) = \{\theta\}$.

If X is a reflexive, strictly convex space and P_C is defined on X, then any selection for P_C is an orthogonal B-operator, in the sense of [10] (see also [3] for more general situations). Moreover, a sequence $\{x_i\}$ such that $x_{i+1} = x_i - P_C(x_i)$ for $i \geq 2$ is then a martingale in that terminology: thus, for example, $P_C(x_i) = x_i - x_{i+1} \underset{i\to\infty}{\to} 0$ if X is uniformly convex (see again [10]).

Many conditions related to approximation problems have been given in terms of norm derivatives; for G an arbitrary subset of X, only the following ones are of interest (see [5]):

$$\tau(x - x_0, x_0 - g) \geq 0 \text{ for every } g \in G \text{ is sufficient that } x_0 \in P_G(x) \tag{2}$$

$$\tau(x - g, g - x_0) \leq 0 \text{ for every } g \in G \text{ is necessary that } x_0 \in P_G(x). \tag{3}$$

If X is an inner product space, then we have also

$$(x - y, x_0 - y_0) \geq 0 \text{ whenever } x_0 \in P_G(x),\ y_0 \in P_G(y). \tag{4}$$

But it is also possible to give a necessary and sufficient condition concerning approximation from closed, convex cones. Namely, we have (see [2], or [4], or [7], p. 362):

<u>Proposition 1</u>. Let C be a convex cone and $x \in X$. Then $x_0 \in C$ is an element of $P_C(x)$ if and only if it satisfies (1), and moreover

$$\tau(x_0 - x, y) \geq 0 \text{ for every } y \in C. \tag{5}$$

Concerning condition (5), see also [1], Lemma 1.

Next result characterizes the elements of best approximation from cones in Hilbert spaces.

Theorem 1. Let X be a pre-hilbert space and C a cone. Then, given $x \in X$ and $x_0 \in C$, we have $x_0 \in P_C(x)$ if and only if x_0 satisfies (1) and

$$(x,y) \leq \| x_0 \| \cdot \| y \| \text{ for every } y \in C. \tag{6}$$

Proof. "Only if" part. Let $x_0 \in P_C(x)$; we already observed that (1) is a necessary condition, thus $(x,x_0) = \| x_0 \|^2$. Condition (6) is trivially true for $y = \theta$; suppose now $y \in C\setminus\{\theta\}$, then also $(y\| x_0 \|)/\| y \| \in C$. We obtain $\| x - x_0 \|^2 \leq \| x - (y\| x_0 \| / \| y \|) \|^2$, which is equivalent to $2(x,x_0) \geq 2(x,(\| x_0 \|)/\| y \|)$; but hten $\| x_0 \|^2 = (x,x_0) \geq (x,y)\| x_0 \| / \| y \|$, that is (6).

"If" part. Assume $x_0 \in C$ satisfy (1) and (6) and $x_0 \notin P_C(x)$. Then there exists $y \in C$ with $\| x - y \| < \| x - x_0 \|$. Hence $y \neq 0$ (since by (1) we have $\| x - x_0 \| \leq \| x \|$) and $\theta \notin P_{C_y}(x)$, where $C_y = \{\lambda y;\ \lambda \geq 0\}$. We can suppose $y \in P_{C_y}(x)$, i.e., $x - y \perp y$. Hence, using (6) $\| y \|^2 = (x,y) \leq \| x_0 \| \cdot \| y \|$ and so $\| y \| \leq \| x_0 \|$. On the other hand the condition $\| x - y \|^2 < \| x - x_0 \|^2$ implies $\| y \|^2 = \| x \|^2 - \| x - y \|^2 > \| x \|^2 - \| x - x_0 \|^2 = \| x_0 \|^2$, so $\| x_0 \| < \| y \|$, a contradiction. Therefore $x_0 \in P_C(x)$.

Remarks. i): since (6) is trivially true for $y = \theta$, we could equivalently ask in (6) that $(x,y) \leq \| x_0 \| \cdot \| y \|$ for every y such that $\| y \| = 1$.

(ii): from the proof of the "if" part of Theorem 1 we can see that $\{x_0,y\} \subset P_C(x)$ implies $(x,x_0) = \| x_0 \| \cdot \| y \| = (x,y)$, so an equality in (6). Conversely, let $(x,y) = \| x_0 \| \cdot \| y \|$ for some $y \notin \{x_0,\theta\}$; then we have also $(x,y') = \| x_0 \| \ \| y' \|$ for $y' = (y/\| y \|)\| x_0 \|$. Hence $\| x - y' \|^2 = \| x \|^2 + \| y' \|^2 - 2\| x_0 \| \cdot \| y' \| = \| x \|^2 - \| x_0 \|^2$. Therefore, if (1) holds, then

$\| x - y' \| = \| x - x_0 \|$.

iii): our condition (6) is formally weaker than (5), therefore the "if" part of Theorem 1 generalizes the corresponding part of known characterizations concerning convex cones, in the case of X a pre-hilbert space.

It seems to be difficult to generalize to general Banach spaces the characterization given by Theorem by using norm derivatives. In fact, the following two examples show that the analogue of (6) is neither necessary, nor sufficient.

Example 5. Let $X = C[0,1]$, $x(t) = 1 = |2t - 1|$, $x_0(t) = 1/2$, $y(t) = 2t - |2t - 1|$, and C the cone generated by x_0 and y. We have $\| x\| = \| y\| = 1$; $\| x_0\| = 1/2$; $x_0 \in P_C(x)$; $-\tau(x,-y) = \tau(x,y) = 1 > \| x_0\| \cdot \| y\| = 1/2$, thus the condition $\| x\| \tau(x,y) \leq \| x_0\| \cdot \| y\|$ is not necessary that $x_0 \in P_C(x)$.

Example 6. Let $X = C[0,1]$, $x(t) = t$, $y(t) = 1$, $x_0(t) = (1 - 2t + |2t - 1|)/2$, and C the cone generated by x_0 and y. We have $\| x\| = \| x_0\| = \| y\| = 1$; the condition (1) (and also (5)) is satisfied, and moreover $\| x\| \tau(x,y) = 1 = \| x_0\| \cdot \| y\|$, but $x_0 \notin P_C(x)$ since $\| x - y/2\| = 1/2 < \| x - x_0\| = 1$; thus the conditions (1) and $\| x\| \tau(x,y) \leq \| x_0\| \cdot \| y\|$ are not sufficient that $x_0 \in P_C(x)$.

We conclude this section recalling that in [8] and in [9], approximation from sets called "pencils", satisfying the following condition (more restrictive than (0)) has been considered:

$$\text{if } y \in C \text{ and } \lambda \in R, \text{ the } \lambda y \in C. \tag{00}$$

Theorem 1 applies also to pencils.

3. CONIC DIFFERENTIALS

In this section, C will denote a closed convex subset of X. For $x \in C$, the support cone $S_C(x)$ to C at x is (by definition) the closure of the convex cone $\cup \{\lambda(C - x); \lambda > 0\}$. Clearly, the set $S_C(x)$ is smallest convex cone S (with vertex at θ) whose translate x + S contains C. This same cone can also be defined in some different ways (see §1 of [1]).

Now we wish to indicate a rather general class of space in which the following result holds

> If C is a nonempty closed convex subset of X, then for each $x \in C$ and any $y \in X$ we have (R)

$$P_C(x + ty) = x + tP_S(y) + o(t), \quad t > 0 \qquad \text{(r)}$$

where $S = S_C(x)$.
This means that $P_S(y)$ is the directional derivative of P_C, at $x \in C$, in the direction y.
It was proved in [6] that (R) holds not only Hilbert spaces (this was already known) but also in a class of spaces including e.g. L_p spaces, $1 < p < \infty$. This fact can be used to apply a "gradient projection method" to some problems of constrained optimization.
Here we want to prove that smoothness of X, assumed to prove (R) in Lemma 3 of [6], is not necessary. We recall the following definition. The space X has property (H) provided

$$\| x_n - x\| \to 0 \text{ whenever } \| x_n\| \to \| x\| \text{ and } x_n \to x \text{ weakly} \qquad \text{(H)}$$

To prove property (R), it is enough to reason for $x = \theta$. In this case, (r) reduces to (set $S = S_C(\theta)$):

$$\frac{P_C(ty)}{t} \to P_S(y) \text{ when } t \to 0^+ \qquad \text{(r')}$$

or, equivalently, to

$$P_{a_n C} \to P_S(y) \quad \text{for any sequence } a_n \to \infty. \qquad \text{(r'')}$$

We state in advance the following (almost trivial) lemma.

Lemma 1. Let $\lim_{n\to\infty} a_n = \infty$. Then, for every $x \in X$, we have $d(x,S) = \lim_{n\to\infty} d(x,a_nC)$.

Proof. Let $x \in X$. Take $\varepsilon > 0$, then choose $s \in S$ such that $d(x,s) < d(x,S) + \varepsilon$. We have $\| s - s'\| < \varepsilon$ for some $s' \in a_nC$, n large enough. Therefore $d(x,a_nC) \le d(x,s') < d(x,S) + 2\varepsilon$, which implies $\lim_{n\to\infty} d(x,a_nC) \le d(x,S)$. The opposite inequality follows trivially since $d(x,S) \le d(x,a_nC)$ for every $n \in N$ ($a_nC \subset S$), so the lemma is proved.

We are now able to prove our result.

Theorem 2. Let C be a closed convex set and $x \in C$ in a reflexive, strictly convex space X with property (H). Then, for any $y \in X$, $P_S(y)$ is the directional derivative of P_C at $x \in C$ in the direction y (i.e., (r") holds).

Proof. Assume again with no loss of generality $x = \theta$. Take a sequence $a_n \to \infty$, then set $C_n = a_n C$ and $S = S_C(\theta)$. Since X is reflexive and strictly convex, for any $n \in N$ there exist a unique best approximation y_n to $y \in X$ from C_n, and a unique best approximation $\overline{y}$ to y from S. The sequence $\{y_n\}$ is bounded, since $d(y,y_n) \leq d(y,y_1)$ for any n. Thus there exists a weakly convergent subsequence $\{y_{n_k}\}$, whose limit y' belongs to S (which is closed and convex, thus weakly closed). Then $d(y',y) \leq \lim_{k\to\infty} d(y_{n_k},y) = \lim_{k\to\infty} d(n_k C,y) = d(S,y)$. Hence, by uniqueness, $\overline{y} = y'$. Thus we have $y_{n_k} \to \overline{y}$, and also

$\lim_{n\to\infty} \| y_{n_k} - y\| = d(y,S) = \| \overline{y} - y\|$. Property (H) then implies

$y_{n_k} - y \to \overline{y} - y$, i.e. $y_{n_k} \to \overline{y}$.

Thus the theorem is proved. In fact, assume that for some $\varepsilon > 0$ and a sequence $\{a_n\}$ we have $\| y_{n'_k} - \overline{y}\| > \varepsilon$ for infinitely many indexes n'_k: the above reasonding would imply a contradiction.

ACKNOWLEDGEMENT

The authors are indebted to G. Godini for many remarks: in particular, she gave us the permission to reproduce her Example 4.

REFERENCES

1. J. P. Aubin, *Lipshitz behaviour of solutions to convex minimization problems*, Math. Oper. Research 9 (1984), 87-111.

2. F. Deutsch, J. H. McCabe and G. M. Phillips, *Some algorithms for computing best approximations from convex cones*, SIAM J. Numer. Anal. 12 (1975), 390-403.

3. G. Godini, *A generalization fo set-valued metric projections*, Mathematica - Rev. Anal Numér. Th. Approx. 12 (1983), 25-44.

4. G. Godini, *Despre cea mai buna aproximare in spatii vectoriale normate prin elemente din conuri convexe*, Stud. Cerc. Mat. 21 (1969), 931-936.

5. P. L. Papini, *Approximation and norm derivatives in real normed spaced*, Resultate Math 5 (1982), 81-94.

6. R. R. Phelps, *Metric projections and the gradient projection method in Banach spaces*, SIAM J. Control Optim., to appear.

7. I. Singer, *Best approximation in normed linear spaces by elements of linear subspaces*, Springer-Verlag, Berlin 1970.

8. S. Srinivasan, Sastry, Sundaram M. A. and K. Viswanath, *Orthogonality in Banach space and approximation theory*, J. Madurai Kamaraj Univ. 8 (1979), 93-98.

9. S. Srinivasan - K. Viswanath, *Coproiminality and biproximinality in Banach spaces*, preprint.

10. F. Sullivan, *A generalization of best approximation operators*, Ann. Mat. Pura Appl. (4) 107 (1975), 245-261.

11. C. Zalinesu, *An algorithm for best approximation by elements of cones in Banach spaces*, Bull. Math. Soc. Sci. Math. R.S. Roumanie 20 (68) (1976), 199-211.

12. E. H. Zarantonello, *L'algébre des projecteurs coniques*, in Analyse convexe et ses applications, 232-242, Springer-Verlag, Berlin 1974.

13. E. H. Zarantonello, *Projections on convex sets and spectral theory*, in Contributions to nonlinear functional analysis ed. by E. H. Zarantonello, 237-424, Academic Press, New York, 1971.

14. E. H. Zarantonello, *Projectors on convex sets in reflexive Banach spaces*, Rev. Un. Mat. Argentina, 29 (1984) 252-269.

INVARIANT MANIFOLD THEOREMS WITH APPLICATIONS

Peter W. Bates
Department of Mathematics
Brigham Young University
Provo, UT 84602

and Christopher K. R. T. Jones
Department of Mathematics
University of Arizona
Tucson, AZ 85721

ABSTRACT. Invariant Manifold Theorems have proved to be very useful in the study of dynamical systems. Here we present a sketch of a geometrical proof of the local stable/center/unstable manifold decomposition theorem for flows or semiflows in Banach space. We distinguish two cases, called dissipative and conservative, anticipating applications to semilinear parabolic and hyperbolic partial differential equations. As examples we show how the stability of the travelling pulse for the FitzHugh-Nagumo equations and the instability of stationary solutions to a nonlinear wave equation follow from this decomposition theorem.

1. INTRODUCTION

When studying the behavior of a dynamical system in the neighborhood of an equilibrium point the first step is to construct the stable, unstable and center manifolds. These are manifolds that are invariant under the flow relative to a neighborhood of the equilibrium point and carry the solutions that decay or grow (or neither) at exponential rates. The existence theory and usefulness of these manifolds has a long history, dating back to Poincare and Hadamard. In finite dimensional space or when the semigroup of the linearized operator is analytic and the nonlinearity is smooth one may refer to Carr [1], Chow and Hale [2] or Henry [4]. The purpose of this paper is to show the existence of these manifolds in the context of semilinear partial differential equations (evolution equations in a Banach space) when the linearized operator generates a C_0-semigroup and the nonlinearity is only Lipschitz continuous. Our approach is geometric and relies on ideas described to us by C. Conley.

Consider the equation

(1.1) $\qquad u_t = Au + f(u)$

where $u \in X$, a Banach space, and t represents time. We shall assume

S. P. Singh (ed.), Nonlinear Functional Analysis and Its Applications, 177–186.

(H1) $A:X \to X$ is a closed, densely defined linear operator that generates a C_0-semigroup on X, call it S(t) (see Pazy [7]).
(H2) The spectrum of A, $\sigma(A) = \sigma_- \cup \sigma_0 \cup \sigma_+$, with

$$\sigma_-=\{\lambda\in \sigma(A): \mathrm{Re}\lambda<0\}, \quad \sigma_0=\{\lambda\in\sigma(A): \mathrm{Re}\lambda=0\}, \quad \sigma_+=\{\lambda\in\sigma(A):\mathrm{Re}\lambda>0\}$$

where σ_-, σ_0 and σ_+ are spectral sets (open and closed subsets of $\sigma(A)$).
(H3) The nonlinearity f(u) is defined on X and is locally Lipschitz. At 0 we have $f(0)=0$ and for all $\varepsilon>0$, there exists a neighborhood $U=U(\varepsilon)$ of 0 such that f has Lipschitz constant ε in U.

Remarks:

(1) The hypothesis (H3) says that f(u) is higher order. To express this in terms of a Lipschitz constant is a fairly weak way to do it. As a consequence we shall only get the weakest results about the smoothness of the invariant manifolds, namely that they are Lipschitz.
(2) In many applications, the manifolds being Lipschitz is sufficient. We shall prove some stability and instability results in this paper for which this suffices.
(3) (H3) is satisfied if $f\in C^1$ and $Df(0) = 0$.

Under the hypothesis (H2), there are invariant (under A) subspaces associated to σ_-, σ_0 and σ_+, call these X^-, X^0, X^+. The association is that

$$\sigma_-= (A|_{X^-}) \qquad \sigma_0= (A|_{X^0}) \qquad \sigma_+= (A|_{X^+}),$$

see Taylor [10]. Set $X^L=X^- \oplus X^0$ and $X^R = X^0 \oplus X^+$.

Let us introduce some notation. With $*= -, 0, +, L, R$, let $\pi^*: X \to X^*$ be the natural projection, $A^*=A|_{X^*}$ and $S^*(t)=S(t)|_{X^*}$. Note that A^* generates $S^*(t)$.

We shall make two further sets of assumptions. The first set we shall call the dissipative case (D).
(D1) $\dim X^0 < +\infty$.
(D2) $\dim X^+ < +\infty$.
(D3) There exists $M > 0$ and $\sigma > 0$ such that

$$\| S^-(t)\| \le Me^{-\sigma t}$$

for all $t > 0$ ($\| \ \|$ here denotes the operator norm). Without loss of generality we may also assume $\| S^+(-t)\| \le Me^{-\sigma t}$ for all $t > 0$.

The second set is called the conservative case (C).
(C1) $\dim X^+ < +\infty$.
(C2) $\dim X^- < +\infty$.
(C3) A generates a C_0-group S(t) and for all $\rho > 0$ there exists $M>0$ such that $\|S^0(t)\| \le Me^{\rho|t|}$ for all t.

Remarks:

(1) Instead of (D3) and (C3) it would be more satisfying to make assumptions about σ_- and σ_0 respectively instead of these estimates. However we need the estimates on the semigroups and unless we make stronger assumptions (e.g. analyticity of the semigroup) the desired spectral mapping property fails. In many applications these growth conditions can be found by considering the solutions to (1.1) directly.
(2) The terms dissipative and conservative do not conform to any standard usage. We use them as they are suggestive of the applications. We shall be interested in the FitzHugh-Nagumo equations which falls under case (D) and a nonlinear wave equation which is (C).

By the assumptions on A and f we get existence and uniqueness of solutions to (1) for small time, see Pazy, p. 185 [7]. If $u_0 \in X$ there is a solution $u(t) \in C([0,T];\, X)$ for some $T>0$ and $u(0)=u_0$. This $u(t)$ gives us the usual conditions for a local semiflow. We shall use the notation

$$\Phi_t(u_0)=u(t).$$

<u>Definition</u>: If $V \subset U$, we say that V is positively invariant relative to U if $\Phi_t(V) \cap U \subset V$ for all $t > 0$, where $\Phi_t(V) = \{\Phi_t(v): v \in V\}$.

We will show that in a neighborhood, U, of 0 there are manifolds which are invariant relative to U:

W^u, the local unstable manifold of 0
W^s, the local stable manifold of 0
W^c, a local center manifold for 0
W^{cu}, a local center-unstable manifold for 0
W^{cs}, a local center-stable manifold for 0.

These manifolds can be used to give a "curvilinear" coordinate system in X in a neighborhood of the equilibrium u=0.

Given a sufficiently small neighborhood, U, of O we define

$$W^u = \{u \in U:\ \Phi_t(u) \text{ exists and lies in U for all } t<0 \text{ and } \Phi_t(u) \to 0 \text{ exponentially as } t \to -\infty\}$$

and

$$W^s = \{u \in U: \Phi_t(u) \in U \text{ for all } t \geq 0 \text{ and } \Phi_t(u) \to 0 \text{ exponentially as } t \to \infty\}$$

The other manifolds are not so succinctly defined, instead their properties will be described in the proofs of the following theorems.

Theorem 1.1. Assume (H1)-(H3) and either (D) or (C). Then there exists a neighborhood U of 0 in X such that

(i) W^s exists as a Lipschitz manifold, invariant relative to U and tangent to X^- at 0.
(ii) In $U \backslash W^s$ there is a Lipschitz manifold, W^{cu}, invariant relative to U and tangent to X^R at 0

Theorem 1.2. Assume (H1)-(H3) and either (D) or (C). Then there exists a neighborhood U of 0 in X such that

(i) W^u exists as a Lipschitz manifold, invariant relative to U and tangent to X^+ at 0.
(ii) In $U \backslash W^u$ there is a Lipschitz manifold, W^{cs}, invariant relative to U and tangent to X^L at 0.

Theorem 1.3. Assume (H1)-(H3) and either (D) or (C). Then there exists a neighborhood U of 0 in X and a Lipschitz manifold, W^c, invariant relative to U and tangent to X^0 at 0. In fact $W^c = W^{cu} \cap W^{cs}$.

The basic ingredients of the proofs of these theorems are Gronwall's Lemma, the Contraction Mapping Theorem and a degree argument. The underlying idea is that certain cones and moving cones are positively invariant due to the difference in growth rates in X^-, X^0 and X^+. For instance, after renorming, we will show that $K = \{(v,w) \in X^- \times X^R\colon |v| \leq |w|\}$ is positively invariant under the flow Φ_t (see Figure 1).

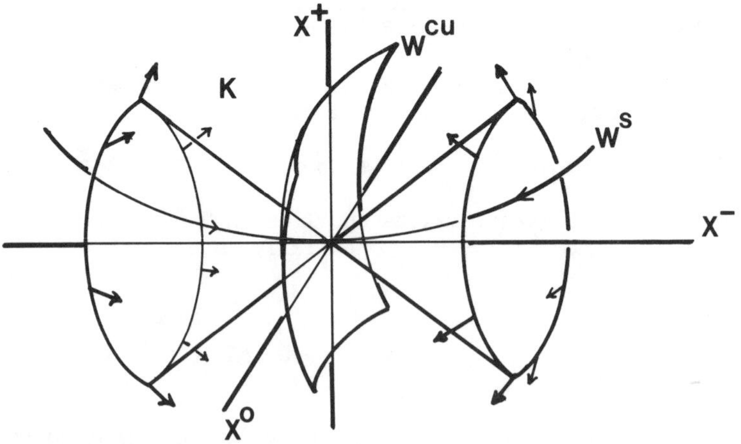

Figure 1.
The flow through ∂K and the local W^s/W^{cu} decomposition of X.

One expects that manifolds over X^R lying in K will be compressed together by Φ_t. This turns out to be true and the forward flow induces a contraction mapping in a certain function space. The fixed point corresponds to W^{cu}. One can also guess that for each $v \in X^-$ there is a point $w(v) \in X^R$ so that $\Phi_t(v,w(v))$ remains in K^c, the complement of K, for all positive time. This is by a modification of the Wazewski Principle applied to $(\{v\} \times X^R) \cap K^c$, noting that ∂K is an exit set for K^c. The set of points (v, w(v)) will form W^s.

2. PROOFS.

Here we shall give the steps to be followed with some of the technical details omitted so that the overall approach is not obscured to the reader. Complete proofs will be given in a forthcoming paper.

We start by proving Theorem 1.1 in case (D) but remark that similar arguments are used in obtaining the other results.

First modify f so that we have global estimates:

Let $g(u) = \phi(u)f(u)$ where

$$\phi(u) = \begin{cases} 1 & \text{for } |u| \leq \delta \\ 2-|u|/\delta & \text{for } \delta < |u| \leq 2\delta \\ 0 & \text{for } 2\delta < |u| \end{cases}$$

With σ given in (D3) choose $\delta > 0$ so small that g has Lipschitz constant $\varepsilon < \sigma/4$.

Consider the equation

$$u_t = Au + g(u) \tag{2.1}$$

and note that solutions to (2.1) agree with solutions to (1.1) in the neighborhood $U \equiv B(0,\delta)$. Now identify X with $X^- \times X^R$ and write (2.1) as a system

$$\begin{aligned} v_t &= A^-v + g^-(v,w) \\ w_t &= A^Rw + g^R(v,w) \end{aligned} \tag{2.2}$$

where $u = (v,w) \in X^- \times X^R$, $g^- = \pi^-g$ and $g^R = \pi^Rg$.

From (D) we have the estimates

$$\|S^-(t)\| \leq Me^{-\sigma t} \text{ for } t > 0 \tag{2.3}$$

and for any $\rho > 0$ there exists $M_1 > 0$ such that

$$\|S^R(t)\| \leq M_1e^{-\rho t} \quad \text{for } t < 0. \tag{2.4}$$

We take ρ so that $\varepsilon < (\sigma-\rho)/4$. Also, following Pazy [7], page 19, we renorm X^- and X^R so that we have (2.3) and (2.4) with $M = M_1 = 1$.

To obtain growth estimates for the components v and w of a solution we will use the following versions of Gronwall's Lemma:

<u>Lemma 2.1</u>. Let $a,b:[0,T]\to[0,\infty)$ be continuous and A and B be positive constants. If

$$a(t) \le a(0) + A\int_0^t a(s)\,ds, \qquad 0 \le t \le T,$$

then

$$a(t) \le a(0)e^{At}, \qquad 0 \le t \le T.$$

If

$$b(t+r) \le b(t) + B\int_r^0 b(t+s)ds, \qquad -T \le -t \le r \le 0,$$

then

$$b(t) \ge b(0)e^{-Bt}, \qquad 0 \le t \le T.$$

Applying the variation of constants formula to (2.2) and using the estimates (2.3) and (2.4) we get

$$|v(t)|e^{\sigma t} \le |v(0)| + \varepsilon\int_0^t (|v(s)|+|w(s)|)e^{\sigma s}\,ds, \quad 0 \le t \tag{2.5}$$

and

$$|w(t+r)|e^{\rho(t+r)} \le |w(t)|e^{\rho t} + \varepsilon\int_r^0 (|v(t+s)|+|w(t+s)|)e^{\rho(t+s)}ds \quad \text{for } -t\le r\le 0. \tag{2.6}$$

Now Lemma 2.1 yields

<u>Lemma 2.2</u>. Let (v,w) be a solution to (2.2). If

$$|w(s)| \le k|v(s)| \quad \text{for} \quad 0 \le s \le t, \tag{2.7}$$

then

$$|v(t)| \le |v(0)|\exp((-\sigma + \varepsilon(1+k))t). \tag{2.8}$$

If

$$|v(s)| \le k|w(s)| \quad \text{for } 0 \le s \le t, \tag{2.9}$$

then

$$|w(t)| \ge |w(0)|\exp((-\rho - \varepsilon(1+k))t). \tag{2.10}$$

These are the growth rates which will give the flow through ∂K depicted in Figure 1. For $\lambda>0$ define $K_\lambda = \{(v,w):\lambda|v| \le |w|\}$.

<u>Lemma 2.3</u>. If $\lambda > 0$ is such that

$$\varepsilon < (\sigma - \rho)/(2 + \lambda + \lambda^{-1}) \tag{2.11}$$

then K_λ is positively invariant.

The proof follows from Lemma 2.2 by noting that in a neighborhood of a point on ∂K we have versions of (2.7) and (2.9). The rapid decay of v and the slow decay of w in this neighborhood, given by (2.8) and (2.10) respectively, gives the result. One can view this as (2.2) defining a vector field which on ∂K_λ points into K_λ. In fact a stronger type of invariance, the positive invariance of moving cones, may be proved in a similar way.

Lemma 2.4. Let λ satisfy (2.11). If u_1 and u_2 are solutions to (2.1) with $u_2(0) \in u_1(0) + K_\lambda$, then $u_2(t) \in u_1(t) + K_\lambda$ for all $t \geq 0$.

We will now sketch the proof of Theorem 1.1 in case (D). To construct W^s we use an idea based on the Wazewski Principle. For each $v_0 \in X^-$ consider the finite dimensional disk $D = \{(v_0, w) : |w| \leq \lambda |v|\}$. For $t>0$ $\Phi_t(\partial D) \subset K_\lambda$ by Lemma 2.3 and a degree-theoretical argument can be used to show that there is at least one point $u_0 \in D$ such that $\Phi_t(u_0) \in K_\lambda^c$ for all $t>0$. Lemma 2.4 is used to show that u_0 is unique, noting that $\Phi_t(u_0)$ remaining in K_λ^c implies that it must tend to zero at a rate of $\exp((-\sigma+\varepsilon(1+\lambda))t)$ and points in $\Phi_t(u_0)+K_\lambda$ cannot approach zero faster than $\exp((-\rho-\varepsilon(1+\lambda^{-1})t)$. The same reasoning shows that the set of these points $\{u_0 : v_0 \in X^-\}$ forms the graph of a Lipschitz function, $h^- : X^- \to X^R$, with Lipschitz constant λ. Because $W^s = \text{graph}(h^-)$ is obtained point by point rather than globally and points in $W^s \cap U$ do not leave U we see that in U, W^s does not depend on the modification of f outside of U. This allows us to conclude that $Dh^-(0) = 0$ since by shrinking U we can let $\lambda \to 0$ while preserving the inequality (2.11).

The construction of W^{cu} relies on the Contraction Mapping Theorem. Fix λ satisfying (2.11) and let $Y = \{h \in C(X^R, X^-) : h(0) = Dh(0) = 0$ and h has Lipschitz constant $\lambda\}$ then Y is a complete metric space with norm $\|h\| = \sup\{|h(w)|/|w| : w \neq 0\}$. For $h \in Y$ let $H = \text{graph}(h)$. By using Lemma 2.4 one shows that $\Phi_t(H)$ is the graph of a function in Y, call this $T_t(h)$. Again using Lemma 2.4 and the relative growth rates of v and w one can show that T_t is a contraction on Y for t sufficiently large. The details are technically complicated and we omit them, however, the idea is that for $h \in Y$, Φ_t tends to stretch its graph in the X^R direction, relative to the flow in the X^- direction. Clearly, the unique fixed point, h_0, of T_t for t large is also the unique fixed point of T_s for all s. We define $W^{cu} = \text{graph}(h_0) \cap U$.

Remark:

Notice that W^{cu} was obtained in a global way and so we must expect that it is dependent upon the modification of f outside of U, as can be demonstrated by example (see e.g. Carr [11]).

The proof of Theorem 1.2 in case (D) proceeds as follows: Identify X with $X^L \times X^+$ and write (2.1) as the system

$$(2.12) \qquad \begin{cases} v_t = A^L v + g^L(v,w) \\ w_t = A^+ w + g^+(v,w) \end{cases}$$

where $(v,w) \in X^L \times X^+$, $g^L = \pi^L g$ and $g^+ = \pi^+ g$. For any $\rho > 0$, from (D) and after renorming we have the estimates

(2.13) $\quad \|S^L(t)\| \leq e^{\rho t} \quad$ for $t > 0$

and

(2.14) $\quad \|S^+(t)\| \leq e^{\sigma t} \quad$ for $t < 0$.

With ρ as before the relative growth rates given by (2.13) and (2.14) lead to a version of Lemma 2.2 with $-\sigma$ replaced by ρ in (2.8) and $-\rho$ replaced by σ in (2.10). Lemma 2.3 and Lemma 2.4 then follow as before since it is the <u>relative</u> rates of growth (or decay) of v and w that is important. The constructions of W^{cs} and W^u are the same as those of W^s and W^{cu}, respectively. The tangency of W^{cs} to X^L at 0 needs a separate argument, however, since the positive evolution of a point on W^{cs} does not necessarily stay in U and we cannot use the previous uniqueness argument (in general W^{cs} is not unique). We omit this rather technical proof but mention that it is similar to showing that W^{cu} is tangent to X^R at 0 (only with time reversed) in that Lemma 2.4 plays the fundamental part.

We can dispose of case (C) rather quickly. Note that since σ^+ is a spectral set in the open right half plane and since $\dim X^+ < \infty$ there exists $\sigma > 0$ such that (after renorming)

$$\|S^+(t)\| \leq e^{\sigma t} \quad \text{for } t < 0.$$

The decomposition of X^L played no part in the above proof and so Theorem 1.2 case (C) is contained in the proof for case (D). To prove Theorem 1.1 in case (C) we simply observe that we may reverse time ($X^+ \leftrightarrow X^-$), since S is a group, and apply Theorem 1.2.

To prove Theorem 1.3 we need some type of Implicit Function Theorem in order to form the intersection $W^{cs} \cap W^{cu}$. Since we do not have smoothness we use the following

<u>Lemma 2.5</u>. Let Z be a Banach space and suppose $F: Z \to Z$ is a contraction. Then I-F is one-to-one and onto.

<u>Proof of Theorem 1.3</u>. We have two globally defined Lipschitz continuous functions $h^0: X^R \to X^-$, from the proof of Theorem 1.1, and $h^-: X^L \to X^+$, from the proof of Theorem 1.2. Each of these can be taken to have Lipschitz constant less than one and so, for fixed $x \in X^0$, the function $F: X^- \times X^+ \to X^- \times X^+$ defined by $F(y,z) = (h_0(x+z),\ h^-(x+y))$ is a contraction. By Lemma 2.5 there is a unique solution $(y,z) = h^c(x)$ to the equation $(y,z) - F(y,z) = 0$, that is, $y = h_0(x+z)$ and $z = h^-(x+y)$. It follows that h^c is Lipschitz continuous and $Dh^c(0) = 0$. This completes the proof of Theorem 1.3.

3. APPLICATIONS

Without giving all the technical details we mention two applications whereby recent important results may be cast in our setting and obtained more simply.

The first concerns the stability of the travelling pulse for the FitzHugh-Nagumo system:

$$(3.1)\qquad \begin{cases} u_t = u_{xx} + f(u) - w \\ w_t = \varepsilon(u - \gamma w) \end{cases} \qquad \text{on } \mathbb{R}\times[0,\infty),$$

where ε and γ are positive but very close to zero. The function f looks like u(1-u)(u-a) for some $a\in(0,1/2)$. One finds a solution $(u,w)(x,t)=(\bar{u},\bar{w})(x-ct)=\bar{U}(x-ct)$, a travelling pulse with speed c, so that $\bar{U}(\pm\infty)=0$ (see Hastings [3]). Changing to coordinates moving with this wave, $\xi=x-ct$, we find $\bar{U}$ is an equilibrium of

$$(3.2)\qquad \begin{cases} u_t=u_{\xi\xi}+cu_\xi + f(u)-w \\ w_t = cw_\xi + \varepsilon(u-\gamma w) \ . \end{cases}$$

We then linearize about $\bar{U}$ and rewrite as

$$(3.3)\qquad \begin{pmatrix} p \\ r \end{pmatrix}_t = A\begin{pmatrix} p \\ r \end{pmatrix} + \begin{pmatrix} f(u+p)-f(u)-Df(u)p \\ 0 \end{pmatrix} \equiv A\begin{pmatrix} p \\ r \end{pmatrix} + g(p,r)$$

Working in the space of bounded uniformly continuous functions on $\mathbb{R}^2$ we find that A generates a C_0-semigroup. For ε small Jones [5] shows that A has a simple eigenvalue at 0 (corresponding to translations of $\bar{U}$) and no spectrum in the right half-plane. Furthermore one can show that for some $\sigma>0$ $\|S^-(t)\| \le Me^{-\sigma t}$ for all $t > 0$. The nonlinear term g is C^1 with $Dg(0)=0$.

Applying Theorem 1.1 case (D), in a neighborhood of $\bar{U}$, the space of bounded uniformly continuous functions on $\mathbb{R}^2$ can be decomposed into an infinite dimensional stable manifold and a one dimensional center manifold (translates of $\bar{U}$) which is attracting. One can conclude that the set of translates of $\bar{U}$ contains the ω-limit set of a neighborhood of $\bar{U}$, and the stability result of Jones [5] is obtained.

The second application concerns the instability of the radially symmetric stationary solutions to the nonlinear wave equation

$$(3.4)\qquad u_{tt} = \Delta u + |u|^\gamma u - m^2u \qquad \text{on } \mathbb{R}^n\times[0,\infty),$$

where $0 < \gamma < 2/(n-2)$ and $m > 0$.

We rewrite (4.4) as a system

$$(3.5)\qquad \begin{cases} u_t = v \\ v_t = \Delta u + f(u) \end{cases}$$

where $f(u) = |u|^{\gamma}u - m^2u$. Linearize at $\tilde{u}$, a radially symmetric stationary solution, to get

$$(3.6)\quad \begin{pmatrix} p \\ q \end{pmatrix}_t = A\begin{pmatrix} p \\ q \end{pmatrix} + \begin{pmatrix} 0 \\ f(p+\tilde{u})-f(\tilde{u})-Df(\tilde{u})p \end{pmatrix} \equiv A\begin{pmatrix} p \\ q \end{pmatrix} + g(p,q).$$

It can be shown that A generates a C_0-group on $H^1 \times L^2$ and in fact that condition (C) holds. Again g is C^1 with $Dg(0)=0$. The fact that $\dim X^+>0$ may be found in the survey article by W. Strauss [9]. Hence, Theorem 1.2 gives a nontrivial unstable manifold emanating from $\tilde{u}$. This yields results like those of Keller [6] and Shatah [8].

REFERENCES

1. J. Carr, Applications of Center Manifold Theory, Appl. Math. Sci. **35**, Springer-Verlag, New York, 1981.
2. S.N. Chow and J.K. Hale, Methods of Bifurcation Theory, Springer Verlag, New York, 1982.
3. S.P. Hastings, 'On the existence of homoclinic and periodic orbits for the FitzHugh-Nagumo equations,' Quart. J. Math., Oxford, **27** (1976), 123-134.
4. D. Henry, Geometric Theory of Semilinear Parabolic Equations, Lecture Notes in Math. **840**, Springer-Verlag, New York, 1981.
5. C.K.R.T. Jones, 'Stability of the travelling wave solution of the FitzHugh Nagumo System,' Trans. Amer. Math. Soc., **286** (1984), 431-469.
6. C. Keller, 'Stable and unstable manifolds for the nonlinear wave equation with dissipation,' J. Diff. Eq., **50** (1983), 330-347.
7. A. Pazy, Semigroups of Linear Operators and Applications to Partial Differential Equations, Appl. Math. Sci. **44**, Springer-Verlag, New York, 1983.
8. J. Shatah, 'Unstable Ground State and Standing Waves of the Nonlinear Klein-Gordon Equation.' Preprint.
9. W.A. Strauss, 'Stable and Unstable States of Nonlinear Wave Equations,' in Nonlinear Partial Differential Equations, Contemporary Mathematics **17**, Amer. Math. Soc., Providence, 1983, pp. 429-441.
10. A.E. Taylor, Introduction to Functional Analysis, Wiley, New York, 1961.

SOME APPLICATIONS OF THE LERAY-SCHAUDER ALTERNATIVE TO DIFFERENTIAL EQUATIONS

R. Bielawski and L. Gorniewicz
Institute of Mathematics
University of Nicholas Copernicus
Chopina 12/18
87 100 Torun
POLAND

Many authors comp. [2], [5], [10], [11] considered differential equations of the following type:

$$x'(t) = f(t, x(t), x'(t)). \tag{1}$$

where $f : [0,a] \times R^n \times R^n \to R^n$ is a continuous map satisfying some suitable assumptions.

To solve (1) or to give the topological characterization of the set of solutions for the Cauchy problem associated with (1), always was used the fixed point theory for condensing-type maps. We want to show that (1) can be reduced very easily to the following:

$$x'(t) = g(t, x(t)), \tag{2}$$

where $g : [0,a] \times R^n \to R^n$ is a continuous map determined by f. To get this we use only the Leray-Schauder alternative for continuous maps from R^n into itself. Moreover, we get that the set of all solutions of (1) is equal to the set of all solutions of (2); so from the well known Peano theorem or from the Aronszajn theorem, in the case of the topological characterization of the set of solutions, we obtain all information about (1).

It is quite interesting that our method works, without any change, for differential inclusions. Therefore we get a generalization of the respective results given in [1], [8], [9] for the following types of differential inclusions:

(i) ordinary differential inclusions,
(ii) hyperbolic differential inclusions,
(iii) eliptic differential inclusions.

Note that even in the case of differential equations from ou results follow some new information .

S. P. Singh (ed.), Nonlinear Functional Analysis and Its Applications, 187–194.

1. THE LERAY-SCHAUDER ALTERNATIVE.

In this section we start from a formulation of the Leray-Schauder alternative [3]) for upper semi continuous (u.s.c.) convex and compact valued multivalued maps from the euclidean n-space R^n into itself. In fact it is true in more general situation [6] or [3]) but for our applications we need only the version presented below.

1.1 Theorem.

(The Leray-Schauder alternative) Let $\psi : R^n \to R^n$ be an u.s.c., convex and compact valued map. We let

$$\varepsilon(\psi) = \{y \in R^n \quad y \in \lambda\psi(y) \quad \text{for some} \quad 0 < \lambda < 1\}.$$

Then either $\varepsilon(\psi)$ is unbounded or ψ has a fixed point, i.e., $y \in \psi(y)$ for some $y \in R^n$.

Before the proof of (1.1) we will prove the following:

1.2 Theorem.

(The non-linear alternative) Let $\psi : K(0,r) \to R^n$ be an u.s.c., convex and compact valued map, where $K(0,r) = \{y \in R^n : \|y\| \leq r\}$. Then ψ has at least one of the following two properties:

(i) ψ has a fixed point,

(ii) there is $y \in \partial K(0,r)$ with $y \in \lambda\psi(y)$ for some $0 < \lambda < 1$,

where $\partial K(0,r)$ denotes the boundary of $K(0,r)$ in R^n.

Proof of (1.2). We can assume, without loss of generality, that ψ is fixed-point free on the boundary $\partial K(0,r)$ of $K(0,r)$ in R^n. Consider a multi-valued homotopy:

$$\chi : K(0,r) \times [0,1] \to R^n$$

given as follows:

$$\chi(y,t) = t \cdot \psi(y), \text{ for each } t \in [0,1] \text{ and } y \in K(0,r).$$

Then χ is a convex and compact valued u.s.c. homotopy joining with the map $\psi_1 : K(0,r) \to R^n$, $\psi_1(y) = \{0\}$, for each y; either this homotopy is fixed-point free on $\partial K(0,r)$ or it is not. If the homotopy is not fixed-point f ee on $\partial K(0,r)$ then we get a point $y \in \partial K(0,r)$ and a real number $0 < \lambda < 1$ such that $y \in \lambda\psi(y)$, so (i) holds. If χ is a fixed-point free on $\partial K(0,r)$ then the topological degree $\deg(\psi)$ of ψ is equal to $\deg(\psi_1)$ of ψ_1 [6], [8] or [9]) and moreover:

$$\deg(\psi) = \deg(\psi_1) \neq 0,$$

because $\deg(\psi_1) \neq 0$. So ψ has a fixed point (comp. [6], [8] or [9]) and the proof is completed.

Proof of (1.1). Assume $\varepsilon(\psi)$ is bounded, and let $K(0,r)$ be a closed ball containing $\varepsilon(\psi)$ in its interior. Then the map $\tilde{\psi} : K(0,r) \to R^n$ to be contraction of ψ to $K(0,r)$ the second property in (1.2). Therefore, in view of (1.2), $\tilde{\psi}$ has a fixed point and the proof is completed.

1.3 Corollary. Let $\psi : R^n \to R^n$ be an u.s.c. convex and compact valued map. Assume further that the following condition is satisfied:

$$\exists 0 \leq \psi \leq 1 \exists \; M \geq 0 \quad \forall y \in R^n \quad \forall u \in \psi(y) : \| u \| \leq M + k \cdot \| y \|. \tag{1.3.1}$$

Then ψ has a fixed point.

Proof. Let $y \in \varepsilon(\psi)$. Then we have $y \in \lambda\psi(y)$ for some $0 < \lambda < 1$. By using (1.3.1) we get:

$$\| y \| \leq \lambda M + \lambda K \; \| y \| \leq M + K \; \| y \|$$

and consequently $\| y \| \leq (1 - K)^{-1} M$, so $\varepsilon(\psi)$ is bounded. Now (1.3) follows from the Leray-Schauder alternative.

Now we are going to formulate some consequences of (1.3) which are important in applications to differential equations.

Let T be a compact metric space. We put $R^{np} = R^n \; \ldots \; R^n$. We need to consider multi-valued maps of the following types:

$$\psi : T \times R^{np} \times R^n \to R^n.$$

To end of this section we will keep the following assumptions:

$$\psi \text{ is u.s.c., convex and compact valued,} \tag{1.4}$$

$$\exists 0 \leq k < 1 \exists \; \alpha, \beta \geq 0 \; \forall t \in T \; \forall x \in R^{np} \; \forall u \in \psi(t,xy) : \| u \| \leq \alpha + \beta \cdot \| x \| + k \cdot \| y \|, \tag{1.5}$$

the set $\{y \in R^n : y \in \psi(t,x,y)\}$ is convex for each $t \in T$ and $x \in R^{np}$. (1.6)

1.7 Proposition.

Assume that $\psi : T \times R^{np} \times R^n \to R^n$ satisfies conditions (1.4), (1.5) and (1.6). Then the map

$$\Psi : T \times R^{np} \to R^n$$

given as follows:

$$\Psi(t,x) = \{y \in R^n;\ y \in \psi(t,x,y)\}$$

is an u.s.c., non-empty, convex and compact valued map which satisfies the following conditions.

$$\forall\, t \in T\ \forall\, x \in R^{np}\ \forall\, y \in \Psi(t,x) : \|y\| \le (1-k)^{-1} (\alpha + \beta \|x\|). \tag{1.7.1}$$

Proof. First observe, that $\Psi(t,x) \neq 0$ because (1.5) implies (1.3.1). By (1.6) we get that $\Psi(t,x)$ is convex and non-empty. Moreover (1.7.1) immediately follows from (1.5). Therefore we have to prove that $\Psi(t,x)$ is compact and that Ψ is u.s.c. We will prove that $\Psi(t,x)$ is compact. Let $\{y_m\}$ be a sequence contained in $\Psi(t,x)$. By (1.7.1) $\{y_m\}$ is a bounded subset of R^n, so there is a subsequence $\{y_{m_1}\}$ of $\{y_m\}$ which converges for some $y_0 \in R^n$. We have:

$$y_{m_1} \in \psi(t,x,y_{m_1}), \text{ for each } 1.$$

Consequently from (1.4) follows that $y_0 \in \psi(t,x,y_0)$ and hence $\Psi(t,x)$ is compact.

Now we will show that Ψ is u.s.c. Let $(t_0,x_0) \in T \times R^{np}$ and let U be an open neighbourhood of $\Psi(t_0,x_0)$ in R^n. We set $\Psi^{-1}(U) = \{(t,x) \in T \times R^{nk};\ \Psi(t,x) \subset U\}$ is an open subset of $T \times R^{np}$. Assume contrary. Then we get a point $(\bar{t},\bar{x})$ in $\Psi^{-1}(U)$ and a sequence $\{t_m,x_m\}$ which converges to $(\bar{t},\bar{x})$ and $(t_m,x_m) \notin \Psi^{-1}(U)$. Let $y_m \in \Psi(t_m,x_m)$ be a point such that $y_m \notin U$. From 1.7.1 we get

$$\|y_m\| \le (1-k)^{-1} (\alpha + \beta \|x_m\|).$$

Because $\lim x_m = \overline{x}$ we can assume, without loss of generality, that the sequence y_m converges to some point $\overline{y} \in R^n$. We have $y_m \in \psi(t_m, x_m, y_m)$ for each m, so from upper semi continuity of ψ we obtain $\overline{y} \in \psi(\overline{t}, \overline{x}, \overline{y},)$, but $\overline{y}_0 \in U$ and $y_m \notin U$ for each m, so we have a contradiction and the proof is completed.

1.8 Remark. Observe that if ψ satisfies the following condition: $\psi(t,x,sy + (1 - s) \cdot \overline{y}) \supset s \cdot \psi(t,x,y) + (1 - s) \cdot \psi(t,x,\overline{y})$, then it implies (.16).

1.9 Remark. Observe that if ψ is a bounded map, then Ψ is a bounded map too.

Assume that $\psi = f$ is a single-valued map.

1.10 Proposition. Assume that $f : T \times R^{np} \to R^n$ is a single-valued map which satisfies (1.4), (1.5) and the following

$$\forall\, t \in T\ \forall\, x \in R^{np}\ \forall\, \overline{y},\ y \in R^n : \langle f(t,x,y) - f(t,x,\overline{y}), y - \overline{y}\rangle \leq k \cdot \| y - y \|^2, \text{ where k is the same as in (1.5)} \tag{1.10.1}$$

and $\langle\ ,\ \rangle$ denotes the inner product in R^n.

Then $\Psi(t,x)$ is a singleton for each $(t,x) \in T \times R^{np}$, so in particular f satisfies (1.6); then we put $\Psi = g$ because Ψ is a single-valued.

Proof. Assume that $y, \overline{y} \in \Psi(t,x)$. By using (1.10.1) we get:

$$\langle y - \overline{y}\ ,\ y - \overline{y}\rangle = \| y - \overline{y} \|^2 \leq k \cdot \| y - \overline{y} \|^2$$

and because $k < 1$ we get that $y = \overline{y}$. The proof is complete.

1.11 Remark. Condition (1.10.1) is well known in theory of differential equations (see [2] or [10] or [11]). Let us remark that the following condition:

$$\forall\, t \in T\ \forall\, x \in R^{np}\ \forall\, y,\ \overline{y} \in R^n : \| f(t,x,y) - f(t,x,\overline{y}) \| \leq k\ \| y - y \| \tag{1.11.1}$$

is stronger Finally observe that (1.11.1) and the following:

$$\alpha,\ \beta \geq 0\ \forall\, t \in T\ \forall\, x \in R^{np} : \| f(t,x,0) \| \leq \alpha + \beta \cdot \| x \| \tag{1.11.2}$$

imply (1.10.1) and (1.5); for details see [5] or [11].

1.12 Proposition. Assume that $f : T \times R^{np} \times R^n \to R^n$ satisfies (1.4), (1.11.1) and (1.11.2). Assume further that f satisfies the Lipschitz condition with respect to the second variable. Then $g : T \times R^{np} \to R^n$ (comp. (1.10)) satisfies the Lipschitz condition with respect to the second variable and is continuous.

Proof. We have:

$$\| g(t,x) - g(t,\bar{x}) \| = \| f(t,x,g(t,x)) - f(t,\bar{x},g(t,\bar{x})) \|$$
$$\leq \| f(t,x,g(t,x)) - f(t,x,g(t,\bar{x})) \| + \| f(t,x,g(t,\bar{x}))$$
$$- f(t,\bar{x},g(t,\bar{x})) \| \leq k \cdot \| g(t,x) - g(t,\bar{x}) \| + L \| x - \bar{x} \| ,$$

and consequently we get:

$$\| g(t,x) - g(t,\bar{x}) \| \leq (1 - k)^{-1} L \| x - \bar{x} \| ;$$

the proof is complete.

2. APPLICATIONS.

Let $\psi : T \times R^{np} \times R^n \to R^n$ be a convex, compact valued map satisfying conditions (1.4), (1.5) and (1.6). We shall consider the following differential inclusions (for details see [1] or [9]).

2.1 Ordinary differential inclusions

$T = [0.a] : x^{(p)}(t) \in \psi(t,x(t),x'(t),\ldots,x^{(p-1)}(t),x^{(p)}(t))$, a.e. $t \in [0,a]$.

2.2 Hyperbolic differential inclusions

$T = [0,a] \times [0,a]$, $p = 1$: $\frac{\partial^2 u}{\partial t \partial s}(s,t) \in \psi(s,t,u(s,t), \frac{\partial^2 u}{\partial t \partial s})$, a.e. $(s,t) \in [0,a] \times [0,a]$.

2.3 differential inclusions

$T = K(0,r) \subset R^n$; $n = 1$, $p = 2$:

$$\Delta(u)(t) \in \psi(t,u(t), D(u)(t),\Delta(u)(t)), \text{ a.e. } t \in K(0,r),$$

where $D(u)(t) = \frac{\partial u}{\partial t_1}(t) + \ldots + \frac{\partial u}{\partial t_n}(t)$ and $\Delta(u) = \sum_{i=1}^{n} (\frac{\partial u}{\partial t_1})^2$ is the Laplace operator.

Let $\Psi : T \times R^{np} \to R^n$ be the multi-valued map associated with ψ in (1.7). Then we can formulate:

$$x^{(p)}(t) \in \Psi(t,x(t),x'(t),\ldots,x^{(p-1)}(t)), \quad \text{a.e. } t \in [0,a]; \tag{2.1.1}$$

$$\frac{\partial^2 u(s,t)}{\partial t \partial s} \in \Psi\,(s,t,u(s,t)), \text{ a.e. } (s,t) \in [0.a] \times [0.a]; \tag{2.2.1}$$

$$\Delta(u)(t) \in \Psi(t,u(t),D(u)(t)], \text{ a.e. } t \in K(0,r). \tag{2.3.1}$$

Denote by $S_i(\psi)$ the set of all solutions of problem (2.i), $i = 1,2,3$ and by $A_i(\Psi)$ the set of all solutions of problem (2.i.1), $i = 1,2,3$.

As an immediate consequence of (1.7) we obtain:

2.4 Theorem.

$S_i(\psi) = A_i(\Psi)$, for each $i = 1,2,3$; in particular problem (2.i) has a solution if and only if problem (2.i.1) has a solution for every $i = 1,2,3$.

Now, it is evident that all results related to the respective differential inclusion (differential equation) of type (2.i.1), $i = 1,2,3$, given in [1,5,6,7,8,9,10,11] can be obtained for the differential inclusion (differential equation) of type (2.i), $i = 1,2,3$; in particular let us note that for problems of type (2.i) the following holds:

1. existence theorems (also for boundary value problems) - for $i = 1,2,3$ (comp. [3] and [9]);
2. the theorem on topological characterization of the set of solutions - for $i = 1,2$ (comp. [7], [8] and [11]);
3. the theorem on upper semicontinuous dependence of the set of solutions on initial values - for $i = 1$ (comp. [2] and [4]);
4. the theorem on existence of periodic solutions - for $i = 1$ (comp. [4]);
5. the Hukuhara theorem - for $i = 1$ (comp. [1]).

REFERENCES

1. J. P. Aubin and A. Cellina, *Differential inclusions*, Lecture Notes in Math.

2. K. Deimling, *Ordinary differential equations in Banach spaces*, Lecture Notes in Math., 596, 1977.

3. J. Dugundji and A. Granas, *Fixed point theory*, Vol. I, PEN, Warszawa, 1982.

4. G. Dylawerski and L. Gorniewicz, *A remark on the Krasnosielski's translation operator along trajectories of ordinary differential equations*, Serdica Bul. J. of Math., 9, 1983, pp. 102-107.

5. K. Goebel, *Grubość zbiorów w przestrzeniach metrycznych i jej zastosowanie w teorii punktów stałych*, Lublin 1970, in Polish.

6. L. Górniewicz and Z. Kucharski, *On k-set contraction pairs*, J. Math. Anal. Appl., 81, 1985.

7. L. Górniewicz and T. Pruszko, *On the set of solutions of the Darboux problem for some hyperbolic equations*, Bull, Acad. Polon Sci., 5-6, 1980, pp. 279-285.

8. J. M. Lasry and R. Robert, *Analyse non linéaire multivoque*, Centre de Recherche de Math. de la Decision, No 7611, Université de Paros - Dauphine.

9. T. Pruszko, *Some applications of the topological degree theory to multi-valued boundary value problem*, Dissertations Math,. 229, 1984, pp. 1-48.

10. B. N. Sadovskij, *Limit-compact and condensing operators*, Uspehi Mat, Nauk, 27, 1971, pp. 81-146, in Russian.

11. R. Schoneberg, *Some applications of the degree theory for semi-condencsing vectorfields*, Preprint No. 203, 1978, University of Bonn.

SEQUENCES OF ITERATES IN LOCALLY CONVEX SPACES

James M. Boyte &
K. L. Singh
Department of Mathematics
Fayetteville State University
Fayetteville, NC 28301

J. H. M. Whitfield
Department of Mathematics
Lakehead University
Thunder Bay, ON, Canada
P7B 5E1

ABSTRACT. Locally convex spaces can be normed over a topological semifield. Using this norm, we prove the strong and weak convergence of sequence of iterates for generalized nonexpansive mappings. Our results generalize those of Browder and Petryshyn, Hicks and Huffman, Kannan, Kirk, Massabo, Opial, Ray and Rhoades, Rhoades and others.

1. PRELIMINARIES

Let Δ be a nonempty set and $R^{\Delta} = \prod_{\alpha\varepsilon\Delta} R_{\alpha}$ be the product of Δ copies of the real line with the product topology. Addition and scalar multiplication in R^{Δ} are defined pointwise by the equations

$$(f + g)(\alpha) = f(\alpha) + g(\alpha),\ (f \cdot g)(\alpha) = (f(\alpha)),\ (g(\alpha)),$$

$$(\lambda f)(\alpha) = \lambda f(\alpha),$$

where $f, g \varepsilon R^{\Delta}$, $\alpha \varepsilon \Delta$ and $\lambda \varepsilon R$. R^{Δ} is called a Tychonoff semifield. A partial ordering is defined by the cone $R_{+}^{\Delta} = \{f : f(\alpha) \geq 0, \alpha \varepsilon \Delta\}$.

A general introduction to the space R^{Δ} may be found in [1]. For $f, g \varepsilon R^{\Delta}$

(1) $f \leq g$ means $f(\alpha) \leq g(\alpha)$ for all $\alpha \varepsilon \Delta$;
(2) $f \lneq g$ means $f \leq g$ and there exists $\alpha \varepsilon \Delta$ with $f(\alpha) < g(\alpha)$;
(3) $f << g$ means $f(\alpha) < g(\alpha)$ for all $\alpha \varepsilon \Delta$.

If E is a real locally convex space whose topology is generated by a family $\{\rho_{\alpha} : \alpha \varepsilon \Delta\}$ of continuous seminorms, then the function $\rho : E \rightarrow R_{+}^{\Delta}$ defined by $[\rho(x)](\alpha) = \rho_{\alpha}(x)$, $x \varepsilon E$, $\alpha \varepsilon \Delta$ satisfies:

(1) $\rho(x) \geq 0$
(2) $\rho(\lambda x) = |\lambda|\ \rho(x)$, and
(3) $\rho(x + y) \leq \rho(x) + \rho(y)$,

S. P. Singh (ed.), Nonlinear Functional Analysis and Its Applications, 195–206.

where "$\leq$" denotes the natural order induced by R_+^Δ. In the case where E is a Hausdorff space (1) becomes

(1)' $\rho(x) = 0$ if and only if $x = 0$, and ρ will be referred to as the corresponding semifield norm. We note that ρ satisfies the asioms of a norm.

The topology t_ρ induced by ρ in E is the original topology, where a t_ρ neighborhood of x is of the form $\Omega(x,U) = \{y : \rho(x - y) \in U\}$, U being a neighborhood of zero in R^Δ.

In this section we prove a theorem for generalized nonexpansive mappings in locally convex spaces which extends results of Kirk [7], Kannan [6], Massabo [9], Ray and Rhoades [11], Naimpally and Singh [10], and others. In section 3, we give an example of nonnormable locally convex space, which demonstrates that our results are indeed extensions of some known results.

Definition 1.1. Let E be a Hausdorff locally convex space whose topology is generated by a family $\{\rho_\alpha, \alpha \in \Delta\}$ of continuous seminorms and let ρ be the corresponding semifield norm. Let C be a subset of E. A mapping $T : C \to C$ is nonexpansive if $\rho(Tx - Ty) \leq \rho(x - y)$ for all $x, y \in C$ and is quasinonexpansive provided that for $p \in C$ with $Tp = p$ we have $\rho(Tx - p) \leq \rho(x - p)$ for all $x \in C$. T is said to satisy condition (A) if $\rho(Tx - Ty) \leq \max\{\rho(x,y), \frac{1}{2}[\rho(x - Tx) + \rho(y,-Ty)], \frac{1}{2}[\rho(x - Ty) + \rho(y - Ty)]\}$ for all $x, y \in C$.

The map T is called generalized nonexpansive if there exists a number k, $0 \leq k < 1$ such that for all $x, y \in C$,
$\rho(Tx - Ty) \leq \max\{\rho(x - y), \frac{1}{2}[\rho(x - Tx) + \rho(y - Ty)], \frac{1}{2}[\rho(x - Ty) + \rho(y - Tx)], k\rho(x - Ty), k\rho(y - Tx)\}$.

Remark 1.1. Clearly any nonexpansive mapping satisfies condition (A), and mappings satisfying condition (A) are generalized nonexpansive. However, the converse is not true as can be seen from the following example.

Example 1.1. Let $M_1 = \{\frac{m}{n}, m = 0, 1, 3, 9, \ldots, n = 1, 4, \ldots, 3k + 1 \ldots\}$ $M_2 = \{\frac{m}{n} : m = 1, 3, 9, 27, \ldots; n = 2, 5, \ldots, 3k + 2, \ldots\}$.

Let $M = M_1 \cup M_2$ with the usual metric. Define $T : M \to M$ by

$$T(x) = \begin{cases} 3x/4 & \text{for } x \in M_1 \\ x/2 & \text{for } x \in M_2. \end{cases}$$

T is a generalized nonexpansive mapping. Indeed, if both x and y are in M_1 or in M_2, then $|(Tx - Ty)| \leq \frac{3}{4}(x - y)$. Now suppose $x \in M_1$ and $y \in M_2$, then for $x > \frac{2}{3}y$, we have $|(Tx - Ty)| = \frac{3}{4}(x - \frac{2y}{3})$

$\leq \frac{3}{4}(x - y/2) = \frac{3}{4}|(x - Ty)|$. For $x < 2/3y$, $(Tx - Ty) = \frac{3}{4}(\frac{2y}{3} - x)$ $\leq \frac{3}{4}|(y - x)|$. Interchanging the roles of x and y we have $(Tx - Ty) \leq \frac{3}{4}|(y - Tx)|$. Letting $k = \frac{4}{5}$, we see that T is a generalized nonexpansive mapping. We show that condition (A) is not satisfied. Take $x = 1$, $y = \frac{3}{5}$. Then $d(Tx,Ty) = \frac{9}{20} \not\leq \max\{\frac{2}{5}, \frac{11}{40}, \frac{17}{40}\} = \max\{d(x,y), \frac{1}{2}[d(x,Ty) + d(y,Ty)], \frac{1}{2}[d(x,Ty) + d(y, Tx)]\}$.

Theorem 1.1. Let E be a locally convex Hausdorff topological vector space and C be a convex subset of E. Let $T : C \to C$ be a generalized nonexpansive mapping and define $S : C \to C$ by

$$S = \alpha_0 I + \alpha_1 T + \alpha_2 T^2 + \ldots + \alpha_k T^k \text{ where } \alpha_i \geq 0,\ \alpha_0 > 0,$$

$$\alpha_1 > 0 \text{ and } \sum_{i=0}^{k} \alpha_i = 1.$$

Then $Sx = x$ if and only if $Tx = x$.

Proof. Clearly, if $Tx = x$, then $Sx = x$ since

$$Sx = (\alpha_0 I + \alpha_1 T + \alpha_2 T^2 + \ldots + \alpha_k T^k)\, x = (\alpha_0 + \alpha_1 + \alpha_2 + \ldots + \alpha_k)x = x.$$

Conversely, let $x \in C$ such that $Sx = x$; that is, such that

$\alpha_0 x + \alpha_1 Tx + \alpha_2 T^2 x + \ldots + \alpha_k T^k_x = x$. Consider the points

$x_0 = T^0x$, $x_1 = Tx_0$, $x_2 = T^2 x_0$, ..., $x_k = T^k x_0 = Tx_{k-1}$, and

the numbers $\rho(x_p - x_q)$, $p,q = 0,1,2, \ldots, k$.

Let $\delta = \max\{\rho(x_p - x_q)\}$ and suppose that $\delta > 0$. Let p be the least positive integer for which $\delta = \rho\,(x_p - x_q)$, $p > q$. Then $q = 0$. For if $q > 0$, then $p > 1$ and we have

$$\begin{aligned}\delta = \rho(x_p - x_q) &= \rho(Tx_{p-1} - Tx_{q-1}) \\ &\leq \max\{\rho(x_{p-1} - x_{q-1}), \tfrac{1}{2}[\rho(x_{p-1} - Tx_{p-1}) \\ &+ \rho(x_{q-1} - Tx_{q-1})], \tfrac{1}{2}[\rho(x_{p-1} - Tx_{q-1}) \\ &+ \rho(x_{q-1} - Tx_{p-1})], \\ &k\,\rho(x_{p-1} - Tx_{q-1}), k\,\rho(x_{q-1} - Tx_{p-1})\} \\ &= \max\{\rho(x_{p-1} - x_{q-1}), \\ &\quad \tfrac{1}{2}[\rho(x_{p-1} - x_p) + \rho(x_{q-1} - x_q)], \\ &\quad \tfrac{1}{2}[\rho(x_{p-1} - x_q) + \rho(x_{q-1} - x_p)], \\ &\quad k\,\rho(x_{p-1} - x_q), k\,\rho(x_{q-1} - x_p)]\} < \delta,\end{aligned}$$

a contradiction. Hence, $q = 0$.

Since $\alpha_1 > 0$, $\alpha_0 < 1$ and $Sx = x$, we obtain $x = \sum_{i=1}^{k} \beta_i Tx^i$, where $\beta_i = \dfrac{\alpha_i}{1 - \alpha_0}$. Note that if $\beta_1 = 1$, then $\alpha_i = 0$ for $i \geq 2$ and $S = \alpha_0 I + \alpha_1 T$, $\alpha_0 + \alpha_1 = 1$. In this case it is easily seen that $Tx = x$. Thus, we may assume that $\beta_1 < 1$. Observe that $\sum_{i=1}^{k} \beta_i = 1$ and $\beta_i \geq 0$, and set $\gamma_i = \dfrac{\beta_i}{1 - \beta_1}$ and $y = \sum_{i=2}^{k} \gamma_1 T^i x$. Then $x = \beta_1 Tx + (1 - \beta_1)y$. Moreover, since $\sum_{i=2}^{k} \gamma_i = 1$, $\gamma_i \geq 0$, and C is convex, it follows that y is in C, and we have

$$\begin{aligned}\rho(T^p x - y) &= \rho\Big(\sum_{i=2}^{k} \gamma_i (T^p x - T^i x)\Big) \\ &\leq \sum_{i=2}^{k} \gamma_i\, \rho(T^p_x - T^i x) = \sum_{i=2}^{k} \gamma_i\, \rho(x_p - x_i)\end{aligned}$$

$$\leq \sum_{i=2}^{k} \gamma_i \ \delta = \delta$$

Consequently,

$$\delta = \rho(x_p - x) = \rho(T_x^p - \beta_1 Tx + (1 - \beta_1)y)$$

$$\leq \beta_1 \ \rho(T_x^p - Tx) \quad + \quad (1 - \beta_1) \ \rho(T_x^p - y)$$

$$< \beta_1 \delta + (1 - \beta_1) \ \delta = \delta.$$

This contradiction yields $\delta = 0$, and $Tx = x$.

Remark 1.2. Theorem 4 [6], Theorem 1 [7], Lemma 4.1 [9], Theorem 2 [10], and Lemma 2.1 [12] are special cases of Theorem 1.1 above.

2. STRONG CONVERGENCE OF THE SEQUENCE $\{S^n x)$ OF ITERATES

In this section we prove a generalization of theorems due to Huffman [4], Naimpally & Singh [10], Ray and Rhoades [11], Rhoades [12], and others using a generalization of the notion of strict convexity in Banach spaces to Hausdorff locally convex linear topological spaces.

Definition 2.1. The Banach space definition of strict convexity can ba extended to a Hausdorff locally convex linear topological space E in several ways:

If $x, y \in E$, $x \neq y$ and $\rho(x) = \rho(y)$, then
$$\rho(\frac{x + y}{2}) \neq \rho(x); \text{ i.e. } \rho(\frac{x + y}{2}) < \rho(x). \qquad \text{(SC1)}$$

If $x, y \in E$, $x \neq y$ and $\rho(x) = \rho(y)$, then
$$\rho(\frac{x + y}{2}) << \rho(x). \qquad \text{(SC2)}$$

If $x, y \in E$, and $\rho(x + y) = \rho(x) + \rho(y)$, then
x and y are not linearly independent. (SC3)

Note that (SC1) is equivalent to: if $x \neq y$ and $\rho(x) = \rho(y)$, then $\rho(\lambda \ x + (1 - \lambda)y) < \rho(x)$ for all λ in (0.1).

Example 2.1. (Huffman [4]). Let $E = C(R)$ be the space of all continuous real-valued functions on the reals. The space E is a locally convex Hausdorff space when given the topology of pointwise convergence generated by the family $\{\rho_\alpha, \alpha \in R\}$ of seminorms, where ρ_α is defined by $\rho_\alpha(f) = | f(\alpha) |$, $f \in E$, $\alpha \in R$. Here E satisfies (SC1), but not (SC 2) or (SC3).

Example 2.2. Let A be a nonempty set and X_α be a strictly convex normed linear space with norm $\|\ \|_\alpha$ for each $\alpha \in A$. Consider the product $\prod_{\alpha\in A} X_\alpha$ and for each α define $\rho_\alpha(x) = \|x_\alpha\|_\alpha$ for each $x = (x_\alpha) \in \prod_{\alpha\in A} X_\alpha$. Then $[\rho_\alpha : \alpha \in A]$ is a family of seminorms on $\prod_{\alpha\in A} X_\alpha$ and therefore generates a locally convex topology (which is the product of topology) on $\prod_{\alpha\in A} X_\alpha$. It is easily seen that $\prod_{\alpha\in A} X_\alpha$ satisfies (SC1) and (SC2), but not (SC3).

Since (SC1) is clearly the weaker of the three conditions, we shall adopt this as the definition of strict convexity in our setting.

The following result will be needed:

Theorem 2.1. (Tarafdar [15]) Let (X,h) be a separated uniform space, T_h be the topology induced by the uniformity h, and $\{\rho_\alpha : \alpha \in \Delta\}$ be the family of semimetrics on X which generates the topology T_h. Let $T\colon X \to X$ be a T_h - continuous mapping. Further suppose that

(i) $F(T) = \{x \in X\colon Tx = x\}$ is nonempty

(ii) for each $x \in X$ with $x \notin F(T)$ and $y \in F(T)$ we have for each $\alpha \in \Delta$, $\rho_\alpha(Tx,y) < \rho_\alpha(x,y)$ if $\rho_\alpha(x,y) \neq 0$ and $\rho_\alpha(Tx,y) = 0$ if $\rho_\alpha(x,y) = 0$.

Then for each x in X, either the sequence (net) $\{T^m x\}$ of iterates has no T_h - convergent subsequence or the T_h - $\lim T^n x$ exists and belongs to F(T).

Theorem 2.2. Let E be a strictly convex quasicomplete Hausdorff locally convex linear topological space. Let K be a closed convex subset of E. Let $T\colon K \to K$ be a continuous generalized nonexpansive mapping such that $T(K) \subseteq K_1$, where K_1 is compact. Define $S\colon K \to K$ as in Theorem 1.1.

Then for each $x \in K$, the sequence $\{S^n x\}$ of iterates converges to a fixed point of T.

Proof. It follows from Tychonoff's theorem that $F(T) \neq 0$. Moreover, an appeal to Theorem 1.1 guarantees the nonemptyness of F(S). Since E is quasicomplete, it follows that the closed convex hull of $T(K) \cup \{x\}$

is compact for each $x \in K$. [8, p.241]. Hence for each $x \in K$, the sequence $\{S^n x\}$ is contained in a compact set and so has a convergent subnet.

Let $y \in F(T)$ and $x \in K - F(T)$. Using the definition of T it can be easily seen that $\rho(Tx - y) \leq \rho(x - y)$.

Let $z = \frac{1}{1 - \alpha_0} [\alpha_1 T(x - y) + \alpha_2(T^2 x - y) + ... + \alpha_k(T^k x - y)]$

Then $\rho(z) \leq \frac{1}{1 - \alpha_0} [\alpha_1 \rho(Tx - y) + \alpha_2 \rho(Tx^2 x - y) + ... + \alpha_k \rho(T^k x - y)]$

$= \rho(x - y)$

Now, $\rho(Sx - y) = \rho[\alpha_0(x - y) + \alpha_1(Tx - y) + \alpha_2(T^2 x - y) + ... + \alpha_k(T^k x - y)]$

$= \rho(\alpha_0(x - y) + (1 - \alpha_0)z)$

If $\rho(x - y) = 0$, then $\rho(Sx - y) = 0$. If $\rho(x - y) \neq 0$, then

$$\rho(Sx - y) = \rho(x - y) \left(\frac{\alpha_0 (x - y)}{\rho(x - y)} + \frac{(1 - \alpha_0)z}{\rho(x - y)}\right).$$

If $\rho(z) < \rho(x - y)$, then

$$\rho(Sx - y) \leq \rho(x - y) \left(\frac{\alpha_0 \rho(x - y)}{\rho(x - y)} + \frac{(1 - \alpha_0)\rho(z)}{\rho(x - y)}\right)$$

$$< \rho(x - y) [\alpha_0 + (1 - \alpha_0)] = \rho(x - y).$$

If $\rho(z) = \rho(x - y)$, then it follows from the strict convexity of E that

$$\rho\left(\frac{\alpha_0 (x - y)}{\rho(x - y)} + \frac{(1 - \alpha_0) z}{\rho(x - y)}\right) < 1$$

Therefore, for any $y \in F(T)$ and $x \notin F(T)$, we obtain

$\rho(Sx - y) < \rho(x - y)$ and the desired conclusion follows from Theorem 2.1.

As immediate corollaries to Theorem 2.2 we have Theorem 4.1 [9], Theorem 8 [10], Theorem 2.9 [11], and Theorem 6 [12].

3. WEAK CONVERGENCE OF SEQUENCES OF ITERATES IN GENERALIZED HILBERT SPACES.

Definition 3.1. A locally convex linear topological space which is complete and whose generating family of seminorms satisfies the parallogram law is called a generalized Hilbert space.

It follows from Corollary 1 [14 p. 102] that nuclear spaces are generalized Hilbert spaces. Other examples of generalized Hilbert spaces which are also F-spaces may be found in [14 p. 106-108].

Definition 3.2. A mapping T froma Hausdorff locally convex linear topological space into itself is said to be asymptotically regular if $\lim_{n\to\infty} \rho(T^{n+1}x - T^{n}x) = 0$ for all $x \in E$.

Theorem 3.1. Let E be a generalized Hilbert space, C be a nonempty convex subset of E, and T: $C \to C$ be a generalized nonexpansive mapping such that $F(T) \neq 0$. Let S: $C \to C$ be a mapping as defined in Theorem 1.1. Then S is asymptotically regular.

Proof. It follows from Theorem 1.1 that $F(T) = F(S)$. Let $x_0 \in C$ and define the sequence $\{x_n\}$ by $x_n = S^n x_0$, $n = 1, 2, \ldots\ldots$ Let $y \in F(T)$. Consider the sequence $\{\rho(x_n - y)\}$.

Now $\rho(x_{n+1} - y) = \rho(Sx_n - y)$

$$\leq \sum_{i=1}^{k} \alpha_i \, \rho(T^i x_n - y).$$

Using the definition of T and $y \in F(T)$ we obtain

$$\rho(T^i n_n - y) \leq \rho(T^{i-1} x_n - y) \leq \ldots \leq \rho(x_n - y).$$

Therefore $\{\rho(x_{m+1} - y)\}$ is a nonincreasing sequence. Let $\lim_{n\to\infty} (x_n - y) = d$. If $d = 0$, the theorem is proved. Assume then that $d > 0$. Now

$$x_{n+1} - y = Sx_n - y = \alpha_0(x_n - y) + (1 - \alpha_0)z_n \quad \text{where}$$

$$z_n = \frac{1}{1 - \alpha_0} \sum_{i=1}^{k} \alpha_i (T^i x_n - y).$$

Since $\rho(T^i x_n - y) \leq \rho(x_n - y)$ and $\sum_{i=1}^{k} \alpha_i = 1$, we have $\lim_{n\to\infty} \rho(z_n) \leq d$.

Also $\lim_{n\to\infty} \rho(x_n - y) - y = d = \lim_{n\to\infty} \rho(x_{n+1} - y)$.

Since E is a generalized Hilbert space we have

$$\rho^2(x_n - y - z_n) = \rho^2(x_n - y) + \rho^2(z_n) - 2\rho\,(x_n - y - z_n).$$

Taking the limit we have $\lim_{n\to\infty} \rho^2(x_n - y - z_n) = 0$, and the result follows.

Remark 3.1. Special cases of Theorem 3.1 appear in the literature as Theorem 6 [5], Theorem 11 [10].

If $\{x_n\}$ is a sequence in E, then by $\underline{\lim}\ \rho(x_n)$ is meant the element R^Δ defined by $\underline{\lim}\ \rho(x_n)(\alpha) = \underline{\lim}\ \rho_\alpha(x_n)$, provided $\underline{\lim}\ \rho_\alpha(x_n)$ is finite for all $\alpha \in \Delta$. The quantity $\lim \overline{\rho(x_n)}$ is similarly defined.

For the proof of our next result we need the following variant of Opial's Lemma.

Lemma 3.1. If in a generalized Hilbert space, the sequence $\{x_n\}$ is weakly convergent to $x_0 \in E$, then for any $x \in E$, $x \neq x_0$

$$\underline{\lim}\ \rho(x_n - x) \geq \underline{\lim}\ \rho(x_n - x_0)$$

Definition 3.3. Let E be a sequentially complete locally convex linear topological space. The mapping S: $E \to E$ is said to be demiclosed if for any sequence $\{x_n\}$ such that $x_n \rightharpoonup x$ and $Sx_n \to y$ we have $y = Sx$.

Theorem 3.2. Let E be a semireflexive generalized Hilbert space, C be a closed, convex, weakly sequentially compact subset of E, and T be a continuous generalized nonexpansive mapping of C into itself. Define the mapping S: $C \to C$ as in Theorem 1.1. If F(T) is nonempty and I-S is demiclosed, then for any $x_0 \in C$, $S^n x_0 \rightharpoonup y \in F(T) = F(S)$.

Proof. Let $y \in F(T)$. Set $x_n = S^n x_0$. Using the definition of T we see that (1) $\rho(x_{n+1} - y) \leq \rho(x_n - y)$. It follows from (1) that $\{\rho_\alpha(x_n - y)\}$ is a nonincreasing sequence of non-negative real numbers for fixed $\alpha \in \Delta$ and fixed $y \in F(T)$. Fixing α define $g_\alpha : F(T) \to R^+$ by $g_\alpha(y) = \lim \rho_\alpha(x_n - y)$. Set $d_\alpha = \inf \{g_\alpha(y): y \in F(T)\}$. Let $\delta > 0$ and set $F_{\delta,\alpha} = \{y \in F(T): g_\alpha(y) \leq d_\alpha + \delta\}$. If $d_\alpha \in \{g_\alpha(y) : y \in F(T)\}$, then there exists $z \in F(T)$ such that $d_\alpha = g_\alpha(z)$ and $z \in F_{\delta,\alpha}$. If $d_\alpha \notin \{g_\alpha(y): y \in F(T)\}$, then $\{g_\alpha(y): y \in F(T)\}$ must be infinite, and by the definition of infimum, there must exist $z_0 \in F(T)$ such that $g_\alpha(z_0) - \delta < d_\alpha$; that is, $z_0 \in F_{\delta,\alpha}$. Hence for each $\delta > 0$, $F_{\delta,\alpha} \neq \emptyset$. Since E is semireflexive and $F_{\delta,\alpha}$ is bounded as a subset of C, $F_{\delta,\alpha}$ is weakly compact. But E is Hausdorff in the weak topology and consequently $F_{\delta,\alpha}$ is weakly closed. Furthermore, $\delta_1 < \delta_2$ implies

$F_{\delta_1,\alpha} \subseteq F_{\delta_2,\alpha}$. Thus the family $\{F_{\delta,\alpha} : \delta > 0\}$ has the finite intersection property and $\bigcap_{\delta>0} F_{\delta,\alpha} = \{y: g_\alpha(y) = d_\alpha\} = F_\alpha \neq \emptyset$. By hypothesis C is weakly sequentially compact and there exists a subsequence $\{x_{n_j}\}$ of $\{x_n\}$ such that x_{n_j} converges to $y_0 \in C$. By Theorem 3.1, S is asymptotically regular. Thus $\lim_j(-x_{n_j+1} + x_{n_j})$

$$= \lim_j (-Sx_0^{n_j+1} + Sx_0^{n_j}) = \lim(-S + I)(Sx_0^{n_j}) = 0.$$

Since by assumption I-S is demiclosed, $(I - S)y_0 = 0$. Hence $y_0 \in F(S) = F(T)$. We shall show that $y_0 \in F_\alpha$ for all $\alpha \in \Delta$. Let $x \in F_\alpha$ and suppose that $x \neq y_0$.

Then by Lemma 3.1,

$$d_\alpha = g_\alpha(x) = \underline{\lim}\, \rho_\alpha(x_{n_j} - x)$$

$$\geq \underline{\lim}\, \rho_\alpha(x_{n_j} - y_0) = g_\alpha(y_0).$$

But, by the definition of d_α, $g_\alpha(x) \leq g_\alpha(y_0)$. Thus, $g_\alpha(y_0) = g_\alpha(x) = d_\alpha$, and $y_0 \in F_\alpha$. Hence $y_0 \in \cap_\alpha F_\alpha = F_0$. Now suppose $y_1 \in F_0$ with $y_1 \neq y_0$. Then $y_1 \in F_\alpha$ for all $\alpha \in \Delta$. By Lemma 3.1 there exists $\beta \in \Delta$ such that

$$g_\beta(y_1) = \underline{\lim}\, \rho_\beta(x_{n_j} - y_1) > \underline{\lim}\, \rho_\beta(x_{n_j} - y_0) = g_\beta(y_0).$$

This contradicts $g_\beta(y_1) = g_\beta(y_0) = d_\alpha$.

Thus $F_0 = \{y_0\}$. We next show that $x_n \rightharpoonup y_0$. Assume, to the contrary that $x_n \not\rightharpoonup y_0$. Then it follows that there exists a continuous linear functional f such that $f(x_n) \not\to f(y_0)$. This in turn implies that there exist an $\varepsilon > 0$ and a subsequence $\{x_{n_k}\}$ of $\{x_n\}$ such that $|f(x_{n_k}) - f(y_0)| \geq \varepsilon$ for k = 1, 2, ... Since C is weakly sequentially compact, there exists a subsequence $\{x_{n_{k_i}}\}$ of $\{x_{n_k}\}$ such that $x_{n_{k_i}} \rightharpoonup z_0 \in C$. By a previous argument $z_0 = y_0$. Hence we have

$f(x_{n_{k_i}}) \to f(y_0)$. Thus there exist an N such that for all $i \geq N$, $|f(x_{n_{k_i}}) - f(y_0)| < \varepsilon$, which contradicts $|f(x_{n_{k_i}}) - f(y_0)| \geq \varepsilon$ for all $x_{n_{k_i}}$. The result follows.

Corollary 3.1. [2, Theorem 8] Let E be a generalized Hilbert space, C be a closed, bounded, convex, weakly sequentially compact subset of E, and T be a nonexpansive self-mapping of C. Define the mapping $S:C \to C$ as in Theorem 1.1. Then for any $x_0 \in C$, $S^n x_0 \to y_0 \in F(T) = F(S)$.

Proof. It follows from a variant of Browder's theorem that $F(T) = F(S) \neq 0$. (Theorem 2 [3]). Since T is nonexpansive S is nonexpansive. Hence I - S is demiclosed. Indeed if $\{x_n\}$ is any sequence in C such that $s_n \to x$ and $x_n - S_{x_n} \to y_0$, then using the nonexpansiveness of S and Lemma 3.1 we have $\underline{\lim}\ \rho(x_n - x_0) \geq \underline{\lim}\ \rho(Sx_n - Sx_0) = \underline{\lim}\ \rho(x_n - y_0 - Sx_0)$

Hence $x_0 - Sx_0 = y_0$.

Remark 3.1. Several other results may also be seen to follow as immdediate corollaries to Theorem 3.2. Included among these are the following:

Huffman & Hicks, Theorem 9 [5] and Singh & Naimpally, Theorem 13 [10].

Finally, we present an example of a nonnormable, locally convex space and a nonexpansive mapping with a fixed point, for which our results are applicable but not those of Browder and Petryshyn, Kannan and Kirk.

Example. Let Ω be an open subset of R^n, and $E = C(\Omega)$ be the space of continuous real valued functions on Ω. Let Δ be the family of all compact subsets of Ω. For $K \in \Delta$, define

$$P_K(f) = \max_{x \in K} |f(x)|, \quad f \in E.$$

The P_K is a seminorm, and the family $\{P_K : K \in \Delta\}$ generates a topology under which E is strictly convex locally convex Frechet space. For a special case, let $\Omega = (-1,1)$ and $E = C(-1,1)$. Let $D = \{f \in E : f : [0, 3/4] \to [0, 3/4]\}$. Then D is compact convex subset. Since E is Frechet space, D is weakly sequentially compact and convex.
Define $T:D \to D$ by $(Tf)(x) = (\sin x)\, f(x)$. Then T is nonexpansive with fixed point $f = 0$ in D.

REFERENCES

1. M. Ja. Antonovskii, V. G. Boltjanskii and T. A. Sarymsakov, *Topological semifields and their applications to general topology*, American Math. Soc. Trans. Ser. 2(106) (1977).

2. F. E. Browder and W. V. Petryshyn, *Construction of fixed points of nonlinear mappings in Hilbert space*, J. Math. Anal. Appl. 20(1967), pp. 197-228.

3. F. E. Browder, *Nonexpansive nonlinear operators in a Banach space*, Proc. Nat. Acad. Sci. U.S.A. 54(1965), 1041-1044.

4. E. W. Huffman, *Strict convexity in locally convex spaces and fixed point theorems*, Math. Japonica, 22(1977), 323-333.

5. E. W. Huffman and T. . Hicks, *Fixed point theorems in generalized Hilbert spaces*, J. Math. Anal. Appl. 64(1978), 562-569.

6. R. Kannan, *Construction of fixed points of a class of nonlinear mappings*, J. Math. Anal. Appl. (1973), 430-438.

7. W. A. Kirk, *On successive approximation for nonexpansive mappings in Banach spaces*, Glasgow Math. J. 12(1971), 6-9.

8. G. Kothe, *Topological Vector spaces*, I. Springer-Verlag, New York (1969).

9. I. Massabo, *On the construction of fixed points for a class of nonlinear mappings*, Boll. U. M. I. 10(1974), 512-528.

10. S. A. Naimpally and K. L. Singh, *Sequence of iterates in locally convex spaces*, Nonlinear Phenomena in Mathematical Sciences, Academic Press (1982), 725-736.

11. Z. Opial, *Nonexpansive and monotone mappings in Banach spaces*, Lecture Note No. 1, Brown University (1967).

12. B. K. Ray and B. E. Rhoades, *A class of fixed point theorems*, math. Sem. Notes, 71(1979), 477-489.

13. B. E. Rhoades, *A fixed point theorem for locally convex linear topological spaces*, Math. Sem. Notes, 5(1977), 413-414.

14. H. H. Schadfer, *Topological Vector Spaces*, Springer-Verlag (1971).

15. E. Tarafdar, *An approach to fixed-point theorems on uniform spaces*, Trans. Amer. Math. Soc. 191(1979), 209-225.

PERIODIC SOLUTIONS OF HAMILTONIAN SYSTEMS: THE CASE OF THE SINGULAR POTENTIAL

A. Capozzi and A. Salvatore
Dipartimento di Matematica
Bari (ITALY)

0. INTRODUCTION AND STATEMENT OF THE RESULT.

Let us consider the Hamiltonian system of 2n ordinary differential equations

$$\begin{cases} \dot{p} = -H_q(p,q) \\ \dot{q} = H_p(p,q) \end{cases} \tag{0.1}$$

where $p, q \in \mathbb{R}^n$, "." denotes $\frac{d}{dt}$, $H \in C^1(\mathbb{R}^{2n}, \mathbb{R})$ and H_q (resp. H_p) is the gradient of H respect to q (resp. p).

In this paper we are interested in the following problem:

$$\begin{cases} \text{Given } T > 0, \text{ to find non constant T-periodic} \\ \text{solutions of (0.1)} \end{cases} \tag{0.2}$$

This problem has been studied by many authors (cf. [1],[2], [7], [20] and their references) under different assumptions on the Hamiltonian H(p,q).

Observe that if we consider a mechanical system with holonomous constraints embedded in a conservative field of forces, then the Hamiltonian H(p,q) has the form

$$H(p,q) = \frac{1}{2}\sum_{i,j=1}^{n} a_{i,j}(q)p_i p_j + \sum_{i=1}^{n} b_i(q)p_i + V(q) \tag{0.3}$$

where $a_{i.j}(q)$, $b_i(q)$ and V(q) are real valued functions and the matrix $\{a_{i,j}(q)\}$ is (uniformly in q) positive definite.

The problem (0.2) with Hamiltonian of the type (0.3) has been studied in [4], [5], [6], [21] in the case when

S. P. Singh (ed.), Nonlinear Functional Analysis and Its Applications, 207–216.

$$V(q) \to \infty \qquad \text{as} \quad |q| \to \infty\,, \tag{0.4}$$

in [10] in the case when

$$V(q) \text{ is "generically" bounded} \tag{0.5}$$

and in many other papers in particular cases (cf. [9], [11], [12], [13], [17], [18], [22] and their references).

In [4], [5], [6], [21] the problem (0.2) with assumption (0.4) has been solved by "direct method" (cf. [2]). More precisely the problem has been reconduced to the problem of the research of critical points of a suitable functional of the form

$$f(u) = \frac{1}{2}(Lu,u) - \psi(u) \qquad u \in H \tag{0.6}$$

where H is an Hilbert space, $L : H \to H$ is a self-adjoint continuous operator and $\psi \in C^1(H,\mathbb{R})$ is a functional such that $\psi' : H \to H$ is a compact operator.

In order to find critical points of (0.6) it has been applied, in a suitable way, an abstract critical point theorem proved in [5] (cf. also [4], [6], [8]). This theorem insures the existence of multiplicity of critical points for a functional of the form (0.6) if such a functional, invariant under the action of a suitable group of transformations, satisfies a weaker version of Palais-Smale condition. Moreover it is required that 0 is a regular value for L or it is an isolated eigenvalue of finite multiplicity of L.

If we consider the problem (0.2) under the assumption (0.5) the above mentioned abstract theorem cannot be applied "directly" to action functional associated to the problem. In fact, because of the boundness of potential, the set of critical points at same critical level could be unbounded and the Palais-Smale condition could not be satisfied. In [10] the problem has been solved by considering the Lagrangian formulation of the problem and by showing that the solutions of (0.1) are the critical points of a suitable restriction of the action functional to a subspace of the Hilbert space, where one usually works.

In this paper we will be concerning in the case when

$$V(q) \text{ is singular}\,,\ a_{i,j}(q) = \text{const and } b_i(q) = 0 \tag{0.7}$$

More precisely we shall consider the equation

$$-\ddot{x} = \nabla\left(-\frac{1}{|x|^2}\right) + e(t) \tag{0.8}$$

where $x(t) : \mathbb{R} \to \mathbb{R}^2$ and $e(t) : \mathbb{R} \to \mathbb{R}^2$.

We shall prove the following theorem:

Theorem 0.9 - For any $T > 0$ and for any T-periodic integrable function $e(t)$, there exists a T-periodic solution of (0.8), which winds around the origin.

Remark 0.10 - The results known to the authors concerning the equation (0.8) are contained in [15], [16].

We refer to these papers also for a detailed discussion.

Remark 0.11 - The proof of the theorem is based on classical variational methods and on the weighted Palais-Smale condition introduced by Benci in [3]. In the same manner it can be proved an analogous theorem in presence of n differential equations.

Moreover the weighted Palais-Smale condition seem to be an useful tool to treat nonlinear problems in presence of singularities.

However we refer to [14] for a more detailed discussion on these arguments and for other results.

1. Preliminaries.

Before proving the theorem we recall some notations and definitions introduced by Benci [3].

Let H be a Hilbert space with norm $\|\cdot\|$ and let Λ be an open set in X. We will denote by $C^n(\Lambda,\mathbb{R})$ the set of n-times Fréchet differentiable functions from Λ to $\mathbb{R}$. If $f \in C^1(\Lambda,\mathbb{R})$ we will denote its Frêchet derivative by f', which can be identified, by inner product, with a function from Λ to $\mathbb{R}$.

Definition 1.1 - A function $\rho : \Lambda \to \mathbb{R}$ is called a weight function for Λ if it satisfies the following assumptions:

$$\rho \in C^1(\Lambda,\mathbb{R}) \tag{1.2}$$

$$\rho(x) > 0 \quad \text{for any} \quad x \in \Lambda \tag{1.3}$$

$$\lim_{x\to\partial\Lambda} \rho(x) = +\infty \tag{1.4}$$

Definition 1.5 - We say that a functional $f \in C^1(\Lambda,\mathbb{R})$ satisfies the weighted Palais-Smale condition (W.P.S.) if there exists a weight function ρ such that for any sequence $\{x_n\} \subset \Lambda$ it happens that:

(WPS1) If $\{\rho(x_n)\}$ and $\{f(x_n)\}$ are bounded and $f'(x_n) \to 0$, then there exists a subsequence converging to $\bar{x} \in \Lambda$.

(WPS2) If $f(x_n)$ is convergent and $\rho(x_n) \to +\infty$, then there exists ν, $\bar{n} > 0$ and a subsequence $\{x'_n\}$ such that

$$\| f'(x'_n) \| \geq \dot{\nu} \| \rho'(x'_n) \| \quad \text{for any } n > \bar{n}.$$

In [3] it has been proved the following lemma:

Lemma 1.6 - Let $f \in C^1(\Lambda, \mathbb{R})$ satisfy (W.P.S.). Suppose that c is not a critical value of f nor an accumulation point of critical values of f.

Then there exist constants $\bar{\varepsilon} > \varepsilon > 0$ and a function $\eta : [0,1] \times \Lambda \to \Lambda$ such that

(a) $\eta(0,x) = x$ for any $x \in \Lambda$

(b) $\eta(1,x) = x$ for any $x \notin \Lambda_{c+\bar{\varepsilon}}/A_{c-\bar{\varepsilon}}$

(c) $\eta(1, A_{c+\varepsilon}) \subset A_{c-\varepsilon}$

where $A_s = \{x \in \Lambda | f(x) < s\}$. Moreover $\bar{\varepsilon}$ can be chosen arbitrarily small.

By lemma 1.6 and by standard arguments it can be easily proved the following lemma:

Lemma 1.7 - Let $f \in C^1(\Lambda, \mathbb{R})$ satisfy (W.P.S.). Suppose that there exists a constant α such that

$$f(x) \geq \alpha \quad \text{for any } x \in \Lambda \tag{1.8}$$

then $c = \inf_\Lambda f$ is a critical value or an accumulation point of critical values of f.

Our aim will be to make a suitable choice of the open set Λ and of the weight function for Λ. After we shall prove that 1.8 is satisfied.

2. Proof of theorem.

If $1 \leq p < \infty$, we set

$$L^p = L^p(S^1, \mathbb{R}^2) = \{u : \mathbb{R} \to \mathbb{R}^2 \,|\, u \text{ } 2\pi\text{-periodic}, \int_0^{2\pi} |u(t)|^p dt < +\infty\}$$

and we denote by $|\cdot|_p$ the norm in L^p and by $(\cdot,\cdot)_2$ the inner product in L^2. We denote by $(\cdot|\cdot)$ the inner product in $\mathbb{R}^2$ and by $|\cdot|$ its norm. Let $H = H^1(S^1, \mathbb{R}^2)$ the Sobolev space obtained by closure of the C^∞ 2π-periodic $\mathbb{R}^2$ valued functions x(t) with respect to the norm

$$\| x \| = [\int_0^{2\pi} (|\dot{x}|^2 + |x|^2) dt]^{\frac{1}{2}}.$$

Obviously H is an Hilbert space and its inner product will be denoted by $((\cdot,\cdot))$. Moreover we denote by $<\cdot,\cdot>$ the duality between H and its dual H^{-1} and we set $|x|_\infty = \max_{[0,2\pi]} |x(t)| \; \forall x \in H$.

For any $x \in H$ we set $\bar{x} = \frac{1}{2\pi}\int_0^{2\pi} x(t)dt$ and $\tilde{x}(t) = x(t) - \bar{x}$.

Moreover we set

$$\Lambda = \{x \in H | x(t) \text{ winds around the origin}\}. \tag{2.1}$$

Obviously Λ is an open set of H, then the critical points of the functional

$$f(x) = \frac{1}{2}\int_0^{2\pi} |\dot{x}|^2 dt + \int_0^{2\pi} \frac{1}{|x|^2} dt - \int_0^{2\pi} (e(t)|x(t))dt \qquad x \in \Lambda \tag{2.2}$$

are T-periodic solutions of (0.1).

Now we define the weight function for Λ by

$$\rho(x) = \int_0^{2\pi} \frac{1}{|x|^2} dt \tag{2.3}$$

Since $x(t) \in H = H^1(S^1,\mathbb{R}^2)$, it is easy to see that $\rho(x)$ satisfies the assumptions (1.2)-(1.4).

Now we will prove that the functional (2.2) satisfies (W.P.S.) with Λ defined by (2.1) and $\rho(x)$ defined by (2.3).

First observe that, given an orbit x(t), which winds around the origin, we have that

$$[\text{arc length } x(t)] > c_1|x|_\infty \tag{2.4}$$

By (2.4) it follows that

$$|\bar{x}| \le c_2|x|_\infty < c_3 \, [\text{arc length } x(t)] < c_4|\dot{x}|_2. \tag{2.5}$$

Now let $\{x_n\} \subset \Lambda$ be a sequence such that

$$\{\rho(x_n)\} \text{ is bounded} \tag{2.6}$$

$$\{f(x_n)\} \text{ is bounded} \tag{2.7}$$

$$f'(x_n) \to 0 \tag{2.8}$$

By (2.6) we have that

$$\{\int_0^{2\pi} \frac{1}{|x_n|^2}\, dt\} \text{ is bounded.} \tag{2.9}$$

By (2.7) we have that

$$\{|\tfrac{1}{2}\int_0^{2\pi} |\dot{x}_n|^2 dt + \int_0^{2\pi} \frac{1}{|x_n|^2} dt - \int_0^{2\pi} (e(t)|x_n(t))dt)\} \text{ is bounded,} \tag{2.10}$$

then by (2.10) it follows that there exist two positive constants k_1 and k_2 such that

$$\tfrac{1}{2}|\dot{x}_n|_2^2 \leq k_1 + \int_0^{2\pi} (e(t)|x_n(t))dt \leq k_1 + \int_0^{2\pi} (\tilde{e}(t)|\tilde{x}_n(t))dt$$

$$+ \int_0^{2\pi} (\bar{e}|\bar{x}_n)dt \leq k_1 + |\tilde{e}(t)|_1 \cdot |\tilde{x}_n(t)|_\infty + 2\pi|\bar{e}| \cdot |\bar{x}_n|$$

$$\leq k_1 + 2\pi|\bar{e}| \cdot |\bar{x}_n| + k_2 \cdot \|\tilde{x}_n(t)\| \tag{2.11}$$

By (2.11) and (2.5) we obtain

$$\|\tilde{x}_n\|^2 \leq k_3 + k_4 \|\tilde{x}_n\| \tag{2.12}$$

with suitable positive constants k_3 and k_4, then

$$\{\tilde{x}_n\} \text{ is bounded} \tag{2.13}$$

Moreover by (2.13) we deduce that arclength $\{x_n(t)\}$ is bounded and since $x_n(t) \in \Lambda$ for any n, then we obtain that

$$\{\bar{x}_n\} \text{ is bounded} \tag{2.14}$$

By (2.13) and (2.14) it follows that

$$\{\|x_n\|\} \text{ is bounded} \tag{2.15}$$

Then there exists a subsequence, still denoted by $\{x_n\}$, such that

$$x_n \to x \text{ weakly in } H^1 \text{ and uniformly.} \tag{2.16}$$

By (2.8) we have that, for any $h \in H^1$

$$\int_0^{2\pi} (\dot{x}_n|\dot{h})dt - 2\int_0^{2\pi} \frac{(x_n|h)}{|x_n|^4} dt - \int_0^{2\pi} (e|h)dt = \varepsilon_n \| h \|,$$

$$\varepsilon_n \to 0 \tag{2.17}$$

By (2.6) x does not belong to $\partial\Lambda$, then by (2.16) and (2.17) it follows that there exists a real constant k_5 such that

$$\int_0^{2\pi} (\dot{x}_n|\dot{h})dt \leq \varepsilon_n \| h \| + k_5 |h|_\infty . \tag{2.18}$$

For $h = x_n - x$ we have that

$$\int_0^{2\pi} (\dot{x}_n|\dot{x}_n - \dot{x})dt \leq \varepsilon_n \| x_n - x \| + k_5 |x_n - x|_\infty \tag{2.19}$$

and therefore

$$\int_0^{2\pi} (\dot{x}_n|\dot{x}_n - \dot{x})dt \to 0. \tag{2.20}$$

Moreover the equality

$$\int_0^{2\pi} |\dot{x}_n - \dot{x}|^2 dt = \int_0^{2\pi} (\dot{x}_n|\dot{x}_n - \dot{x})dt - \int_0^{2\pi} (\dot{x}|\dot{x}_n - \dot{x})dt \tag{2.21}$$

implies that

$$\dot{x}_n \to \dot{x} \quad \text{in} \quad L^2 \tag{2.22}$$

By (2.16) and (2.22) we easily get that

$$x_n \to x \text{ strongly in } H^1.$$

Then (WPS_1) is satisfied.

Now let $\{x_n\} \subset \Lambda$ be a sequence such that

$$f(x_n) \to c \tag{2.23}$$

$$\rho(x_n) \to +\infty \tag{2.24}$$

By (2.23) we have that there exist k_6, $k_7 > 0$ such that

$$|\dot{x}_n|_2^2 \leq k_6 + 2\int_0^{2\pi} (e(t)|x_n(t))dt \leq k_6 + k_7 \cdot \|\tilde{x}_n(t)\|, \tag{2.25}$$

then, also in this case, we obtain that

$$\| x_n \| \text{ is bounded} \tag{2.26}$$

By (2.26) and (2.23) it follows that

$$\int_0^{2\pi} \frac{1}{|x_n|^2} dt \text{ is bounded} \tag{2.27}$$

which contradicts (2.24). Then there are no sequence satisfying (2.23), (2.24) and (WPS) is completely satisfied.

Now consider the functional (2.2). By (2.5) we have that

$$f(x) \geq \tfrac{1}{2} \int_0^{2\pi} |\dot{x}|^2 dt - \int_0^{2\pi} (e(t)|\tilde{x}(t))dt \geq c_1 \| \tilde{x} \|^2$$

$$- c^2 \| \tilde{x} \| \geq \alpha \tag{2.28}$$

with suitable positive constants c_1, c_2 and α.

Since $f(x)$ verifies (W.P.S.), by (2.28) and by lemma 1.7 the conclusion of theorem (0.9) follows.

References

1. A. Ambrosetti, *Recent Advances in the Study of the Existence of Periodic Orbits of Hamiltonian systems*, Advances in Hamiltonian systems, Annals of the Ceremade, Birkhauser (1983).

2. V. Benci, *The Direct Method in the Study of Periodic Solutions of Hamiltonian System with Prescribed Period*, Ceremade, Birkhauser, (1983), 23-42.

3. V. Benci, *Normal Modes of Lagrangian Systems Constrained in a Potential Well*, Ann. Inst. H. Poincare, 1, 5 (1984), 379-400.

4. V. Benci, A. Capozzi, D. Fortunato, *Periodic Solutions for a Class of Hamiltontian Systems*, Lecture Notes in Mathematics, Springer Verlag, 964 (1982), 86-94.

5. V. Benci, A. Capozzi, D. Fortunato, *Periodic Solutions of Hamiltonian Systems of Prescribed Period*, Math. Res. Center, University of Wisconsin-Madison, Technical Summary Report n. 2508 (1983).

6. V. Benci, A. Capozzi, C. Fortunato, *Periodic Solutions of Hamiltonian Systems with Superquadratic Potential*, to appear on Ann. Mat. Pura e App.

7. H. Berestycki, *Solutions Periodiques de Systèmes Hamiltoniens*, Séminaire Bourbaki, 35e année, 1982/83, n.603.

8. A. Capozzi, D. Fortunato, *An Abstract Critical Point Theorem for Strongly Indefinite Functionals*, Proc. of Symposia in Pure Math., 44, (1985).

9. A. Capozzi, D. Fortunato, A. Salvatore, *Periodic Solutions of Dynamical Systems*, Atti del VII Congresso AIMETA, 1, (1984), 9-16.

10. A. Capozzi, D. Fortunato, A. Salvatore, *Periodic Solutions of Lagrangian Systems with Bounded Potential*, Preprint.

11. A. Capozzi, A. Salvatore, *Periodic Solutions for Nonlinear Problems with Strong Resonance at Infinity*, Comm. Math. Un. Car., 23, 3 (1982), 415-425.

12. A. Capozzi, A. Salvatore, *Sull'equazione* $Lu=\nabla V(u)$, Atti del convegno "Problemi Differenziali e teoria dei punti critici", ed. Pitagora, (1984), 41-63.

13. A. Capozzi, A. Salvatore, *Nonlinear Problems with Strong Resonance at Infinity: An Abstract Theorem and Applications*, Proc. R. Soc. Edinb., 99 A, (1985), 333-345.

14. A. Capozzi, C. Greco, A. Salvatore, *Lagrangian Systems in Presence of Singularities*, in preparation.

15. W.B. Gordon, *Conservative Dynamical Systems Involving Strong Forces*, Trans. Amer. Math. Soc., 204, (1975), 113-135.

16. W.B. Gordon, *A Minimizing Property of Keplerian Orbits*, Am. J. of Math., 99, 5, (1977), 961-971.

17. J. Mawhin, *Periodic Oscillations of Forced Pendulum-like Equations*, Lecture Notes in Math., Springer-Verlag, 964, (1982), 458-476.

18. J. Mawhin, M. Willem, *Multiple Solutions of the Periodic Boundary Value Problem for Some Forced Pendulum-type Equations*, J. Diff. Eq. 52, (1984), 264-287.

19. P.H. Rabinowitz, *Periodic Solutions of Hamiltonian Systems*, Comm. Pure Appl. Math., 31 (1978), 157-184.

20. P.H. Rabinowitz, *Periodic Solutions of Hamiltonian Systems: A Survey*, SIAM J. Math. Anal., 13 (1982), 343-352.

21. A. Salvatore, *Periodic Solutions of Hamiltonian Systems with a Subquadratic Potential*, B.U.M.I. (c), 1 (1984), 393-406.

22. M. Willem, *Oscillations Forcées de Systèmes Hamiltonians*, Publications Semin. Analyse non Linéaire, Univ. Besancon (1981).

Work supported by Ministero P.I. (40%, 60%) and by G.N.A.F.A. of C.N.R.

OSCILLATIONS ET ANALYSE NON LINEAIRE: PROPRIETES DES PULSATIONS DES SOLUTIONS PERIODIQUES (CYCLES) DE CERTAINES EQUATION DIFFERENTIELLES AUTONOMES NON LINEAIRES. APPLICATION DE LA THEORIE DU DEGRE DE LERAY SCHAUDER.

Robert Faure
Université des Sciences et Techniques de Lille I
U.E.R. de Mathématiques Pures et Appliquées
Service de Mécanique, bâtiment M 3
59655 VILLENEUVE D'ASCQ CEDEX

In this paper we establish the fundamental properties of periodical solutions of non linear differential equations (cycles).
If the pulsation ω tends to zero, the parameter $|\lambda|$ increases indefinitely, and reciprocally. With the norm's properties we can use of Leray Schauder's degree theory for the existence of the periodical aolution.

Dans l'article cité en référence (I), nous exposions une méthode d'application de Théorème de Leray Schauder pour établir l'existence de cycles d'équations différentielles schématisées par $Lx = \lambda F(x, x')$ $\lambda \in R$. Lx opérateur différentiel linéaire, $F(x, x')$ fonction non linéaire de l'inconnu x et de sa dérivée x', λ paramètre.

Nous y démontrions en particulier les résultats 1°, 2°, 3°, 4° qui suivent et nous associons la pulsation inconnue ω avec $\omega T = 2\pi$, T période aux autres inconnues x. Nous complétons dans ce qui suit résultats de (I).

On considère l'équation différentielle (E) $y'' + \omega_1^2 y = \lambda P(y')$, λ paramètre supposé ici borné inférieurement et supérieurement $0 < \lambda_0 \leq \lambda \leq \lambda_1$ λ_0 et λ_1 constantes positives. Les résultats porteront sur la valeur absolue $|\lambda|$ de λ. $P(y')$ Polynôme de degré impair en y' à coefficients constants $a_{2n+1} \neq 0$ $a_1 \neq 0$ $P(y') = a_{2n+1} y'^{2n+1} + \ldots + a_1 y'$.

On choisit ici $a_1 > 0$.

T est la période d'une solution périodique non nulle; on suppose essentiellement l'existence de celles-ci. ω est la pulsation $\omega T = 2\pi$.

Par des changements de variables et de foncitions on remème (E) à (E') : $\omega^2 y'' + \omega_1^2 y = \lambda \omega P(y')$: la période est alors 2π. On a déjà obtenu les résultats suivants : $\omega_1 \geq \omega$; si $\lambda \to \infty$ $\omega \to 0$. Examinons le cas $\omega \to 0$ qui reste ouvert.

I) On a établi les résultats suivants (Réf. I) avec K_i; $i = 1, 2, \ldots 5$ constantes positive

S. P. Singh (ed.), Nonlinear Functional Analysis and Its Applications, 217–221.

1°) par multiplication de (E') par y' et intégration entre 0 et 2π on a $\int_0^{2\pi} P(y')\, y'\, dt = 0$ il en résulte que $\int_0^{2\pi} y'^{2n+2}\, dt \leq 2\pi\, \ell^{2n+2}$ et $\text{Max}\, |y'| > m$ ou ℓ et m sont des constantes positives indépendantes de λ et de ω.

2°) Si $2\pi\, \alpha = \int_0^{2\pi} y(s)\, ds$, on tire de (E') : $|\alpha| \leq K_1\, \omega$ car $2\pi\, \omega_1^2\, \alpha = \lambda\omega \int_0^{2\pi} P(y')\, ds$, <u>$\lambda$ borné</u>, t_0 définie par $y(t_0) = \alpha$; d'où puisque $y(t) = y_0 + \int_{t_0}^{t} y'(s)\, ds$; vu le 1°) on a $\sup\limits_{\omega,\lambda} \sup\limits_{t} |y'| < +\infty$.

3°) Si on recherche le maximum de $|y'|$ on a $\lambda\, \omega\, P(y') = \omega_1^2\, y$, d'où $\text{Max}\, |y'| \leq K_2\, \omega^{-\frac{1}{2n+1}}$ avec $2n + 1 \geq 3$.

4°) En multipliant par y' et en intégrant entre les valeurs t_0 et t_1, avec $y(t_0) = \alpha$ on tire de (E') : $|y(t)| \leq K_3\, \omega^{1/2}$ ceci pour $|\lambda| < \lambda_1$.

5°) On déduit de ce qui précède et qui est fondamental. Si le nombre de changement de signe de y' sur une période est fini; puisque $\left|\int_{t_1}^{t_2} y'(s)\, ds\right| = \left|y(t_2) - y(t_1)\right| \leqq K_3\, \omega^{1/2}$ et que $\text{Max}\, |y'| \leq K_4\, \omega^{-1/3}$; on a $\int_0^{2\pi} y'^2\, dt \leq K_5\, \omega^{1/6}$, ceci parce que $|\lambda|$ est borné. $y'(s)$ définie à une translation près de la variable s converge alors vers zéro dans $L_2(0,2\pi)$ lorsque $\omega \to 0$.

Nous allons vérifier que le nombre de zéros de y est borné sur $[0,2\pi]$.

II) Nous représentons la solution de l'équation différentielle dans le plan des phases; $M(x,z)$ est le point courant de $\Gamma(\omega)$ trajectoire périodique on a le système d'équation

$$x' = z \tag{1}$$

$$\omega^2\, z' = -\omega_1^2\, x + \lambda\, \omega\, P(z) \tag{2}$$

ainsi que $\omega^2 z'' = -\omega_1^2 z + \lambda\, \omega\, P'(z)\, z'$. (3)

Il résulte de ce système que tout point correspondant à un extremum de z est pour $z > 0$ un maximum et pour $z < 0$ un minimum.

On désigne par A et C les points situés sur l'axe des x : $A(x_1 < 0, 0)$: $C(x_2 > 0, 0)$. B correspond au maximum positif de z. C au minimum négatif de z. La trajectoire dont la durée de trajet est 2π est parcourue dans le sens ABCD. Il résulte de la structure de la courbe que z ne s'annule que deux fois sur $[0,2\pi]$ on suppose que pour $t = 0$ le point M se trouve en A. Il résulte de (II) que toute solution périodique de (E') converge dans $L_2(0,2\pi)$ vers zéro quand $\omega \to 0$. Nous pouvons appliquer le Théorème d'Egoroff.

(III) Nous devons envisager pour $\Gamma(\omega)$ les trois possibilités suivantes:

1°) le maximum et le minimum de $z = y'$ tendent tous deux vers zéro avec ω - cette propriété est en contradiction avec le 1°) de I).

2°) Les valeurs absolues de minimum et du maximum sont toutes deux bornées inférieurement par un nombre positif C^2. C constante

Considérons alors la branche croissante $z = y' \leq 0$ de $\overline{ABCD}$ soit $\overline{DA}$ et choisissons trois nombres α, β, γ avec $0 < \alpha < \beta < \gamma < C^2$ et tels que P(z) soit croissante sur $[-\gamma, \gamma]$ ceci est compatible avec l'hypothèse $a_1 > 0$. Désignons par t_1, t_2, t_3 les valeurs de t correspondant aux points M_1, M_2, M_3 d'ordonnées respectives $-\gamma$, $-\beta$, $-\alpha$, t est croissant sur $\overline{DAC}$.

Remarquons que sur $\overline{DA}$:

1°) $y' = z$ et croissant.

2°) $x = y$ est décroissant.

3°) l'équation (3) avec $y' = z$;

$\omega^2 y''' + \omega_1^2 y' = \lambda\, \omega\, P'(y')\, y''$ (4) montre que y'' étant positif, y' négatif, y''' est positif.

Alors on voit que pour $t > t_3$ on a K étant une constante positive $\omega\, y''(t) > \omega\, y''(t_3) > \lambda\, (P(y'(t_3)) - P(y'(t_2)) = \lambda K$ (5) $y''(t_3) \to \infty$ avec ω^{-1}.

Considérons maintenant les valeurs de y'' pour $t \geq 0$ on a $y'''(0) = \lambda\, a_1\, \omega_1^{-2}$; il en résulte que le maximum de y'' sera atteint par la valeur de y' satisfaisant à $\omega_1^2 y' = \lambda\, \omega\, P'(y')\, y''$ qui vu (5) entraine $\omega_1^2 y' > \lambda^2 K\, P'(y')$ (6) pour $\lambda \in [\lambda_0\ \lambda_1]$; ces valeurs de y' sont donc dans leur ensemble borné inférieurement par un nombre b positif.

Soit J le point de la trajectoire pour lequel ce maximum est réalisé, sur la branche $\overline{M_3J}$ la durée de trajet est égal au plus a $(y'(J) + \alpha)y''(t_3)^{-1}$ et tend donc vers zéro avec ω.

On pourra pour la partie décroissante $\overline{BCD}$ de la trajectoire raisonner de la même manière. On définira sur la branche $\overline{BC}$ trois point M_1', M_2', M_3' de coordonnées respectifs γ, β, α; on montre que dans ce cas aussi une branche $\overline{M'J'}$ aura également une durée de trajet tendant vers zéro avec ω.

Considérons maintenant l'ensemble E_ω des valeurs de t pour lesquels $|y'| \geq \text{Min}\ (\gamma, b, b')$ ou b' (J') est analogue de b(J). Pour ces valeurs de t il ne pourra y avoir convergence uniforme de $|y'|$ vers zéro; nous pouvons donc affirmer d'après le théorème d'Egoroff que Mes (t, $M \in E_\omega$) peut être prise pour ω suffisamment petit inférieure à un nombre η suffisamment petit. La durée totale de parcours de la courbe Γ (ω) sera alors inférieure à 2π pour ω suffisamment petir, ceci puisque $|\lambda|$ est bornée, alors qu'elle est supposée égale à 2π. λ ne peut être borné supérieurement.

3°) Envisageons : l'hypothèse suivante : le maximum de y', $y' > 0$ tend vers zéro avec ω mais le minimum de y' est borné inférieurement en valeur absolue.

L'inégalité donnant la valeur de y'(J) : $\omega_1^2\ y' \geq \lambda\ K\ P'(y')$ montre que cette hypothèse est à rejeter.

Nous pouvons donc affirmer dans tous les cas.

Théorème. Pour toute équation (E) lorsque $|\lambda| \to \infty$ $\omega \to 0$ et réciproquement si $\omega \to 0$ $|\lambda| \to \infty$ pour toute solutions périodeque non nulle. Si $\omega \to \omega_1$, $\lambda \to 0$, si $\lambda \to 0$, $\omega \to \omega_1$. On a toujours $\omega \leq \omega_1$.

APPLICATION DE LA METHODE DU DEGRE DE LERAY SCHAUDER

Si nour revenons maintenant à l'application de la méthode du degré. Nous reprenons l'équation (E') multipliée par y" et intégrons entre $[0,2\pi]$ on a :

$$\omega^2 \int_0^{2\pi} y''^2\ dt - \omega_1^2 \int_0^{2\pi} y'^2\ dt = 0 \tag{7}$$

de ce fait $\omega\ \|\ y''\ \|_H = (2\pi^{-1} \int_0^{2\pi} y''^2\ dt)^{1/2}\ \omega$, vu le 1° est borné supérieurement par une constant : $\|\ y''\ \|_H \leq A_1\ \omega^{-1}$, A_i constantes positives i = 1, 2, 3.

Si maintenant on considère $||y'||_B = \sum_{n=-\infty}^{n=+\infty} |\alpha_n|$; $y' = \sum_{n=-\infty}^{n=+\infty} \alpha_n e^{nit}$,

$n \neq 0$ on a $||y'||_B \leq A_2 \, ||y''||_H \leq A_3 \, \omega^{-1}$.

De ce fait et compte tenu résultats 1°, 2°, 2°, 4°, 5° pour $0 < \alpha < \omega < \beta \leq \omega_1$; $|\lambda|$ est borné supérieurement, il en est de même de $||y'||_B$; α, β sont des constantes arbitraires.

La norme $||\ \ ||_B$ étant pourvue d'une algèbre de Banach, le polynome $P(y')$ sera borné à l'aide de $||y'||_B$ dans l'espace (B) des séries de période 2π absolument convergentes.

Nous pouvons alors comme dans (I) écrire l'équation (E') sous la forme

$$\left.\begin{array}{l} Y' = T(\lambda, y') \\ \quad = \lambda \end{array}\right\} S \quad \text{ou } Y = y', \quad = \lambda. \quad \lambda \in R.$$

T est une transformation dans (B), λ et x' étant associées comme inconnues. La transformation S étant complètement continue dans $B \times R$, on pourra appliquer la méthode du degré de Leray Schauder en utilisant les propriétés si $\lambda \to 0$ $\omega \to \omega_1$ si $\omega \to \omega_1$, $\lambda \to 0$, avec $\omega \leq \omega_1$ en prenant alors ω comme paramètre au lieu de λ.

Remarque. Les équations (E) étudiées ci-dessus sont importantes en Electronique et Biophysique; si par contre on considère l'équation de Duffing utilisée en Mécanique $x'' + \omega_1^2 x + P(x) = 0$ $P(x)$ polynome impair à coefficients positifs si K est le paramètre Energie on a ici si $K \to \infty$, $\omega \to \infty$ et réciproquement avec $\omega \geq \omega_1$, si $K \to 0$, $\omega \to \omega_1$.

REFERENCE

1. Robert Faure, *Existence et comportement de cycles (solutions périodiques non excitées) dans certains systèmes mécanique et électrique*, SIAM Journal of Control 1972.

A FIXED POINT THEOREM FOR TWO COMMUTING MAPPINGS

B. Fisher and S. Sessa
University of Leicester
Department of Mathematics
Leicester LE1 7RH
ENGLAND

S. Sessa
University of Naples
Institute of Mathematics
80134 Naples
Italy

ABSTRACT. Let S and T be two mappings of a complete metric space (X,d) into itself satisfying the inequality

$$d(Sx,Ty) \leqslant c.\max\{d(x,y),d(x,Sx),d(y,Ty),d(x,Ty),d(y,Sx)\}$$

for all x, y in X, where $0 \leqslant c < 1$. Suppose further that for some particular x in X

$$\sup\{d(S^{r+1}x,S^rx),d(T^{r+1}x,T^rx) : r = 0,1,2, \ldots\} < \infty.$$

It is proved that S and T have a unique common fixed point z and that z is the unique common fixed point of S and T.

1. INTRODUCTION

Let S and T be two mappings of a complete metric space (X,d) into itself satisfying the inequality

$$d(Sx,Ty) \leqslant c.\max\{d(x,y),d(x,Sx),d(y,Ty),d(x,Ty), d(y,Sx)\} \quad (1)$$

for all x, y in X, where $0 \leqslant c < 1$.

Following an open question of Ćirić [1], it is well known that in general S and T do not have a common fixed point, as is shown in the following example of Fisher [6]:

<u>Example 1.1</u>: Let $X = \{x,y,z,w\}$ be a finite set with metric d defined by

$$d(x,x) = d(y,y) = d(z,z) = d(w,w) = 0,$$
$$d(x,z) = d(x,w) = d(y,z) = d(y,w) = 1,$$
$$d(x,y) = d(z,w) = 2.$$

Clearly (X,d) is a complete metric space and if S and T are defined by

$$Sx = y, \quad Sy = Sz = Sw = x,$$
$$Tx = Ty = Tw = z, \quad Tz = w$$

then it is easily seen that inequality (1) is satisfied with $c = \frac{1}{2}$, but S and T have no fixed points.

An extensive literature exists about common fixed point theorems of two mappings. We refer the reader to the papers of Rhoades [13] where a multitude of contractive conditions involving two mappings

S. P. Singh (ed.), Nonlinear Functional Analysis and Its Applications, 223–227.

are compared. Other results are established by Fisher [4], [9], Meade and Singh [12] and Wong [14].

As pointed out by Rhoades [13], one needs in addition to inequality (1), a further condition either on the space or on the mappings in order to guarantee a common fixed point of S and T. From this point of view, Fisher [2] was able to prove the following theorem:

THEOREM 1.1. Let S and T be commuting mappings of a complete metric space (X,d) into itself satisfying inequality (1) for all x, y in X, where $0 \leqslant c < 1$. Suppose further that for some particular x in X

$$\sup\{d(S^{r+1}T^nx,S^rT^nx),d(S^rT^{n+1}x,S^rT^nx) : r,\ n = 0,1,2,\ \ldots\} = M < \infty. \qquad (2)$$

Then S and T have a unique common fixed point z. Further z is the unique fixed point of S and T.

2. MAIN RESULT

We now prove the following generalization of theorem 1.1.

THEOREM 2.1. Let S and T be commuting mappings of a complete metric space (X,d) into itself satisfying inequality (1) for all x, y in X, where $0 \leqslant c < 1$. Suppose further that for some particular x in X

$$\sup\{d(S^{r+1}x,S^rx),d(T^{r+1}x,T^rx) : r = 0,1,2,\ \ldots\} = L < \infty. \qquad (3)$$

Then S and T have a unique common fixed point z. Further z is the unique fixed point of S and T.

PROOF. Using inequality (1) we have

$$\begin{aligned} d(S^rx,Tx) &\leqslant c.\max\{d(S^{r-1}x,x),d(S^{r-1}x,S^rx),d(x,Tx),\\ &\qquad d(S^{r-1}x,Tx),d(x,S^rx)\}\\ &\leqslant c.\max\{d(S^{r-1}x,S^rx) + d(S^rx,Tx) + d(Tx,x),L,\\ &\qquad d(S^{r-1}x,S^rx) + d(S^rx,Tx),d(x,Tx) + d(Tx,S^rx)\}\\ &\leqslant c[2L + d(S^rx,Tx)] \end{aligned}$$

and so

$$d(S^rx,Tx) \leqslant 2Lc/(1 - c)$$

for $r = 1,2,\ \ldots$. It follows that the set

$$\{S^rx : r = 0,1,2,\ \ldots\}$$

is bounded.

We can prove similarly that the set

$$\{T^rx : r = 0,1,2,\ \ldots\}$$

is also bounded and it follows that

$$\sup\{d(S^rx,S^nx),d(T^rx,T^nx) : r,\ n = 0,1,2,\ \ldots\} = K < \infty. \qquad (4)$$

Let us now suppose that the set

$$A = \{S^{n-r}T^r x : 0 \leqslant r \leqslant n \,;\, n = 0,1,2, \ldots\}$$

is unbounded. Then there exist integers r and n, with $r < n$, such that

$$d(S^{n-r}T^r x, Tx) > \max\{Kc/(1-c), K\} \tag{5}$$

and

$$d(S^{n-r}T^r x, Tx) > \max\{d(S^{m-i}T^i x, Tx), d(S^{n-j}T^j x, Tx) : 0 \leqslant i \leqslant m \,;\, 0 \leqslant m < n \,;\, 0 \leqslant j < r\}. \tag{6}$$

Now choose an integer k such that

$$c^k.\max\{dS^i T^{i'} x, T^j S^{j'} x) : 0 \leqslant i,i',j,j' < n\} < d(S^{n-r}T^r x, Tx). \tag{7}$$

Using inequality (1) we have

$$\begin{aligned} d(S^{n-r}T^r x, Tx) &\leqslant c.\max\{d(S^{n-r-1}T^r x, x), d(S^{n-r-1}T^r x, S^{n-r}T^r x), \\ &\qquad d(x, Tx), d(S^{n-r-1}T^r x, Tx), d(x, S^{n-r}T^r x)\} \\ &\leqslant c.\max\{d(S^{n-r-1}T^r x, Tx) + d(Tx, x), d(S^{n-r-1}T^r x, S^{n-r}T^r x), \\ &\qquad d(x, Tx), d(S^{n-r-1}T^r x, Tx), d(x, Tx) + d(Tx, S^{n-r}T^r x)\} \\ &\leqslant c.\max\{d(S^{n-r}T^r x, Tx) + K, d(S^{n-r-1}T^r x, S^{n-r}T^r x)\} \end{aligned}$$

because of (4) and (6).

If

$$d(S^{n-r}T^r x, Tx) \leqslant c[d(S^{n-r}T^r x, Tx) + K]$$

then

$$d(S^{n-r}T^r x, Tx) \leqslant Kc/(1-c),$$

contradicting inequality (4). We must therefore have

$$d(S^{n-r}T^r x, Tx) \leqslant c\,d(S^{n-r-1}T^r x, S^{n-r}T^r x), \tag{8}$$

where it follows that $r > 0$, otherwise inequality (5) would again be contradicted. Inequality (1) can therefore be applied to the right-hand side of inequality (8) to give terms of the form

$$d(S^i T^j x, T^{i'} S^{j'} x),$$

where $i, i' \leqslant n - r$ and $j, j' \leqslant r$. Inequality (1) can be applied to these and resulting terms either indefinitely or until terms of the form

$$d(S^i x, S^{i'} x), \quad d(T^j x, T^{j'} x), \quad d(S^i T^j x, x)$$

are obtained, where again $i, i' \leqslant n - r$ and $j, j' \leqslant r$. Terms obtained after k applications of inequality (1) can be omitted because of inequality (7) and terms of the form

$$d(S^i x, S^{i'} x), \quad d(T^j x, T^{j'} x)$$

can be omitted because of (4). We must therefore have

$$\begin{aligned} d(S^{n-r}T^r x, Tx) &\leqslant c.\max\{d(S^i T^j x, x) : 0 \leqslant i \leqslant n-r ; \\ &\qquad 0 \leqslant j \leqslant r\} \\ &\leqslant c.\max\{d(S^i T^j x, Tx) + d(Tx, x) : 0 \leqslant i \leqslant \\ &\qquad n-r ; 0 \leqslant j \leqslant r\} \\ &\leqslant c[d(S^{n-r}T^r x, Tx) + K], \end{aligned}$$

because of (4) and (6), again leading to a contradiction of inequality (5).

The set A must therefore be bounded and so condition (2) must hold. The conditions of theorem 1.1 therefore hold and the result of theorem 2.1 follows.

We note that it is not known if condition (3) is necessary in theorem 2.1. Further note that example 1.1 shows that the commutativity of the mappings S and T is a necessary condition in theorem 2.1.

Relaxing inequality (1) with

$$d(Sx,Ty) < \max\{d(x,y), d(x,Sx), d(y,Ty), d(x,Ty), d(y,Sx) \qquad (9)$$

where x, y are in X, Fisher [3] proved the following result:

THEOREM 2.2. Let S and T be mappings of a compact metric space (X,d) into itself satisfying inequality (9) for all x, y in X for which the right hand side of inequality (9) is positive. If S and T commute and if S and T are continuous, then S and T have a unique common fixed point z. Further, z is the unique fixed point of S and T.

The same result holds if one assumes the continuity of the mapping ST instead of the continuity of both S and T, see [8].

We point out that in theorem 2.2 the compactness of the space is a necessary condition. Indeed, consider the following

Example 2.1: Let $X = [1,\infty)$ with the euclidean metric and let $Sx = 2x$ and $Tx = 7x$ for all x in X. Clearly S commutes with T and they are both continuous. We have

$$d(Sx,Ty) = \begin{cases} 2x - 7y < 2x - y = d(Sx,y) & \text{if } 2x \geqslant 7y > y, \\ 7y - 2x < 7y - x = d(x,Ty) & \text{if } 7y > 2x > x \end{cases}$$

for all x, y in X for which the right hand side of inequality (9) is positive. Then all the assumptions of theorem 2.2 are satisfied except the compactness of X but S and T have no fixed points.

Other results on common fixed points in compact metric spaces can be found in Fisher [5], [7], Fisher and Sessa [10] and Kasahara and Rhoades [11].

REFERENCES

1. L.B. Ćirić, 'On common fixed points in uniform spaces', Publ. Inst. Math. (Beograd), 24(38)(1978), 39-43.
2. B. Fisher, 'A common fixed point theorem for commuting mappings', Math. Sem. Notes, Kobe Univ., 7(1979), 297-200.

3. B. Fisher, 'Results on common fixed points on bounded metric spaces', Math. Sem. Notes, Kobe Univ., 7(1979), 73-80.
4. B. Fisher, 'Results on common fixed points on complete metric spaces', Glasgow Math. J., 21(1980), 165-167.
5. B. Fisher, 'Common fixed points of commuting mappings', Bull. Inst. Math. Acad. Sinica, 9(1981), 399-406.
6. B. Fisher, 'A fixed point theorem for commuting mappings', Bull. Malaysian Math. Soc., (2) 5(1982), 65-67.
7. B. Fisher, 'Common fixed points of four mappings', Bull. Inst. Math. Acad. Sinica, 11(1983), 103-113.
8. B. Fisher, 'A common fixed point theorem for four mappings on a compact metric space', Bull. Inst. Math. Acad. Sinica, 12(1984), 249-252.
9. B. Fisher,'A common fixed point theorem', Publ. Math. Debrecen, to appear.
10. B. Fisher and S. Sessa, 'On fixed points of weakly commuting mappings in compact metric spaces', Jñānābha, to appear.
11. S. Kasahara and B.E. Rhoades, 'Common fixed point theorems in compact metric spaces', Math. Japon., 23(1978), 227-229.
12. B.A. Meade and S.P. Singh, 'On common fixed point theorems', Bull. Austral. Math. Soc., 16(1977), 49-53.
13. B.E. Rhoades,'A comparison of various definitions of contractive mappings', Trans. Amer. Math. Soc., 226(1977), 257-290.
14. C.S. Wong, 'Fixed point theorems for generalized nonexpansive mappings', J. Austral. Math. Soc., 18(1974), 265-276.

NONLINEAR ELLIPTIC PROBLEMS INVOLVING CRITICAL SOBOLEV EXPONENT IN THE CASE OF SYMMETRICAL DOMAINS

Donato FORTUNATO and Enrico JANNELLI
Dipartimento di Matematica - Università di Bari
Via G. Fortunato - 70125 BARI
ITALY

ABSTRACT. We consider the boundary value problem

$$(*) \qquad -\Delta u - \lambda u - u|u|^{2^*-2} = 0 \qquad u \in H^1_0(\Omega)$$

where $\Omega \subset \mathbb{R}^n$ is a bounded domain, $n \geq 4$, $2^* = 2n/(n-2)$ is the critical exponent for the Sobolev embedding and λ is a real positive parameter. We state some theorems which ensure the existence of infinitely many solutions of $(*)$ when Ω exhibits suitable simmetries.

1. THE MAIN RESULTS

There are many interesting problems in differential geometry and mathematical physics which are described by means of nonlinear elliptic equations involving critical Sobolev exponents: for instance, the Yamabe problem, the Yang - Mills equation, the Rellich conjecture and so on.

For these problems, the standard tools used in the theory of nonlinear elliptic problems cannot be directly applied, since a sort of "lack of compactness" occours (for an extensive treatment of this subject see [2,3]).

In this lecture we shall be concerned with a "model equation" which includes the main difficulties which characterize the above problems. More precisely, we shall consider the following equation:

$$(1) \qquad -\Delta u - \lambda u - u|u|^{2^*-2} = 0 \qquad u \in H^1_0(\Omega)$$

where λ is a real parameter, Ω is a bounded domain of $\mathbb{R}^n$, $n \geq 3$, and $2^* = 2n/(n-2)$. We look for non trivial solutions of this problem.

The solutions of (1) are the critical points of the C^1 functional on $H^1_0(\Omega)$ defined by

$$(2) \qquad f_\lambda(u) = \frac{1}{2}\int_\Omega ((\nabla u)^2 - \lambda|u|^2)\,dx - \frac{1}{2^*}\int_\Omega |u|^{2^*}dx \quad .$$

Since the embedding $H^1_0(\Omega) \hookrightarrow L^{2^*}(\Omega)$ is not compact, the fun

S. P. Singh (ed.), Nonlinear Functional Analysis and Its Applications, 229–233.

ctional f_λ does not satisfy the Palais - Smale condition (P - S) in all $\mathbb{R}$, i.e. there exists a sequence $\{u_n\} \subset H_0^1(\Omega)$ satisfying the following properties (see [5])

$$(3) \qquad f_\lambda'(u_n) \to 0 \quad \text{in } H^{-1}(\Omega)$$

$$(4) \qquad \{u_n\} \rightharpoonup 0 \quad \text{in } H_0^1(\Omega)$$

$$(5) \qquad f_\lambda(u_n) \to \frac{1}{n} S^{n/2}$$

where S is the best constant for the Sobolev embedding $H_0^1(\Omega) \hookrightarrow L^{2^*}(\Omega)$.

Moreover, Pohozaev ([9]) has shown that, if Ω is starshaped and $\lambda \leq 0$, problem (1) has only the trivial solution $u \equiv 0$; therefore, we are led to consider the case $\lambda > 0$.

Brezis and Nirenberg have proved in [2] that problem (1) has a positive solution for $0 < \lambda < \lambda$ if the dimension n is greater or equal than 4, where λ_1 is the first eigenvalue of $-\Delta$ on Ω; in the case $n = 3$ they have proved that there exists $\lambda^* \in]0,\lambda_1[$ such that problem (1) has a positive solution for $\lambda^* < \lambda < \lambda_1$. Moreover, if Ω is a ball in $\mathbb{R}^3$, (1) has a positive solution if and only if $\lambda \in]\lambda_1/4 , \lambda_1[$.

If $\lambda \geq \lambda_1$, problem (1) has no positive solutions (this fact can be seen at once by multiplying equation (1) by the first eigenfunction of $-\Delta$ and by integrating on Ω); nevertheless, one can look for changing sign solutions.

In [5] it has been proved that there exists $\delta > 0$ (δ depending only on Ω and n) such that, for any $\lambda \in]\lambda_j - \delta, \lambda_j[$ (λ_j being any eigenvalue of $-\Delta$), problem (1) has at least m_j (pairs of) non trivial solutions, m_j being the multiplicity of λ_j .

On the other hand, in [4] it has been proved that, if $n \geq 4$, (1) has at least one pair of non trivial solutions for any $\lambda > 0$.

We want to point out that the energies (i.e. the critical values of f_λ) of the solutions u found by means of the above results belong to the interval $]0, \frac{1}{n} S^{n/2}[$; this is due to the fact that the functional f_λ satisfies the (P - S) condition in the range $]0, \frac{1}{n} S^{n/2}[$ (see [2,5]), which means, as is well known, that

$$(6) \qquad \begin{cases} \text{for any } c \in]0, \frac{1}{n} S^{n/2}[\text{ , for any } \{u_n\} \subset H_0^1(\Omega) \text{ such} \\ \text{that } f_\lambda(u_n) \to c \text{ and } f_\lambda'(u_n) \to 0 \text{ there exists a stron-} \\ \text{gly convergent (in } H_0^1(\Omega)\text{) subsequence of } \{u_n\} . \end{cases}$$

However, on the analogy of the "subcritical case"

$$(7) \qquad -\Delta u - \lambda u - u|u|^{p-2} = 0 \qquad u \in H_0^1(\Omega)$$

with $2 < p < 2^*$ (see [1]), we could expect infinitely many solutions for problem (1) when $\lambda > 0$ and $n \geq 4$ (the case $n = 3$ is much more delicate; see [2]), but, as far as we know, such a result has not yet been obtained; nevertheless, we shall show that, for a suitable class of domains Ω, problem (1) has infinitely many solutions of arbitrarily large energy.

More precisely, the following theorem holds (see [7]):

THEOREM 1

Let $n \geq 4$, $\lambda > 0$ and suppose that Ω is a "cylinder" in $\mathbb{R}^n$ defined by

$$\Omega = \omega \times]a,b[$$

where ω is an open bounded subset of $\mathbb{R}^{n-1}$ and $a,b \in \mathbb{R}$, with $a < b$.

Then, for any $R > 0$ there exists a solution u_R of (1) such that $f_\lambda(u_R) \geq R$.

In order to state an analogous result for domain which possess a "rotational symmetry" we need to introduce some definitions.

Let V be a 2-dimensional subspace of $\mathbb{R}^n$ and $V^\perp$ be its orthogonal complement. We shall denote by y_1 and y_2 the vectors of V and $V^\perp$ respectively. Moreover, let Σ_V be the group of the rotations of V. Any $\sigma \in \Sigma_V$ induces in a natural way a map $\gamma_\sigma : \mathbb{R}^n \to \mathbb{R}^n$ defined by

$$\gamma_\sigma(y_1, y_2) = (\sigma(y_1), y_2)$$

Roughly speaking, γ_σ corresponds to a "rotation around $V^\perp$".

Keeping these notations in mind, we give the following

DEFINITION

We say that an open subset $\Omega \subset \mathbb{R}^n$ has a rotational symmetry if there exists a 2-dimensional subspace V of $\mathbb{R}^n$ such that $\gamma_\sigma(\Omega) = \Omega$ for any $\sigma \in \Sigma_V$.

Obviously the unit ball in $\mathbb{R}^n$ is the simplest example of a domain with rotational symmetry.

Now we can state the following theorem (see [7]):

THEOREM 2

Let $n \geq 4$, $\lambda > 0$ and suppose that Ω is a bounded domain in $\mathbb{R}^n$ having a rotational symmetry in the sense of the above definition.

Then, for any $R > 0$ there exists a solution u_R of (1) such that $f_\lambda(u_R) \geq R$.

A result related to Theorem 2 has been obtained by G. Cerami, S. Solimini and M. Struwe (this volume).

2. SKETCH OF THE PROOFS

The proofs of Theorems 1 and 2 are contained in [7]. In this lecture we shall give only a brief sketch of the proof of Theorem 1.

Without loss of generality we can consider

$$\Omega = \omega \times]0,\pi[$$

where ω is an open bounded subset of $\mathbb{R}^{n-1}$.

The eigenvalues of the operator $-\Delta$ in Ω are

(8) $\lambda_{jk} = \mu_j + k^2 \qquad j, k \in \mathbb{N}$

where μ_j are the eigenvalues of $-\Delta$ on Ω. The corresponding eigenfunctions are

(9) $e_{jk} = v_j(x) \cdot \sin kt \qquad \forall\, x \in \omega,\ \forall\, t \in]0,\pi[$.

Let u_{jk} be the Fourier coefficients of u along e_{jk}. For any $m \in \mathbb{N}$ we set

(10) $V_m = \left\{ u \in H_0^1(\Omega) \mid u_{jk} = 0 \text{ if } k/m \notin \mathbb{N} \right\}$

(obviously $V_1 = H_0^1(\Omega)$).

Let us define

(11) $S_m = \inf \left\{ \|u\|^2 / |u|_{2^*}^2,\ u \in V_m \setminus \{0\} \right\}$.

Clearly, S_m is the best constant for the embedding $V_m \hookrightarrow L^{2^*}(\Omega)$, and $S_1 = S$.

Finally, we denote by $f_{\lambda|V_m}$ the restriction of f_λ to V_m and by $f'_{\lambda|V_m}$ its Frechét derivative.

Our proof consists of the following steps:

<u>1st step</u>: $S_m = m^{2/n} \cdot S$ for any $m \in \mathbb{N}$

<u>2nd step</u>: $f_{\lambda|V_m}$ satisfies the (P - S) condition in $\left]0, \frac{1}{n} S_m^{n/2}\right[$.

<u>3rd step</u>: using the previous step and the mountain pass lemma ([1]), we show that for any R > 0 there exists a critical point u_R of $f_{\lambda|V_m}$, whose energy is greater or equal than R (m is a suitable positive integer depending on R).

<u>4th step</u>: u_R is a critical point of f_λ , hence is a non trivial solution of problem (1).

The proof of Theorem 2 is based upon the same kind of arguments here exposed; we only remark that, in the case of domains with rotational symmetry (see the above definition), we choose the subspaces V_m in a different way, accordingly to this symmetry.

Summing up, the basic concept of our theorems is the following: the symmetry of the domain Ω allows us to restrict problem (1) to a suitable class of subspaces V_m of $H_0^1(\Omega)$, in which we gain more and more compactness, as m increases, for the functional f_λ (see steps 1 and 2). We think that this simple idea could be applied to other situations, in which the lack of compacteness can be overcome by means of restrictions to suitable subspaces (in this context see also [6]).

REFERENCES

[1] A. Ambrosetti, P.H. Rabinowitz, 'Dual variational methods in critical point theory and applications', J. Funct. Anal., 14, 1973, pp. 349 - 381.

[2] H. Brezis, L. Nirenberg, 'Positive solutions of nonlinear elliptic equations involving critical Sobolev exponents', Comm. Pure Appl. Math., 36, 1983

[3] H. Brezis, 'Some variational problems with lack of compactness', to appear on the Proceedings of the Berkeley Symp. on Nonlinear Functional Analysis.

[4] A. Capozzi, D. Fortunato, G. Palmieri, 'An existence result for nonlinear elliptic problems involving critical Sobolev exponent', to appear on Ann. I.H.P. Analyse Nonlinéaire.

[5] G. Cerami, D. Fortunato, M. Struwe, 'Bifurcation and multiplicity results for nonlinear elliptic problems involving critical Sobolev exponents', Ann. I.H.P. Analyse Nonlinéaire, 1, 1984, pp. 341 - 350.

[6] J.M. Coron, 'Periodic solutions of a nonlinear wave equation without assumptions of monotonicity', Math. Annalen, 262, 1983, pp. 273 - 285.

[7] D. Fortunato, E. Jannelli, 'Infinitely many solutions for some nonlinear elliptic problems in symmetrical domains', to appear.

[8] P.L. Lions, 'The concentration - compacteness principle in the calculus of variations, the limit case', Part I and II, to appear.

[9] S.J. Pohozaev, 'Eigenfunctions of the equation $\Delta u + \lambda f(u) = 0$', Soviet Math. Doklady, 6, 1965, pp. 1408 - 1411.

PERIODIC SOLUTIONS OF PENDULUM LIKE THIRD ORDER DIFFERENTIAL EQUATIONS

G. Fournier
University of Sherbrooke
Department of Mathematics and Computer Sciences
Sherbrooke Québec
Canada J1K 2R1

R. Iannacci
University of Calabria
Department of Mathematics
87036 Rende(Cosenza)
Italy

J. Mawhin
Catholic University of Louvain
Mathematics Institute
Chemin du Cyclotron, 2
B-1348 Louvain-la-Neuve
Belgium

ABSTRACT. In this paper, we intend to show that some techniques used to study the problem of the existence of periodic solutions of the forced pendulum equation can be used to study similar third order equations, and similarly higher order equations, to get almost all the previously obtained results.

1. INTRODUCTION

Many papers have been devoted to the study of periodic solutions, in particular to the study of periodic solutions of the forced pendulum equation.

In this paper, we want to show that the methods of G. Fournier-J. Mawhin [1] can be adapted to study periodic solutions of higher order equations in order to obtain some of the results contained in that paper. We shall limit ourselves to the study of third order equations of the form

$$x''' + ax'' + bx' + A \sin x = e \qquad (1.1)$$
$$\text{with } x''(0) = x''(T),\ x'(0) = x'(T) \text{ and } x(0) = x(T)$$

where A, a and b are constants and $e(0) = e(T)$ and e is continuous.

S. P. Singh (ed.), Nonlinear Functional Analysis and Its Applications, 235–239.

The specific question we will ask ourselves is: given a function $\tilde{e}$ of mean value zero, for which constant $\bar{e}$ does (1.1) have a solution with $e = \tilde{e} + \bar{e}$. Finally note that if x is a solution of (1.1) then $x + 2k\pi$ is also a solution of (1.1) for any integer k.

2. PRELIMINARIES

Let us introduce some notations: for any $x: \mathcal{R} \to \mathcal{R}$ in $L_1([0, T])$ denote

$$\bar{x} = T^{-1} \int_0^T x(t)\, dt \qquad \text{and} \qquad \tilde{x}(t) = x(t) - \bar{x}.$$

Furthermore denote, for x and y elements of $L_2([0, T])$,

$$\langle x, y\rangle = T^{-1} \int_0^T x(t)y(t)\, dt$$

and denote $\|x\| = (\langle x, x\rangle)^{\frac{1}{2}}$. We shall also use the Wirtinger inequality: $\omega\|\tilde{x}\| \le \|x'\|$ and the Sobolev inequality: $\|\tilde{x}\|_C \le \pi\omega^{-1}\sqrt{3}\,\|x'\|$, where $\|\ \|_C$ is the sup norm (see [1]) for C^1-functions satisfying $x(0) = x(T)$, where ω is defined by $\omega = 2\pi T^{-1}$.

Finally we shall need the following lemma in order to obtain our a priori bounds. See [1].

Lemma 2.1: Let x be a t-periodic, n times differentiable map such that $x^{(n)}$ is in $L_2([0, T])$, we have the following for any $0 \le i, j \le n$,

$$\langle x^{(i)}, x^{(j)}\rangle = \begin{cases} 0 & \text{if } i \ne j \pmod 2 \\ (-1)^{|i-j|/2}\|x^{((i+j)/2)}\|^2 & \text{if } i = j \pmod 2 \end{cases}$$

3. A PRIORI BOUNDS

For our purposes, we need a priori bounds for slightly more general type of equations:

$$\begin{aligned} &x''' + ax'' + bx' = g(x, t) + c(x) \\ &x''(0) = x''(T),\ x'(0) = x'(T) \text{ and } x(0) = x(T) \end{aligned} \tag{3.1}$$

where g is continuous and T-periodic in t and $c: C([0, T], \mathcal{R}) \to \mathcal{R}$ is continuous.

Proposition 3.2: Let x be a solution of (3.1), if

$$\|\sup_x g(x, t)\| \le A' \tag{3.2.1}$$

and

$$\omega^2 + a^2 > b \tag{3.2.2}$$

then $\|\tilde{x}\|$, $\|x'\|$, $\|x''\|$, $\|x'''\|$, $\|\tilde{x}\|_C$, $\|x'\|_C$, $\|x''\|_C$ are bounded by constants depending only on ω, A', a and b, and

$$\|\tilde{x}\| \le A'\omega^{-1}(\omega^4 + \omega^2(a^2-b) + b^2)^{-\frac{1}{2}}$$

and

$$\|x'''\| \le \omega^2 A'(\omega^2 + a^2 - b)^{-1}.$$

Proof: Multiply both sides of the equation by $T^{-1}(x''' + ax'' + bx')$ and integrate over [0, T] to get, using the Schwartz inequality,

$$\|x''' + ax'' + bx'\|^2 \le \|x''' + ax'' + bx'\| \; \|g(x, t)\|.$$

That is, by simplification, $\|x''' + ax'' + bx'\| \le A'$. Now using (2.1), we get that

$$(\|x'''\|^2 + (a^2 - b)\|x''\|^2 + b^2\|x'\|^2)^{\frac{1}{2}} \le A'.$$

Using Wirtinger inequality twice, one gets

$$((1 - \omega^{-2}(b-a^2))\|x'''\|^2 + b^2\|x'\|^2)^{\frac{1}{2}} \le A'$$

and

$$(\omega^4 + (a^2 - b)\omega^2 + b^2)^{\frac{1}{2}}\|x'\| \le A'$$

so $\|x'''\| \le \omega^2 A'(\omega^2 + a^2 - b)^{-1}$ and $\|\tilde{x}\| \le \|x'\|\omega^{-1} \le A'\omega^{-1}(\omega^4 + \omega^2(a^2 - b) + b^2)^{\frac{1}{2}}$. Thus, by the Sobolev and the Wirtinger inequalities, we get that $\|\tilde{x}\|$, $\|x'\|$, $\|x''\|$ $\|x'''\|$ are bounded by constants depending only on A', a, b and ω; and so $\|\tilde{x}\|_C$, $\|x'\|_C$, $\|x''\|_C$ are bounded by the same type of constants. □

4. EXISTENCE THEOREM

Now by the well known Leray-Schauder homotopy continuation method since $\tilde{e}$ is of mean value zero, using the bounds given by (3.2) with $A' = \|\tilde{e}\|$, for (3.1) with $g(x, t) = \tilde{e}$ and $c(x) = 0$, one can show that there is exactly one solution E of (3.1) with $g(x, t) + c = \tilde{e}$ such that $\bar{E} = 0$. We have uniqueness because any solution of (3.2) with $g(x, t) + c(x) = 0$ must satisfy the bound $\|\tilde{x}\|_C = 0$.

Define $y = x - E$ then (1.1) becomes

$$y''' + ay'' + by' + A\sin(y + E) = \bar{e} \qquad (4.1)$$
$$y''(0) = y''(T),\ y'(0) = y'(T) \text{ and } y(0) = y(T)$$

Looking at (4.1) one sees that to have a solution y of (4.1), we would need that

$$A\,\overline{\sin(y + E)} = \bar{e} \qquad (4.2)$$

so let us modify (4.1) into

$$y''' + ay'' + by' = A\,\overline{\sin(y + E)} - A\sin(y + E)$$
$$y''(0) = y''(T),\ y'(0) = y'(T) \text{ and } y(0) = y(T). \tag{4.3}$$

Now again, since the right member of (4.2) is always of mean value zero, using homotopy continuation methods, one can show that for any $\lambda \in \mathcal{R}$, there exists y_λ a solution of (4.2). First we use (3.2) with $c(y) = A\,\overline{\sin(y + E)}$ and $g(y, t) = A\sin(y + E)$, taking $A' = |A|$, to get a priori bounds for $\|\tilde{y}\|$, $\|y'\|$ etc... . Secondly, we use (3.2) with $g(x, t) = g(t) = A\,\overline{\sin(y + E)}$ and $c(x) = A\sin(y + E)$ to get for each y a solution $x = \phi_\lambda(y)$ of (3.2) with mean value λ by the homotopy continuation method and also we get, by (3.2), a priori bounds for all the $\phi_\lambda(y)$; this enables us to prove that $\phi_\lambda(y)$ is completely continuous as a function of λ and y. That is, by homotopy continuation, we get the following theorem.

Theorem 4.4: Assume (3.2.2). Let $r, s \in \mathcal{R}$, then there exists K a continuum of solutions of (4.3) such that

$$r \le \overline{x} \le s \quad \text{for any } x \in K \tag{4.4.1}$$

and

$$\text{there exists } x_\lambda \in K \text{ with } \overline{x_\lambda} = \lambda \text{ for all } r \le \lambda \le s. \tag{4.4.2}$$

5. PARTIAL SOLUTION OF OUR SPECIFIC PROBLEM

By (4.2), we need only estimate the possible values of $A\,\overline{\sin(y + E)}$ for a solution y of (4.3). But we can estimate $\sup_\lambda \overline{\sin(\lambda + E)}$ and $|\overline{\sin(y + E)} - \overline{\sin(\lambda + E)}|$ if $\overline{y} = \lambda$. Thus we have bounds for our $\sup A\,\overline{\sin(y + E)}$ if $\overline{y} = \lambda$. The final argument is to vary λ continuously. In fact, by (3.2), $|\overline{\sin(y + E)} - \overline{\sin(\lambda + E)}| \le |\overline{\tilde{y}}| \le \|\tilde{y}\| \le |A|\omega^{-1}(\omega^4 + \omega^2(a^2 - b) + b^2)^{-\frac{1}{2}}$ and since $\overline{\sin(\lambda + E)} = \sin\lambda\,\overline{\cos E} + \cos\lambda\,\overline{\sin E}$, $\sup_\lambda \overline{\sin(\lambda + E)} = \delta_{\tilde{e}} = ((\overline{\sin E})^2 + (\overline{\cos E})^2)^{\frac{1}{2}}$ and $\inf_\lambda \overline{\sin(\lambda + E)} = -\delta_{\tilde{e}}$. Thus $|A|\delta_{\tilde{e}} - A^2\omega^{-1}(\omega^4 + \omega^2(a^2 - b) + b^2)^{-\frac{1}{2}} \le \sup \overline{\sin(y + E)} \le |A|\delta_{\tilde{e}} + A^2\omega^{-1}(\omega^4 + \omega^2(a^2 - b) + b^2)^{-\frac{1}{2}}$ and $-|A|\delta_{\tilde{e}} - A^2\omega^{-1}(\omega^4 + \omega^2(a^2 - b) + b^2)^{-\frac{1}{2}} \le \inf \overline{\sin(y + E)} \le -|A|\delta_{\tilde{e}} + A^2\omega^{-1}(\omega^4 + \omega^2(a^2 - b) + b^2)^{-\frac{1}{2}}$.

Finally, by the periodicity of the sin, let λ_1 be such that $\overline{\sin(\lambda_1 + E)} = \delta_{\tilde{e}}$ and $\overline{\sin(\lambda_1 + \pi + E)} = -\delta_{\tilde{e}}$; by (4.4), let K be a continuum of solutions given for $r = \lambda_1$ and $s = \lambda_1 + \pi$; then since $\overline{\sin(y+E)}$ is a continuous function of y, the image of K is an interval of $\mathcal{R}$ containing the interval [-d, d], where

$$d = |A|\delta_{\tilde{e}} - A^2\omega^{-1}(\omega^4 + \omega^2(a^2 - b) + b^2)^{-\frac{1}{2}}.$$

Repeating the process with $r = \lambda_1 + \pi$ and $s = \lambda_1 + 2\pi$, we get the following theorem.

<u>Theorem 5.1</u>: Assume (3.2.2), fix $\tilde{e}$ and assume that

$$\delta_{\tilde{e}} > |A|\omega^{-1}(\omega^4 + \omega^2(a^2 - b) + b^2)^{-\frac{1}{2}} \tag{5.1.1}$$

then (1.1) has a solution for $e = \tilde{e} + \bar{e}$ for any $\bar{e}$ contained in the interval [-d, d] and for $\bar{e}$ in the interior of this interval, there exists at least two solutions the difference of which is not a multiple of 2π.

REFERENCES

1. G. Fournier et J. Mawhin, 'On periodic solutions of forced pendulum-like equations' to appear.

2. J. Mawhin, 'L_2-Estimates and Periodic Solutions of Some Nonlinear Differential Equations', *Bolletino U.M.I.*, (4) **10** (1974), 341-352.

DOUBLE RESONANCE AT THE FIRST AND SECOND EIGENVALUES FOR THE NONLINEAR HEAT EQUATION.

Maria do Rosário Grossinho
C.M.A.F.
Av.Prof.Gama Pinto, 2
1699 Lisboa Codex
Portugal

ABSTRACT. We prove the existence of generalized time-periodic solutions of the nonlinear heat equation satisfying periodic and boundary conditions when resonance involving the first and second eigenvalues is allowed. We use Leray-Schauder's degree, the regularity of the inverse operator and the imbedding theorems for anisotropic Sobolev spaces.

0. INTRODUCTION

In this paper we are concerned with the existence of time periodic solutions of the nonlinear heat equation satisfying periodic and boundary conditions when resonance involving the first and second eigenvalues is allowed.

Let $g = g(t,x,s)$ be a function of class C^2 defined in $\mathbb{R} \times [0,\pi] \times \mathbb{R}$, 2π-periodic in t and such that $g(t,0,0)=0=g(t,\pi,0)$. Set $J=[0,2\pi] \times [0,\pi]$ and suppose moreover that g and its partial derivatives satisfy linear growth and bounding conditions of the following type : there exist constants $C > 0$, $K > 0$ and a real valued function $b \in L^2(J)$ such that

$$|g(t,x,s)| + |\frac{\partial g}{\partial t}(t,x,s)| + |\frac{\partial g}{\partial x}(t,x,s)| + |\frac{\partial^2 g}{\partial x^2}(t,x,s)| < C|s| + b(t,x) \qquad (g_1)$$

$$|\frac{\partial g}{\partial s}(t,x,s)| + |\frac{\partial^2 g}{\partial s^2}(t,x,s)| + |\frac{\partial^2 g}{\partial s\, \partial x}(t,x,s)| < K \qquad (g_2)$$

and also

$$g(t,x,s)s > 0 \quad \text{if} \quad s \neq 0. \qquad (g_3)$$

Denote by $H(J)$ the space of all functions $u(t,x)$ which are 2π-periodic in t and $u_{|J} \in L^2(J)$.

S. P. Singh (ed.), Nonlinear Functional Analysis and Its Applications, 241–252.

Let

$$H^1(J) = \{u \in H: u_t, u_x \in H\} \quad \text{and}$$

$$H^{1,2}(J) = \{u \in H^1(J): u_{xx} \in H\} .$$

If we define

$$\|u\|_{H^1} = \left[\int_0^{2\pi}\int_0^{\pi}\left[u^2(t,x) + u_t^2(t,x) + u_x^2(t,x)\right]dxdt\right]^{\frac{1}{2}}$$

$$\|u\|_{H^{1,2}} = \left[\int_0^{2\pi}\int_0^{\pi}(u^2(t,x)+u_t^2(t,x)+u_x^2(t,x) + \right.$$

$$\left. + u_{xx}^2(t,x)dxdt\right]^{\frac{1}{2}}$$

then $H^1(J)$ and $H^{1,2}(J)$ are Banach spaces (with respect to the norms just defined).

Denote by $H_o^1(J)$ the closure in $H^1(J)$ of the space of all real functions $u(t,x)$ on J which are infinitely continuously differentiable and 2π-periodic in t and satisfy $u(t,0) = u(t,\pi) = 0$ for $t \in \mathbb{R}$.

Let $h \in H^{1,2}(J) \cap H_o^1(J)$ and consider the problem

$$\begin{cases} -u_t(t,x)+u_{xx}(t,x)+g(t,x,u(t,x)) = h(t,x) & (t,x) \in J \\ u(t,0) = u(t,\pi) = 0 & t \in [0,2\pi] \\ u(0,X) = u(2\pi,x) & x \in [0,\pi] \end{cases} \qquad (0.1)$$

A weak solution of the problem (0.1) is a function $u \in H^{1,2}(J) \cap H_o^1(J)$ which satisfies

$$-u_t(b,x)+u_{xx}(t,x)+g(t,x,u(t,x)) = h(t,x) \ . \qquad (0.2)$$

Defining in $H^{1,2}(J) \cap H_o^1(J)$ the linear operator

$$Lu = u_t - u_{xx}$$

the eigenvalues of L are the numbers m^2 $(m \in \mathbb{N})$.

In [3] it has been proved that the problem (0.1) has at least a weak solution if

$$m^2 \leq \gamma(t,x) \leq \liminf_{|u|\to\infty} \frac{g(t,x,u)}{u} \leq$$

$$\leq \limsup_{|u|\to\infty} \frac{g(t,x,u)}{u} \leq \Gamma(t,x) \leq (m+1)^2$$

hold uniformly for a.e. $(t,x) \in J$, for some $m \in \mathbb{N}$, where $\gamma, \Gamma \in L^\infty(J)$ and satisfy

$$\int_J (\gamma(t,x)-m^2) \sin^2 mx \, dxdt > 0 \qquad (\gamma)$$

and

$$\int_J ((m+1)^2 - \Gamma(t,x)) \sin^2 (m+1)x \, dxdt > 0. \qquad (\Gamma)$$

In this work, we present a result for the case m=1, in which (γ) or (Γ) (or even both) may be false, although there are some restrictions on h.

More precisely, we prove the existence of a weak solution of (0.1) assuming that the following limits exist

$$\beta_{\pm}(t,x) = \lim_{u\to\pm\infty} \frac{g(t,x,u)}{u}$$

(and afterwards under more general hypothesis in which the existence of these limits is not required) and satisfy a double resonance condition of the form

$$1 \leq \beta_{\pm}(t,x) \leq 4 \qquad (\beta)$$

Example : Let α,Θ,ϕ, and ψ be functions defined on $\mathbb{R}$ such that

$$\alpha,\psi,\phi \in C^2(\mathbb{R}), \qquad \Theta \in C^1(\mathbb{R})$$

$$\alpha',\alpha'',\psi',\psi'' \in L^\infty(\mathbb{R})$$

$$\alpha(s)s > 0, \quad \psi(s)s > 0 \quad \text{if } s\neq 0$$

$$\Theta \text{ is } 2\pi\text{-periodic}$$

$$\Theta,\phi > 0$$

$$\lim_{u\to+\infty} \frac{\alpha(u)}{u} = C_1 \qquad \lim_{u\to-\infty} \frac{\alpha(u)}{u} = C_2$$

$$\lim_{u\to+\infty} \frac{\psi(u)}{u} = C_3 \qquad \lim_{u\to-\infty} \frac{\psi(u)}{u} = C_4$$

where C_1, C_2, C_3, $C_4 \in \mathbb{R}$
Then $g(t,x,u) = \phi(\sin x)\psi(u) + \Theta(t)\alpha(u) + u$ is 2π-periodic in t, satisfies (g_1), (g_2) and (g_3) and $g(t,0,0)=0=g(t,\pi,0)$.

If $C_1=C_2=C_3=C_4=0$ then

$$\beta_{\pm}(b,x) \equiv 1 \equiv \gamma(t,x)$$

so (β) holds but (γ) is false.

If $\phi \equiv 1$, $C_1=C_2=C_4=0$ and $C_3=3$

then

$$\beta_+(t,x) \equiv 4 \equiv \Gamma(t,x)$$

and

$$\beta_-(t,x) \equiv 1 \equiv \gamma(t,x)$$

so (β) holds but (γ) and (Γ) are both false.

Note that problem (0.1) is equivalent to

$$\begin{aligned} &-u_t(t,x)+u_{xx}(t,x)+u(t,x)+g(t,x,u(t,x))=h(t,x) && (t,x) \in J \\ &u(t,0) = u(t,\pi) = 0 && t \in [0,2\pi] \qquad (0.3) \\ &u(0,x) = u(2\pi,x) && x \in [0,\pi] \end{aligned}$$

where the new function g (which we still denote by the same symbol) satisfies the hypothesis of the previous one.

We shall study (0.3) assuming that the following limits exist

$$\beta_\pm(t,x) = \lim_{u\to\pm\infty} \frac{g(t,x,u)}{u}$$

and satisfy the resonance condition

$$0 \leq \beta_\pm(t,x) \leq 3 \qquad (\beta^*)$$

(which is equivalent to (β)).

REMARK : If there exist subsets of J of positive measure where $\beta_+(t,x)$ and $\beta_-(t,x)$ are not zero, problem (0.3) falls in the abstract framework of [1]. We, however, allow these limits to be identically zero (even both).

1. PRELIMINARIES

Set $U=H^{1,2}(J) \cap H^1_0(J)$ and define the linear operator

$A: D(A) \subset H \to H$ by
$U = D(A)$ and
$Au = -u_t+u_{xx}+u$ under boundary and periodic conditions (0.3).

If $u \in U$, then u has the Fourier series

$$u(t,x) = \sum_{\substack{k \in Z \\ n \in \mathbb{N}}} u_{kn} e^{ikt} \sin nx$$

The nullspace of A, N(A), is the subspace N(A)=sp(sin x) and the range of A is $R(A) = N(A)^{\perp}$. A has an inverse $A^{-1}:R(A) \to$ $\to R(A)$ which is compact.

The eigenvalues of A are the numbers of the form $\lambda_n=-n^2+1$, $n \in \mathbb{N}$, and then condition (0.4) can be written in the following way

$$-\lambda_1=0 \leq \beta_\pm(t,x) \leq -\lambda_2 .$$

Denote by P and Q the orthogonal projections of H onto N(A) and R(A), respectively, and by G the Nemytskii operator associated to g.

LEMMA 1: Let A be the operator defined before
(i) Then $\langle Au, Au+3u\rangle \geq 0 \quad \forall\, u \in U$
(ii) If $\langle Au, Au+3u\rangle = 0$ for some $u \in U$
then $u \in N(A) \oplus N(A+3I)$.

Proof: Using the Fourier series of u (real) we have

$$\langle Au,Au+3u\rangle=\langle\Sigma(-ik-n^2+1)u_{kn}e^{ikt}\sin n x,\Sigma(-ik-n^2+4)u_{kn}e^{ikt}\sin nx\rangle$$
$$=\pi^2\Sigma\left[k^2+(-n^2+1)(-n^2+4)\right]\,|u_{kn}|^2$$

which implies $\langle Au,Au+3u\rangle \geq 0$, equility holding if and only if k=0 and n=1 or n=2, that is, if and only if $u \in N(A) \oplus N(A+3I)$.

2. EXISTENCE RESULT

In the following we denote by $C^{k,p}(J)$ the Banach space of all functions $u:J \to \mathbb{R}$ that are 2π-periodic in t and such that $D_t^\alpha D_x^\beta u$ is continuous for any $(\alpha,\beta) \in \mathbb{N}^2$ such that $\alpha=0$ and $\beta \leq p$ or $\alpha \leq k$ and $\beta=0$, with the natural norm.

THEOREM 1 : Let g satisfy the previous hypothesis and suppose that

$$0 \leq \beta_\pm(t,x) \leq 3 \tag{2.1}$$

and

$$\int_{v>0}(3-\beta_+) + \int_{v<0}(3-\beta_-) > 0 \qquad \forall\, v \in N(A+3I),\ v \neq 0. \tag{2.2}$$

Then problem (0.3) has a solution $u \in U$ for every $h \in H^{1,2}(J) \cap H_0^1(J)$ satisfying the following condition

$$\int_J g_-\sin x < (h,\sin x) < \int_J g_+\sin x \tag{h_1}$$

where $g_\pm(t,x)=\liminf_{u\to\pm\infty} g(t,x,u)$. Moreover $u \in C^{1,2}(J)$.

Proof : Writing the equation of (0.3) in the following way

$$Au + Gu = h$$

it turns equivalent to the equation

$$(A+P)u + (Gu-Pu-h) = 0$$

or, taking into account that A+P is invertible,

$$u + (A+P)^{-1}(Gu-Pu-h) = 0.$$

For $t \in [0,1]$ and $u \in H$ we define the homotopy mapping

$$\mathcal{H}(t,u) = u+t(A+P)^{-1}(Gu-Pu-h). \tag{2.3}$$

It is clear that, for each $t \in [0,1]$, $\mathcal{H}(t,.)$ is a compact perturbation of the identity on account of the compactness of $(A+P)^{-1}$.

We claim that there exists $R > 0$ such that

$$\mathcal{H}(t,u) \neq 0 \qquad \forall\, t \in [0,1] \qquad \forall\, u \in \partial B_R \tag{2.4}$$

Let us assume for the moment that (2.4) holds and finish the proof.

Using the homotopy invariance of Leray-Schauder degree we have

$$\deg(\mathcal{H}(1,.),B_R,0)=\deg(\mathcal{H}(0,.),B_R,0)=\deg(I,B_R,0)=1.$$

Since $\mathcal{H}(1,.)=0$ is equivalent to our original equation, this implies the existence of a solution of (0.3). Relying on the regularity of the inverse operator and using an embedding theorem for anisotropic Sobolev spaces (see [4]) it is easy to see that $u \in C^{1,2}(J)$.

Let us now estabilish (2.4). The arguments we use were suggested by [1] .

<u>LEMMA 2</u>: There exists a constant $R > 0$ such that $\mathcal{H}(t,u) \neq 0$ for any $t \in [0,1]$ and $u \in \partial B_R$.

<u>Proof</u> : We argue by contradiction. Suppose for each $n \in \mathbb{N}$ we find $t_n \in [0,1]$ and $u_n \in \partial B_R \cap D(A)$ such that

$$\mathcal{H}(t_n,u_n)=0 \qquad \text{and} \qquad \|u_n\| = n. \tag{2.5}$$

Putting $v_n = \frac{u_n}{n}$, $\|v_n\| = 1$, and passing

to subsequences we may assume that

$$t_n \to t^* \qquad \text{and} \qquad v_n \rightharpoonup v \in H.$$

We can even show that $v_n \to v$ in H-norm. In fact, dividing (2.5) by n, we get

$$v_n + t_n(A+P)^{-1} \left(\frac{Gu_n}{n} - P\, v_n - \frac{h}{n}\right) = 0. \tag{2.6}$$

Observing that $\frac{Gu_n}{n}$ is bounded in H on account of (g_1) and using the fact that $(A+P)^{-1}$ is compact we derive (passing to a subsequence)

$$v_n \to v \quad \text{in } H \text{ and a.e. in } J. \tag{2.7}$$

Then it is easy to see (writing $\frac{g(t,x,u_n)}{n} = \frac{g(t,x,u_n)}{u_n} v_n$ if $v \neq 0$ and directly from condition (g_1) if $v=0$) that

$$\frac{g(t,x,u_n)}{n} \xrightarrow[a.e]{} \begin{cases} \beta_+ v & \text{if } v > 0 \\ \beta_- v & \text{if } v < 0 \\ 0 & \text{if } v = 0 \end{cases} \tag{2.8}$$

and, as $\left\| \frac{g(t,x,u_n)}{n} \right\|_H < M$,

by Lebesgue's theorem we have

$$\frac{Gu_n}{n} \longrightarrow \chi_v v \text{ in H-norm,} \tag{2.9}$$

where $\chi_v = \begin{cases} \beta_+ & \text{if } v > 0 \\ \beta_- & \text{if } v < 0 \\ 0 & \text{if } v = 0 \end{cases}$

Passing to the limit in (2.6) and applying (A+P) to the equation we obtain

$$(A+P)v + t^*(\chi_v v - Pv) = 0$$

which is equivalent to

$$Av = -t^* \chi_v v - (1-t^*)Pv. \tag{2.10}$$

By lemma 1. i) we derive

$$t^{*2}\|\chi_v v\|^2 - \langle t^*\chi_v v, 3v\rangle + 2\langle t^*\chi_v v, (1-t^*)Pv\rangle + \tag{2.11}$$
$$+ (1-t^*)^2 \|Pv\|^2 - (1-t^*)3 \|Pv\|^2 \geq 0.$$

But making the inner product with Pv in (2.10) it is easy to see that

$$2 \langle t^*\chi_v v, (1-t^*)Pv\rangle + (1-t^*)^2 \|Pv\|^2 \leq 0$$

and then, from (2.11), we get

$$t^{*2}\|\chi_v v\|^2 \geq \langle t^*\chi_v v, 3v\rangle ,$$

that is,

$$\int_J [(t^*\chi_v)^2 - 3t^*\chi_v] v^2 \geq 0.$$

Therefore, as $0 \leq \chi_v \leq 3$ and $\|v\| = 1$, it follows that $t^*=0$ or $t^*=1$ and

$$\langle Av, Av + 3v\rangle = 0. \tag{2.12}$$

If $t^*=0$, (2.10) implies $v=0$ which contradicts the fact that $\|v\|=1$. Then $t^*=1$ and (2.10) becomes

$$Av = -\chi_v v. \tag{2.13}$$

From (2.12), and applying Lemma 1. ii), we get

$$v \in N(A) \oplus N(A+3I) .$$

Writing $v=d_1 \sin x + d_2 \sin 2x$ it follows from (2.13) that

$$3\, d_2 \sin 2x = \chi_v(t,x)\, v(x)$$

or, in an equivalent way.

$$(3-\chi_v(t,x))d_2 \sin 2x = \chi_v(t,x) d_1 \sin x . \tag{2.14}$$

Then

$$\int_0^\pi (3-\chi_v(t,x))d_2 d_1 \sin 2x \sin x\, dx = \int_0^\pi \chi_v(t,x) d_1^2 \sin^2 x\, dx \geq 0$$

which implies

$$\int_0^\pi \chi_v(t,x) d_2 d_1 \sin 2x \sin x\, dx \leq 0. \tag{2.15}$$

From (2.14) we can also derive

$$\int_0^\pi \chi_v(t,x) d_1 d_2 \sin x \sin 2x\, dx = \int_0^\pi (3-\chi_v(t,x)) d_2^2 \sin^2 2x\, dx \geq 0 \tag{2.16}$$

hence, by (2.15),

$$\int_0^\pi \chi_v(t,x) d_1 d_2 \sin x \sin 2x\, dx = 0$$

and then, by (2.16),

$$\int_0^\pi (3-\chi_v(t,x)) d_2^2 \sin^2 2x\, dx = 0 .$$

Therefore if $d_2 \neq 0$, we deduce

$$\chi_v(t,x) = 3 \quad \text{a.e. in } J$$

and, by (2.13), $v \in N(A+3I)$, which yields a contradiction to (2.2).

Thus $d_2=0$ and $v \in N(A)$, that is, v is of the form v=d sin x. We claim that (passing to a subsequence if necessary) there exists n_0 such that

$$v_n(t,x)v(x)>0 \ \forall\ n > n_0, \quad \text{a.e. in } J\ . \tag{2.17}$$

Let us assume for the moment that (2.17) holds.
From (2.6) we derive

$$(A+P)v_n + t_n \left(\frac{Gu_n}{n} - Pv_n - \frac{h}{n}\right) = 0. \tag{2.18}$$

Taking the inner product of (2.18) with v(=d sin x) and observing that $(Pv_n,v) \to \|v\|^2 = 1$ we get by (2.17) and (g_3)

$$0 \leq \int_J d\ Gu_n \sin x \leq \int_J d\ h \sin x \quad \text{for large } n$$

and then, applying Fatou's lemma,

$$\int_J g_- \sin x \geq (h,\sin x) \quad \text{if } d<0$$

or

$$\int_J g_+ \sin x \leq (h,\sin x) \quad \text{if } d>0$$

which yields a contradiction to (h_1).
Now it remains to establish (2.17).

<u>LEMMA 3</u>: Let (v_n) and v be as in the proof of Lemma 2. Then (passing to a subsequence) there exists n_0 such that

$$v_n(t,x)\ v(x) > 0 \quad \forall\ n > n_0, \text{ a.e. in } J.$$

<u>Proof</u>: In what follows we represent by the same symbol K several constants.

Let us rewrite (2.18) as

$$(A+P)v_n = -t_n \left(\frac{Gu_n}{n} - Pv_n - \frac{h}{n}\right)$$

Observing that the second member is bounded in $H=L^2(J)$ and $(A+P)^{-1}$ is continuous (since A^{-1} is continuous and ker A has finite dimension, see [4]) we conclude that there exists $K>0$ such that

$$\|v_n\|_{H^{1,2}} < K \qquad \forall\ n \in \mathbb{N} \tag{2.19}$$

We can even prove that a subsequence of v_n converges to v in $C^{0,1}$-norm. In order to show it let us first see that

$$\frac{Gu_n}{n} \in H^{1,2}(J) \quad \text{and} \quad \left\|\frac{Gu_n}{n}\right\|_{H^{1,2}} < K. \tag{2.20}$$

In fact it is easy to establish

$$\left|\frac{\partial}{\partial t}\frac{Gu_n}{n}\right| \leq \left|\frac{g_t}{n}\right| + \left|g_s v_{n_t}\right| \leq c\left|\frac{u_n}{n}\right| + K\left|v_{n_t}\right| + b(t,x) \qquad (2.21)$$

$$\left|\frac{\partial}{\partial x}\frac{Gu_n}{n}\right| \leq \left|\frac{g_x}{n}\right| + \left|g_s\right|\left|v_{n_x}\right| \leq c\left|\frac{u_n}{n}\right| + K\left|v_{n_x}\right| + b(t,x) \qquad (2.22)$$

$$\left|\frac{\partial^2}{\partial x^2}\frac{Gu_n}{n}\right| \leq \left|\frac{g_{xx}}{n}\right| + 2\left|g_{xs}\right|\left|v_{n_x}\right| + \left|g_{ss}\right|\left|v_{n_x}^2\right| + \left|g_s\right|\left|v_{n_{xx}}\right| \leq$$

$$\leq c\left|\frac{u_n}{n}\right| + 2K\left|v_{n_x}\right| + K\left|v_{n_x}^2\right| + K\left|v_{n_{xx}}\right| + b(t,x) \qquad (2.23)$$

where $g_t = \frac{\partial g}{\partial t}$, $g_{xs} = \frac{\partial^2 g}{\partial x \partial s}$, etc ... (2.23)

Since $\|v_n\|_{H^{1,2}} < K$, we infer that

$$\|v_{n_t}\|_H < K, \quad \|v_{n_x}\|_H < K \text{ and } \|v_{n_{xx}}\|_H < K \qquad (2.24)$$

and thus, by (2.21) and (2.22),

$$\left\|\frac{\partial}{\partial t}\frac{Gu_n}{n}\right\|_H < K \text{ and } \left\|\frac{\partial}{\partial x}\frac{Gu_n}{n}\right\|_H < K \qquad (2.25)$$

From (2.19) and using embedding theorems for anisotropic Sobolev speces (see [4]), we can say that $v_{n_x} \in L^6(J)$ and $\|v_{n_x}\|_{L^6} < K$ which implies

$$v_{n_x}^2 \in L^2(J) \quad \text{and} \quad \|v_{n_x}^2\|_H < K . \qquad (2.26)$$

Thus by (2.23), (2.24) and (2.26), $\left\|\frac{\partial^2}{\partial x^2}\frac{Gu_n}{n}\right\|_H < K$ which, together with (2.25), implies (2.20).

Hence, as $h \in H^{1,2}(J)$, by (2.19), (2.20) and relying on the regularity of the inverse operator (see [4]), we get v_n bounded in $C^{1,2}(J)$ and, passing to a subsequence,

$$v_n \to v \quad \text{in} \quad C^{0,1}(J). \qquad (2.27)$$

Observing that v has a definite sign in J, $\left|\frac{\partial v}{\partial n}\right| \geq \varepsilon > 0$ in $\{0,\pi\} \times [0,2\pi]$ and that v does not depend on t, we conclude by (2.27) that the result holds.

3. A MORE GENERAL RESULT

In theorem 1 we required the existence of the limits

$$\lim_{u\to\pm\infty} \frac{g(t,x,u)}{u}$$

We can state, however, a more general result. Let us define

$$\alpha_{\pm}(t,x) = \liminf_{u\to\pm\infty} \frac{g(t,x,u)}{u}$$

$$\beta_{\pm}(t,x) = \limsup_{u\to\pm\infty} \frac{g(t,x,u)}{u} .$$

Then we have

THEOREM 2 : Let g satisfy the same hypothesis as in theorem 1 (except (2.1) and (2.2)) and suppose that

$$0 \le \alpha_+(t,x) \le \beta_+(t,x) \le 3, \quad 0 \le \alpha_-(t,x) \le \beta_-(t,x) \le 3 \tag{3.1}$$

and

$$\int_{v>0} (3-\beta_+) + \int_{v<0} (3-\beta_-) > 0 \qquad \forall\, v \in N(A+3I),\ v \neq 0 . \tag{3.2}$$

Then problem (0.3) has a solution $u \in U$ for every $h \in H^{1,2}(J) \cap H^1_0(J)$ such that

$$\int_J g_- \sin x < (h, \sin x) < \int_J g_+ \sin x \tag{h_2}$$

where $g_{\pm}(t,x) = \liminf_{u\to\pm\infty} g(t,x,u)$.

Proof : The proof is very similar to the proof of theorem 1 (we even use similar notations to emphasize the analogy) . In fact we can argue in the same way up to (2.7) (including it). There we introduce the following function

$$\chi_v(t,x) = \begin{cases} \dfrac{\chi(t,x)}{v(t,x)} & \text{if } v \neq 0 \\ 0 & \text{if } v = 0 \end{cases}$$

whese $\chi(t,x)$ denotes the weak limit in H of $\frac{Gu_n}{n}$.

Passing to the limit in (2.6) and applying (A+P) to the equation we obtain (analogously to (2.10))

$$Av = -t^* \chi_v v + (1-t^*)Pv \tag{3.3}$$

It is easy to see (writing $\frac{g(t,x,u_n)}{n} = \frac{g(t,x,u_n)}{u_n} v_n$ if $v\neq 0$)

that

$$\alpha_+(t,x) \leq \chi_v(t,x) \leq \beta_+(t,x) \quad \text{a.e. in } \{v>0\} \tag{3.4}$$

$$\alpha_-(t,x) \leq \chi_v(t,x) \leq \beta_-(t,x) \quad \text{a.e. in } \{v<0\} . \tag{3.5}$$

Using these inequalities, lemma 1.1 and (3.3) we can deduce (as in theorem 1) that

$$V \in N(A) \oplus N(A+3I)$$

Writing $v = d_1 \sin x + d_2 \sin 2x$, supposing $d_2 \neq 0$ and arguing as in theorem 1 we derive

$$t^* = 1 \quad \text{and} \quad \chi_v(t,x) = 3 \quad \text{a.e. in } J$$

Then, from (3.3), $v \in N(A+3I)$ and from (3.4) and (3.5)

$$\beta_+(t,x) = 3 \quad \text{a.e. in } \{v>0\}$$

$$\beta_-(t,x) = 3 \quad \text{a.e. in } \{v<0\}$$

which yields a contradiction to (3.2).

Hence $d_2=0$, $v \in N(A)$ and the proof follows in the same way as in theorem 1 after (2.17)

REMARKS 1. Arguing in an analogous way as in Theorem 4 of |1| we can also prove a result similar to that one about "crossing of eigenvalues" using in the adequate step our lemma 3.

2. A generalization of these results involving any pair of consecutive eigenvalues was established by M.N. Nkashama and the author.

R E F E R E N C E S

[1] *H.Berestycki, D.C.De Figueiredo* : 'Double resonance in semilinear elliptic problems'. Comm. in Partial Diff.Eq., 6(1), 91-120 (1981)

[2] *M.N.Nkashama, M.Willem* : 'Periodic solutions of the boundary value problem for the nonlinear heat equation', Bull.Austral.Math.Soc. 29 (1984), 99-110

[3] *M.N.Nkashama, M.Willem* : 'Time-Periodic solutions of Boundary Value Problems for Nonlinear Heat, Telegraph and Beam Equations', Institut de Mathematique Pure et Appliquée, Université Catholique de Louvain, Belgique. Rapport nº54 - October 1984

[4] *O.Vejvoda* and al : 'Partial differential equations : time-periodic solutions', Martinus Nijhoff Publishers, 1982.

GENERALISED RIEMANN INVARIANTS

A. M. Grundland
Department of Mathematics and Statistics
Memorial University of Newfoundland
St. John's, Newfoundland
A1C 5S7, Canada

ABSTRACT. In this paper a new method of constructing solutions for nonlinear and nonelliptic systems of P.D.E.'s and especially nonhomogeneous ones, is presented. A generalization of the Riemann invariant methods for the case of nonhomogeneous systems has been formulated. Classes of solutions being nonlinear superpositions of solutions of the nonhomogeneous system are studied. The necessary and sufficient conditions for the existence of such solutions are discussed and theorems useful for constructing these solutions are given. These theoretical considerations are illustrated by the examples appearing in various branches of mathematical physics. New classes of solutions of the field equation are obtained.

1. Introduction

The present paper is a continuation of previous papers[1,2] concerning the Riemann invariant methods for quasilinear nonhomogeneous systems of P.D.E.'s. The starting point for the following considerations is the algebraization of these systems done previously and the involutivity conditions found for the case of nonlinear superpositions of Riemann waves[3,4].

In the former analysis quasilinear systems with the coefficients dependent only on the unknown functions were studied. The general case will not be considered, namely when the coefficients are functions of dependent and independent variables as well. Existence conditions will be discussed for elementary solutions - called simple states - for nonhomogeneous systems of this type.

In the main part of this paper the problem of superpositions of elementary solutions of nonhomogeneous systems of P.D.E.'s will be investigated. In previous papers[1-4] superpositions of a single state with many Riemann waves (elementary solutions of homogeneous system) were studied. Now the rules of superpositions for many simple states will be presented. For sake of simplicity we conduct our analysis for

S. P. Singh (ed.), Nonlinear Functional Analysis and Its Applications, 253–276.

the case of systems with coefficients dependent only on the unknown functions.

Let us consider a first order nonelliptic quasilinear system of P.D.E.'s with many independent variables

$$\sum_{j=1}^{\ell} \sum_{\mu=1}^{n} a_j^{s\mu} (x, u)\frac{\partial u^j}{\partial x^\mu} = b^s (x, u), \quad s = 1, \dots , m \qquad (1.1)$$

(m is the number of equations) where the matrix $a_j^{s\mu}$ and the vector b^s are given functions of $n + \ell$ variables; $x = (x^1, \dots , x^m) \in \mathbb{R}^n$ are the independent variables and $u = (u^1, \dots , u^\ell) \in \mathbb{R}^\ell$ are unknown functions defined on an open subset $\mathcal{D} \subset \mathbb{R}^n$. According to traditional terminology the Euclidean space $E = \mathbb{R}^n$ (the space of independent variables) is called the physical space and the space $H = \mathbb{R}^\ell$ (the space of values of dependent variables) is called the hodograph space.

For simplicity we assume that all considered functions, maps and manifolds are of the class C^∞. All our considerations are of a local character. In more rigorous formulation there should be used germs of functions (mappings, manifolds) instead of functions (mappings, manifolds).

We introduce now the basic notions and definitions used in the course of our analysis. We shall deal only with the properties of the system (1.1) connected with the motion of a simple integral element[3,5,6].

Definition 1: A matrix L_μ^j defined on the domain $W \subset E \times H$ is called a simple integral element if the following conditions

$$\text{(i)} \quad a_j^{s\mu}(x_o, u_o)\, L_\mu^j = b^s(x_o, u_o)$$
$$\text{(ii)} \quad \operatorname{rank} \| L_\mu^j (x_o, u_o)\| = 1 \qquad (1.2)$$

are satisfied at the given point $(x_o, u_o) \in W$.

The matrix $L = \left(\frac{\partial u^j}{\partial x^\mu}\right)$ is a matrix of the tangent mapping

$$du(x): \ E \to T_u H, \text{ where } E \ni (\delta x^\mu) \to (\delta u^j) = \left(\frac{\partial u^j}{\partial x^\mu}\, \delta x^\mu\right) \in T_u H$$

So $du(x)$ determines an element of the linear space $L(E, T_u H)$ which is identified with the tensor product $T_u H \otimes E^*$ (where E^* is the dual to E). In order to determine a simple integral element L we seek

$\gamma \in T_u H$ and $\lambda \in E^*$ such that

$$\frac{\partial u^j}{\partial x^\mu} = \gamma^j(x, u)\ \lambda_\mu\ (x, u) \tag{1.3}$$

where

$$a_j^{s\mu}\ \gamma^j \lambda_\mu = b^s \tag{1.4}$$

The necessary and sufficient condition for the existence of a nonzero solution γ of Eqs (1.4) is

$$\text{rank}\ |\ a_j^{s\mu}\ \lambda_\mu,\ b^s| = \text{rank}\ |\ a_j^{s\mu}\ \lambda_\mu|\ . \tag{1.5}$$

If the covector $\lambda \epsilon E^*$ satisfies the relation (1.5), then there exists a vector $\gamma \epsilon T_u H$ satisfying relation (1.4) . Therefore, the vector γ is the function of the variable (x, u, λ) i.e. $E \times H \times E^* \ni (x, u, \lambda) \to \gamma(x, u, \lambda) \in T_u H$.

Eq: (1.3) can also be written as the differential forms

$$du^j\ (x) = \gamma^j(x, u) \otimes \lambda\ (x, u),\ (\lambda = \lambda_\mu\ dx^\mu) \tag{1.6}$$

the problem being to find functions $u(x)$ whose differential forms are given by (1.6). Suppose that the simple elements $\gamma^j \lambda_\mu$ are continuously differentiable in some domain $\mathcal{D}$ of the variables x and u. The Eqs. (1.3) or (1.6) are said to be completely integrable if for any set of values x_o and u_o in $\mathcal{D}$ for which the functions γ^j and λ_μ are analytic there exists one and only one set of ℓ functions $u(x)$ which satisfy Eq. (1.3) and take on the initial values

$$u(x_o) = u_o \quad \text{for} \quad x = x_o\ .$$

It was proved[3] that Eqs. (1.3) are completely integrable in $\mathcal{D}$ if and only if the compatibility conditions (i.e. the symmetry of second derivatives, Schwarz Lemma)

$$\lambda \otimes d_x\lambda + d_x\gamma \otimes \lambda + \gamma \otimes \lambda \wedge \lambda,_\gamma = 0 \text{ modulo } (1.6) \tag{1.7}$$

where:

$$d_x\lambda = \frac{\partial \lambda_\mu}{\partial x^\nu}\ dx^\nu \wedge dx^\mu,\ d_x\gamma = \frac{\partial \nu}{\partial x^\nu} \otimes dx^\nu,\ \lambda,_\gamma = \gamma^i\ \frac{\partial \lambda}{\partial u^i} \tag{1.8}$$

are satisfied identically in the u's and x's. It was also proved[7] that for any $(x_o, u_o) \in \mathcal{D}$ there exists a neighborhood $|x - x_o| < r$ of x_o in which (1.3) determines a unique solution $u(x)$ of the class C^2 such that $u(x_o) = u_o$. The determination of $u(x)$ therefore reduces to solving the system of O.D.E.'s

$$\frac{du^j}{dt} = \gamma^j(u, x(t)) \sum_{\mu=1}^{n} \lambda_\mu(u, x(t)) \frac{dx^\mu}{dt} \tag{1.9}$$

with the initial condition

$$u(o) = u_o \quad \text{and} \quad x(o) = x_o \tag{1.10}$$

Here $x = x(t) \in C^2$ is a line joining the point x_o and the point x and therefore $u = u^i(x(t))$. Thus, a solution $u(x)$ of a completely integrable system (1.3) depends only on the $\ell + n$ arbitrary constants $u_o^1, \ldots, u_o^\ell$ and $x_o^1, \ldots, x_o^n$.

Let us now consider a first order overdetermined linear homogeneous system of P.D.E.'s of the form

$$L_s v: \sum_{\alpha=0}^{n} B^{s\alpha}(y) \frac{\partial v}{\partial y^\alpha} = 0, \; y := (x^\mu, R) \in E \times \mathbb{R}^1. \tag{1.11}$$

$$s \in \{1, \ldots, m\}$$

and

$$B^{s\mu}(y) := \sum_{j=1}^{\ell} a_j^{s\mu}(f(R), x) \frac{df^j}{dR}, \tag{1.12}$$

$$B^{so}(y) := b^s(f(R)), x), \mu \in \{1, \ldots, n\}$$

with one unknown function $v(y)$. In this paper the coefficients $B^{s\alpha}$ are assumed to be sufficiently smooth functions of y.

The algorithm for investigating the compatibility conditions for the system of the type (1.11) is well known[8,9]. It reduces to successively forming the so-called Poisson brackets, i.e.

$$[L_s, L_t] := L_s L_t - L_t L_s = \sum_{\alpha=0}^{n} B^{st\alpha} \frac{\partial}{\partial y^\alpha} \text{ for } s,t \in \{1, \ldots, m\} \tag{1.13}$$

where

$$B^{st\alpha} = \sum_{\beta=0}^{n} \left(B^{s\beta} \frac{\partial B^{t\alpha}}{\partial y^\beta} - B^{t\beta} \frac{\partial B^{s\alpha}}{\partial y^\beta} \right)$$

$[L_s, L_t]$ is often called the commutator of the linear operators L_s and L_t .

If the function

$$v = v(y) \in C^2(\mathbb{R}^{n+1}) \tag{1.14}$$

is a solution of Eqs. (1.11), then it also satisfies the first order linear homogeneous

$$[L_s, L_t] = 0 \quad \text{for} \quad s, t \in \{1, \dots, m\} \tag{1.15}$$

By adjoining the Eqs. (1.15) for all $s, t = 1, \dots, m$ to the system (1.11) we obtain a first order system of equations of the same type as the original system - the so-called extended system. This sequence of operations is often called a prolongation.
After a finite number of prolongations a linear system is obtained for which the adjoining of Poisson brackets yields no new equations, i.e. the commutators of the differential operators of the system are linear combinations of these operators. Such systems are called complete. Thus by definition, the system (1.11) is called complete if

$$[L_s, L_t] = \sum_{\alpha=0}^{n} C^\alpha_{st}(y)\, L_\alpha \quad \text{for } s, t \in \{1, \dots, m\} \tag{1.16}$$

Suppose that the rank of the matrix $\| B^{s\alpha}(y) \|$ in the domain $\mathcal{D} \subset E \times \mathbb{R}^1$ is equal to m . For a complete system (1.11) there are two possible cases:

(i) If $m = n + 1$ then the system (1.11) admits only the trival solution $v = \text{const}$.

(ii) If $m < n + 1$ then it is possible to show[9] that the system (1.11) by means of a change of variables reduces to a singular linear homogeneous equation for a single unknown function of $n - m + 2$ arguments $z_1, \dots, z_{n-m+2}$. Hence, the general solution of a complete system (1.11) in this case depends on one arbitrary function of $n + 1 - m$ arguments.

Thus, the investigation of the compatibility of the system (1.11) consists in extending it to a complete system and in computing the rank of the matrix of coefficients of the complete system.

An interesting problem is the following: Find the class of exact solutions of the form

$$u(x) = f(R(x)) \tag{1.17}$$

for which the functions $x \to R(x) \in \mathbb{R}^1$ and $R \to f(R) \in H$ satisfy identically the equations (1.1) i.e.

$$a_j^{s\mu}(f(R), x) \frac{df^j(R)}{dR} \frac{\partial R}{\partial x^\mu} = b^s(f(R), x), \; s \in \{1, \dots, m\} \tag{1.18}$$

Assume that the matrix

$$\| c_j^s \; (f(R), x) \| := \| a_j^{s\mu} \; (f(R), x) \frac{\partial R}{\partial x^\mu} \|$$

is nonsingular in the neighbourhood of $(x_o, u_o) \in E \times H$. Defined matrix $\| C_s^j \; (f(R), x) \|$ to be inverse matrix (i.e. $C_s^j c_k^s = \delta_k^j$). Let $R \to f(R) \in H$ be a curve of the system

$$\frac{df^j}{dR} = D^j(f(R), x) \text{ where}$$

$$D^j(f(R), x) = \sum_{s=1}^{m} C_s^j(f(R), x) \; b^s(f(R), x) \tag{1.19}$$

Note that if $\frac{\partial R}{\partial x^\mu}$ can be expressed as a function of $R(x)$ and if the coefficients D^j do not depend on x then the problem (1.19) reduces to O.D.E.'s for the function f, i.e.

$$\frac{df^j}{dR} = D^j(f, R). \tag{1.20}$$

The following theorems can now be proved:

Theorem 1: Let $R \to f(R)$ be an integral curve of the class C^1 in the hodograph space H given by condition (1.19). Let us assume, that the linear homogeneous system of equations (1.11) has a solution $v = v(x, R)$.
If the system given in the implicit form:

$$\begin{cases} u = f(R) & (1.21a) \\ v(x, R) = 0 \quad \text{where} \quad \dfrac{\partial v}{\partial R} \neq 0 & (1.21b) \end{cases}$$

can be uniquely solved with respect to the variables $u = u(x)$ and

$R = R(x)$, then the function $u(x)$ is an exact solution of the nonhomogeneous system (1.1).

Proof: From the condition for local solvability of Eq. (1.21) the equation $v(x, R) = 0$ can be solved with respect to the variable R if

$$\frac{\partial v}{\partial x^\mu} = -\frac{\partial v}{\partial R}\frac{\partial R}{\partial x^\mu} .$$

Inserting the above equation into Eq. (1.11) implies

$$B^{s\mu}(R, x)\frac{\partial R}{\partial x^\mu} = B^{so}(R, x) . \tag{1.22}$$

Hence Eqs (1.18) hold. This completes the proof, since in the presence of Eq. (1.21a) we have

$$\frac{\partial u^j}{\partial x^\mu} = \frac{df^j}{dR}\frac{\partial R}{\partial x^\mu} \tag{1.23}$$

Q.E.D.

Remark: If the number of equations are equal to the number of independent variables (i.e. $m = n$) and matrix $\| B^{s\mu}(x, R)\|$ is an invertible one then Eq. (1.22) can be written in the form

$$\frac{\partial R}{\partial x^\mu} = \lambda_\mu(x, R), \text{ where } \lambda_\mu : = \sum_{s=1}^{m} B^{so}B_{s\mu}, \; B_{s\mu}B^{s\rho} = \delta^\rho_\mu . \tag{1.24}$$

Eqs. (1.11) and (1.23) imply

$$\frac{dR}{dt} = \sum_{\mu=1}^{m} \lambda_\mu(x(t), R)\frac{dx_\mu}{dt} \tag{1.25}$$

with the initial condition $R(o) = R^o$ and $x(o) = x_o$. From Eq. (1.24) we have

$$B^{s\mu}(x, R)\,\lambda_\mu(x, R) = B^{so}(x, R) \tag{1.26}$$

which gives Eq. (1.9). The solution of the completely integrable system (1.24) contains arbitrary integration constants. Thus in this case both methods are equivalent.

Theorem 2: Let $u = f(R)$ be a curve of the class C^1 in the hodograph space $\mathcal{H}$. Let $\lambda^o = (\lambda^o_\mu) \in E^*$ be a constant covector and let $\xi(R) \in \mathbb{R}^1$ be a variable such that

$$\text{(i)}\quad \xi(R)\, a_j^{s\mu}(x, f(R))\, \frac{df^j}{dR}\, \lambda_\mu^o = b^s(x, f(R)) \tag{1.27}$$

If $\psi(\Omega)$ is the solution of the O.D.E. of the form

$$\text{(ii)}\quad \frac{d\psi}{d\Omega} = \xi(\psi)\ , \quad \text{where} \quad \Omega = \lambda_\mu^o\, x^\mu \tag{1.28}$$

then

$$\begin{cases} u(x) = f(R(x)) \\ R(x) = \psi(\lambda_\mu^o\, x^\mu) \end{cases} \tag{1.29}$$

is a solution of the nonhomogeneous system (1.1).

This solution is called a simple state[1].

Proof: Differentiating (1.29) we get

$$\frac{\partial u^j}{\partial x^\mu} = \frac{df^j}{dR}\, \frac{d\psi}{d\Omega}\, \lambda_\mu^o = \frac{df^j}{dR}\, \xi(R)\, \lambda_\mu^o\ .$$

Hence Eq. (1.1) implies

$$a_j^{s\mu}(x, u)\frac{\partial u^j}{\partial x^\mu} = a_j^{s\mu}(x, f(R))\, \frac{df^j}{dR}\, \xi(R)\lambda_\mu^o = b^s(x, f(R)) = b^s(x,u)$$

Q.E.D.

2. COMPATIBILITY CONDITIONS FOR THE CASE OF MANY SIMPLE STATES

We consider now an automorphic nonhomogeneous system (1.1). Its coefficients depend only on ℓ unknown functions $u^1, \ldots, u^\ell$. So we have here

$$\sum_{j=1}^{\ell} \sum_{\mu=1}^{m} a_j^{s\mu}(u^1, \ldots, u^\ell)\, \frac{\partial u^j}{\partial x^\mu} = b^s(u^1, \ldots, u^\ell)\quad s \in \{1, \ldots, m\} \tag{2.1}$$

In order to extend the class of solutions obtained (1.6) let us consider the solutions of the system (2.1) in which the tangent mapping $du(x)\colon E \to T_u H$ is of the rank 2. In other words we are looking for solutions of the form

$$du(x) = \gamma_1 \otimes \lambda^1 + \gamma_2 \otimes \lambda^2\ , \quad \lambda^1 \wedge \lambda^2 \neq 0,\ \lambda^s = \lambda_\mu^s\, dx^\mu \in E\ ,$$
$$s = 1, 2, \tag{2.2}$$

where γ_1 and γ_2 are linearly independent vectors and the following conditions should be satisfied.

$$a_j^{s\mu}(\gamma_i^j \lambda_\mu + \gamma_2^j \lambda_\mu^2) = b^s, \ s \in \{1, \dots, m\} \tag{2.3}$$

These conditions (2.3) constitute a system of m equations for 2ℓ unknown functions γ_1 and γ_2. So according to the Kronecker-Cappela theorem the necessary and sufficient condition for existence of non-zero solutions γ_s of the Eqs. (2.3) is

$$\operatorname{rank}|a_j^{s\mu}\lambda_\mu^1, a_j^{s\mu}\lambda_\mu^2, b^s| = \operatorname{rank}|a_j^{s\mu}\lambda_\mu^1, a_j^{s\mu}\lambda_\mu^2| \tag{2.4}$$

while the condition for the homogeneous system is given by

$$\operatorname{rank}|a_j^{s\mu}\lambda_\mu^1, a_j^{s\mu}\lambda_\mu^2| < \ell, \ b^s \equiv 0 \tag{2.5}$$

In general we look for solutions of the rank k, for which the total derivative $du(x)$ is the sum of homogeneous and nonhomogeneous simple elements, i.e.

$$du(x) = \sum_{a=1}^{p} \xi^a \gamma_a \otimes \lambda^a + \sum_{t=1}^{q} \zeta^t \gamma_{ot} \otimes \lambda^{ot}, \ p + q \le (\ell, m) \tag{2.6}$$

where

$$a_j^{s\mu}\Big(\sum_{a=1}^{p} \xi^a \gamma_a \otimes \lambda^a\Big) = 0, \ a_j^{s\mu}\Big(\sum_{t=1}^{q} \zeta^t \gamma_{ot} \otimes \lambda^{ot}\Big) = b^s$$

$$\lambda^{r_1} \wedge \lambda^{r_2} \wedge \lambda^{r_3} \neq 0 \text{ for } r_1 < r_2 < r_3$$

and $\gamma_{01}, \dots, \gamma_{0q}, \gamma_1, \dots, \gamma_p$ are linearly independent vectors.

The solutions of the homogeneous system (2.6) (when $\zeta^t = 0$) representing nonlinear superposition of Riemann waves i.e. double waves and in general k-waves were studied by many authors, e.g. M. Burnat[11], A. Jeffrey[12], Z. Peradzynski[13], B. Rozdestvenskii and N. Janeko[3]. The class of solutions representing nonlinear superposition of Riemann waves in the nonhomogeneous system (2.6) (when $q = 1$) were also studied - by the author of this paper and the involutivity conditions for this case were studied in detail in the papers[1-4,14].

Now we would like to present a new class of solutions which are the nonlinear superpositions of the elementary solutions of the type (1.28). We propose the form of solutions of the system (2.1) for which the total derivative $du(x)$ is given by

$$du(x) = \sum_{t=1}^{k} \zeta^t \gamma_t \otimes \lambda^t, \quad \sum_{t=1}^{k} \zeta^t = 1, \ k \leq \ell$$
$$a_j^{s\mu} \gamma_t^j \lambda_\mu^t = b^s \quad t \in \{1, \ldots , k\} \tag{2.7}$$

where the quantities $\zeta^t \not\equiv 0$ are treated as variables dependent on x. Suppose for example, that the covectors $\gamma^1, \ldots , \gamma^k$ are constant one forms such that

$$\lambda^{t_1} \wedge \lambda^{t_2} \wedge \lambda^{t_3} \neq 0 \quad \text{for} \quad t_1 < t_1 < t_3 . \tag{2.8}$$

The existence of solutions of type (2.7) needs some conditions called involutivity conditions[15]. Namely, closing (2.7) by exterior differentiation we obtain

$$\sum_{t=1}^{k} \{\gamma_t \otimes d\zeta^t \wedge \lambda^t + \zeta^t \, d\gamma_t \wedge \lambda^t\} = 0, \quad \sum_{t=1}^{k} d\zeta^t = 0 \tag{2.9}$$

which should be satisfied modulo (2.7). Using (2.7) we have

$$d\gamma_t = \sum_{i=1}^{\ell} \frac{\partial \gamma_t}{\partial u^i} du^i = \sum_{s=1}^{k} \zeta^s \gamma_{t,\gamma_s} \otimes \lambda^s$$

where we denote by $x,y := \sum_{i=1}^{\ell} y^i \frac{\partial X}{\partial u^i}$ which is the directional derivative.
Inserting the above relations into the prolonged system (2.9) we get

$$\sum_{t=1}^{k} \{\gamma_t \otimes d\zeta^t \wedge \lambda^t + \tfrac{1}{2} \sum_{s=1}^{k} \zeta^t \zeta^s [\gamma_t, \gamma_s] \otimes \lambda^t \wedge \lambda^s\} = 0 ,$$
$$\sum_{t=1}^{k} d\zeta^t = 0 \tag{2.10}$$

where $[\gamma_t , \gamma_s]$ denotes the commutator of the fields γ_t , γ_s , i.e.

$$[\gamma_t, \gamma_s] = \sum_{i=1}^{\ell} \left(\gamma_s^i \frac{\partial \gamma_t}{\partial u^i} - \gamma_t^i \frac{\partial \gamma_s}{\partial u^i}\right)$$

Let A be an annihilator of the fields $\gamma_1, \ldots , \gamma_k$, i.e.

$$< w_1, \gamma_t > = 0 \text{ for } w_1 \in A = \text{An}\{\gamma_1, \ldots, \gamma_k\},$$

$$t \in \{1, \ldots, k\}$$

where by the paranthesis $< w_1, \gamma_t >$ we denote the contraction of the covector $w_1 \in T^*_u H$ with the vector $\gamma_t \in T_u H$. Multiplying scalarly the Eq. (2.10) by the one-form $w_1 \in A$ we obtain

$$\sum_{s,t=1}^{\ell} \zeta^t \zeta^s < w_1, [\gamma_t, \gamma_s] > \lambda^t \wedge \lambda^s = 0 \tag{2.11}$$

We are looking for conditions of integrability such that the system (2.10) does not lead to algebraic constraints on the coefficients ζ^s. This requires that the coefficient of successive powers of ζ^s in the Eq. (2.11) vanish. So by virtue of Eq. (2.8) and of the property of the annihilator A, the above equations are equivalent to the following

$$[\gamma_t, \gamma_s] = \sum_{p=1}^{k} C^p_{st} \gamma_p \tag{2.12}$$

The above means that the commutator of the fields γ_t and γ_s is a linear combination (not necessarily with constant coefficients $C^p_{st}(u)$) of these fields. So the assumptions of the Frobenius theorem[7] are satisfied. Thus, at every point u_o of the hodograph space H there exists a tangent surface denoted by $G \subset H$, spanned by the fields $\gamma_1, \ldots, \gamma_k$ passing through the point u_o.

Let now w_2 be the covectors in the space $T^*_u H$ such that

$$< \overset{r}{w}_2, \gamma_t > = \overset{r}{\gamma}_t \qquad r, t \in \{1, \ldots, k\}$$

$$< \overset{r}{w}_2, \gamma_q > = 0 \qquad q \in \{k+1, \ldots, \ell\}$$

Inserting (2.12) into the Eq. (2.10) and multiplying by $\overset{r}{w}_2$ we get

$$d\zeta^r \wedge \lambda^r + \tfrac{1}{2} \sum_{t,s=1}^{k} \zeta^t \zeta^s C^p_{st} < w_2, \gamma_p > \lambda^t \wedge \lambda^s = 0 \tag{2.13}$$

We require that the system (2.12) form an involutive system. Then according to[15,16] there can not be any algebraic restrictions on the coefficients ζ^s. Thus multiplied externally by λ^r, the system (2.13) takes the form

$$\sum_{t,s=1}^{k} \zeta^t \zeta^s C^p_{st} < \overset{r}{w}_2, \gamma_p > \lambda^t \wedge \lambda^s \wedge \lambda^r = 0 \qquad (2.14)$$

So we require that the coefficients of the respective powers of ζ^s in the Eq. (2.14) disappear. So we have $C^r_{ts} = 0$ for $t \neq s \neq r \in \{1, \ldots, k\}$. Hence from Eq. (2.12) we obtain that the commutators for all vector fields γ_t, γ_s are linear combinations of these fields, i.e.

$$[\gamma_t, \gamma_s] = C^t_{ts}\gamma_t + C^s_{st}\gamma_s \quad s \neq t \in \{1, \ldots, k\} \qquad (2.15)$$

(no summation convention). This proves that the solutions of the postulated form (2.7) can be written in Riemann invariants. Inserting (2.15) into the Eq. (2.10) and together with the assumption that $\gamma_1, \ldots, \gamma_k$ are linearly independent, leads to the Pfaffian system

$$d\zeta^t \wedge \lambda^t + C^t_{ts}\zeta^s\zeta^t \lambda^t \wedge \lambda^s = 0, \quad \sum_{t=1}^{k} d\zeta^t = 0, \; s \neq t. \qquad (2.16)$$

Using the Cartan Lemma we can prove, that there exists functions μ^t such that

$$d\zeta^t = \mu^t\lambda^t + C^t_{ts}\zeta^t\zeta^s\lambda^s, \quad \sum_{t=1}^{k} d\zeta^t = 0, \; s \neq t \in \{1,\ldots,k\} \qquad (2.17)$$

But adding Eqs. (2.17) and taking into account the assumption that the covectors $\lambda^1, \ldots, \lambda^k$ are linearly independent we have

$$\mu^t = \zeta^s\zeta^t C^s_{st}, \; s \neq t \qquad (2.18)$$

Inserting Eq. (2.18) into Eq. (2.17) we get

$$d\zeta^t = \zeta^t\zeta^s\{C^s_{st}\lambda^t + C^t_{ts}\lambda^s\} \quad s \neq t \qquad (2.19)$$

Closing Eq. (2.19) by exterior differentiation we obtain the equations

$$dC^t_{ts} \wedge \lambda^t + dC^s_{st} \wedge \lambda^s = 0 \qquad (2.20)$$

Which should be satisfied modulo Eq. (2.7). Using Eq. (2.7) we have

$$dC^t_{ts} = C^t_{ts,u_i} du^i = \sum_{p=1}^{k} \zeta^p C^t_{ts,\gamma_p} \lambda^p \quad s \in \{1, \ldots, k\} \tag{2.21}$$

We insert Eq. (2.21) into Eq. (2.20) and then multiply the result externally by λ^s. Since the covectors $\lambda^p \wedge \lambda^t \wedge \lambda^s$ are linearly independent (2.8) and there cannot be any algebraic restrictions on the coefficients ζ^s we obtain

$$C^t_{ts,\gamma_s} = 0 \quad s \neq t \tag{2.22}$$

It can be shown according to[15,17] that the conditions found (2.15) and (2.22) guarantee the existence of solutions of the system (2.7). They demand that the set of solutions of the system (2.7) depends only on k arbitrary analytical functions of one variable.

3. NONLINEAR SUPERPOSITIONS OF SIMPLE STATES IN QUASILINEAR SYSTEMS

We are looking for a class of solutions of the Eqs. (2.1) being the nonlinear superpositions of the elementary solutions (1.29) which can be written in Riemann invariants. This method introduces a set of new dependent variables (called latter Riemann invariants) which remain constant along certain noncharacteristic curves of the system (2.1). The methods presented here reduce the number of dependent variables for some problems, so that the process of solving the original system of equations is simplified. Let the number of equations in (2.1) be equal to the number of dependent variables (i.e. $m = \ell$). Suppose that $\lambda^1, \ldots, \lambda^k \in E^*$ are constant linear independent covectors such that the matrices

$$\| c^{st}_j(u) \| := \| \sum_{\mu=1}^{n} a^{s\mu}_j(u)\, \lambda^t_\mu \| \, , \quad t \in \{1, \ldots, k\} \tag{3.1}$$

are nonsingular ones in the neighbourhood of a point $u_0 \in H$. Let matrices $\| c^j_{st}(u) \|$ be the inverse matrices. Thus we get the following.

Theorem 3: Let us assume that $\xi^t \equiv 0$, $t = 1, \ldots, k$ are arbitrary functions of class C^1 of the variable $R = (R^1, \ldots, R^k)$ and

(i) $\mathbb{R}^k \ni R \to f(R) \in H$ is the integral surface of P.D.E.'s

$$\frac{\partial f^j}{\partial R^t} = \frac{1}{k\xi^t} \sum_{s=1}^{\ell} c^j_{st}(f)\, b^s(f) \tag{3.2}$$

satisfying an initial condition $f(0) = u_o$.

(ii) $\mathbb{R}^1 \ni \Omega^t \to R^t(\Omega^t) \in \mathbb{R}^1$ are solutions of the ordinary differential equations

$$\frac{dR^t}{d\Omega^t} = \xi^t(R^1, \ldots, R^k), \text{ where } \Omega^t = \lambda^t_\mu u^\mu \tag{3.3}$$

With the initial conditions $R^t(0) = 0$.
Then the function

$$u(x) := f(R^1(\lambda^1_\mu x^\mu), \ldots, R^k(\lambda^k_\mu x^\mu)) \tag{3.4}$$

is the exact solution of the system (2.1) satisfying the initial condition $u(o) = u_o$.

Proof: Indeed, inserting (3.4) into Eq. (2.1) and using assumptions (i) and (ii) we get

$$\sum_{j=1}^{\ell} \sum_{\mu=1}^{n} a^{s\mu}_j(u) \frac{\partial u^j}{\partial x^\mu} = \sum_{j=1}^{\ell} \sum_{\mu=1}^{n} \sum_{t=1}^{k} a^{s\mu}_j (f(R)) \frac{\partial f^j(R)}{\partial R^t} \cdot$$

$$\cdot \frac{dR^t(\lambda^t_\mu x^\mu)}{d\Omega^t} \lambda^t_\mu = \sum_{j=1}^{\ell} \sum_{t=1}^{k} C^{st}_j(f(R)) \frac{\partial f^j(R)}{\partial R^t} \xi^t(R)$$

$$= \sum_{t=1}^{k} \frac{1}{k\xi^t(R)} b^s(f(R))\xi^t(R) = b^s(f(R)) = b^s(u) .$$

Let us notice that Riemann invariants in this case are the following functions

$$E \ni x \to R^t(\lambda^t_\mu x^\mu) \in \mathbb{R}^1 . \tag{3.5}$$

Q.E.D.

4. THE EXAMPLES OF APPLICATIONS

Let us consider the equation of the motion for the scalar ϕ^6 - theory

$$g^{\mu\nu} \frac{\partial^2\phi}{\partial x^\mu \partial x^\nu} + b(\phi) = 0 \ , \text{ where } b(\phi) = 6A_6\phi^5 + 4A_4\phi^3 + 3A_3\phi^2 + 2A_2\phi + A_1 \qquad (4.1)$$

where ϕ is the unknown function (defined on an open subset $D \subset E$), $g^{\mu\nu}$ is the metric tensor in Minkowski space and A_1, A_2, A_3, A_4, A_6 are the arbitrary constants.

With our method, we reduce Eq. (4.1) to the first-order P.D.E.'s

$$g^{\mu\nu} \frac{\partial\phi_\mu}{\partial x^\nu} + b(\phi) = 0, \ \frac{\partial\phi_\mu}{\partial x^\nu} - \frac{\partial\phi_\mu}{\partial x^\mu} = 0 \ , \ \frac{\partial\phi}{\partial x^\mu} = \phi_\mu \ , \qquad (\mu,\nu = 0, 1, 2, 3) \qquad (4.2)$$

The above system has eleven equations for five unknown functions ϕ_μ and ϕ. The elementary solution

$$u^j(x) = f^j(R(x)) \equiv (\phi_\mu (R(x)), \phi(R(x))), \ (j = 1, \dots , 5)$$

of the nonhomogeneous system (4.2) is determined (according to the equation (1.18)) in the form

$$g^{\mu\nu} \dot{\phi}_\mu \frac{\partial R}{\partial x^\nu} + b(\phi) = 0, \ \dot{\phi}_\mu \frac{\partial R}{\partial x^\nu} - \dot{\phi}_\nu \frac{\partial R}{\partial x^\mu} = 0 \ , \ \dot{\phi} \frac{\partial R}{\partial x^\mu} = \phi_\mu \qquad (4.3)$$

where we have denoted $\frac{d\phi}{dR} =: \dot{\phi}$. In this case the linear homogeneous system (1.11) take the form

$$g^{\mu\nu} \dot{\phi}_\mu \frac{\partial v}{\partial x^\mu} - b(\phi) \frac{\partial v}{\partial R} = 0$$

$$\dot{\phi}_\mu \frac{\partial v}{\partial x^\nu} - \phi_\nu \frac{\partial v}{\partial x^\mu} = 0 \qquad (4.4)$$

$$\dot{\phi} \frac{\partial v}{\partial x^\mu} + \phi_\mu \frac{\partial v}{\partial R} = 0 \ .$$

The rank of the coefficients matrix of the system (4.4) is equal to seven. So we get

$$\dot{\phi}^{-1} g^{\mu\mu} \phi_\mu \dot{\phi}_\mu - b(\phi) = 0 , \quad \dot{\phi}_\nu \phi_\mu - \phi_\nu \dot{\phi}_\mu = 0 \tag{4.5}$$

From these equations we find out

$$\phi_0 = \varepsilon \left[\sum_{k=0}^{6} a_k \phi^k \right]^{\frac{1}{2}} =: \psi(\phi), \quad \varepsilon = \pm 1,$$

$$\phi_s = c_s \psi(\phi), \quad (s = 1, 2, 3) \tag{4.6}$$

where $a_o := \frac{A_o}{2g}$, c_s are arbitrary integration constants and $a_p = \frac{A_p}{2g}$ (p = 1, ... , 6), $g = 1 - \sum_{s=1}^{3} c_s^2 \neq 0$. Then Eqs. (4.4) reduce to the system of four equations with one unknown function v of five arguments (x^μ , R).

$$\dot{\phi}(R) \frac{\partial v}{\partial x^0} + \psi(\phi(R)) \frac{\partial v}{\partial R} = 0 ,$$

$$\dot{\phi}(R) \frac{\partial v}{\partial x^s} + c_s \psi(\phi(R)) \frac{\partial v}{\partial R} = 0, \quad (s = 1, 2, 3) \tag{4.7}$$

where $\dot{\phi}(R)$ is an arbitrary function. Let us investigate now the existence conditions for the system (4.7). According to the previous considerations for the linear operators

$$L_0 = \dot{\phi}(R) \frac{\partial}{\partial x^0} + \psi(\phi) \frac{\partial}{\partial R}$$

$$L_s = \dot{\phi}(R) \frac{\partial}{\partial x^s} + c_s \psi(\phi) \frac{\partial}{\partial R} \tag{4.8}$$

we form the commutators (1.13)

$$[L_0, L_s] = \psi \ddot{\phi} \left(\frac{\partial}{\partial x^s} - c_s \frac{\partial}{\partial x^0} \right) = \frac{\psi \ddot{\phi}}{\dot{\phi}} (L_s - c_s L_0)$$

$$[L_s, L_r] = 0 \tag{4.9}$$

The above equations satisfy conditions (1.16), so the system (4.7) is an involutive one. In this case (since $m < n+1$) the system (4.7) by means of a change of variables reduces to one linear homogeneous equation for one unknown function v of one variable. Hence the integral trajectories of the fields L_0 and L_s are given respectively by

$$\frac{dx^0}{dr} = \dot{\phi}(R) \qquad \frac{dx^0}{ds} = 0$$

$$\frac{dx^s}{dr} = 0 \qquad \frac{dx^s}{ds} = \dot{\phi}(R) \tag{4.10}$$

$$\frac{dR}{dr} = \psi(\phi(R)) \qquad \frac{dR}{ds} = c_s \ \psi(\phi(R))$$

From these equations we find the general integral of the system (4.7) which can be expressed in the implicit form

$$v(x^\mu, R) = v(\lambda_\mu x^\mu - \kappa(R)),$$
$$\lambda_\mu x^\mu = x^0 + c_1 x^1 + c_2 x^2 + c_3 x^3 \tag{4.11}$$

where

$$\kappa(R) = \int_0^R \frac{\dot{\phi}(r)dr}{\psi(\phi(r))} \tag{4.12}$$

Inserting (4.6) into the above equation we obtain an elliptic integral

$$\kappa(\phi) = \varepsilon \int_0^\phi \frac{dt}{[\sum_{k=0}^{6} A_k \ t^k]^{\frac{1}{2}}} \tag{4.13}$$

Thus according to the Theorem 1, the relation

$$v(\lambda_\mu x^\mu - \kappa(R)) = 0 \tag{4.14}$$

determines the Riemann invariant R of system (4.2), so Eq. (4.14) can be uniquely solved with respect to the invariant R .

$$\lambda_\mu x^\mu - \kappa(R) = -R_0 = \text{const. where } R_0 = -v^{-1}(0)$$

So we get

$$R = \kappa^{-1}(\lambda_\mu x^\mu + R_0) \tag{4.15}$$

where by virtue of Eq. (4.13) the function κ^{-1} can be expressed by the Weierstrass P-function[18], satisfying the equation

$$P'^2 = 4P^3 - g_2 P - g_3 \tag{4.16}$$

where the invariants

$$\begin{aligned} g_2(w_1, w_2) &= 60 \sum_{m,m'}{}' (m w_1 + m' w_2)^{-4} \\ g_3(w_1, w_2) &= 140 \sum_{m,m'}{}' (m w_1 + m' w_2)^{-6} \end{aligned} \tag{4.17}$$

are homogeneous functions of the periods w_1 and w_2 of the -4th and -6th order.
Integral (4.13) can be reduced to an elliptic one by a change of variables. There are essentially two different cases to consider.

$$\text{(i)} \quad \phi = \pm \left(y - \frac{A_4}{4A_6} \right)^{\frac{1}{2}} \tag{4.18}$$

or

$$\text{(ii)} \quad \phi = \pm \left(y - \frac{3A_6}{4A_4 A_0} \right)^{-\frac{1}{2}} \tag{4.19}$$

(i): In the first case the values of the invariants g_2 and g_3 are given by

$$g_2(w_1, w_2) = a_4 - 4a_1 a_3 + 3a_3 =: q \tag{4.20}$$

$$g_3(w_1, w_2) = \det \begin{vmatrix} 1 & a_1 & a_2 \\ a_1 & a_2 & a_3 \\ a_2 & a_3 & a_4 \end{vmatrix} =: p$$

where

$$a_1 = \frac{A_4}{2\sqrt{2A_6}} , \quad a_2 = \frac{A_3}{3} , \quad a_3 = A_2\sqrt{\frac{A_6}{2}} \quad a_4 = 4A_6A_0$$

The existence of periods w_1 and w_2 is guaranteed by the following fact[19]. For every real number c, the equation

$$J(\tau) := \frac{g_2^3}{g_2^3 - 27g_3^2} = c, \text{ where } \tau := w_1/w_2 \tag{4.21}$$

(where $J(\tau)$ is the module function of τ) possesses exactly one root in the fundamental region of the modular group. Thus, taking $c := \frac{q^3}{q^3 - 27p^2}$ we can obtain the ratio τ. If $g_2 = q \neq 0$ then from the homogenity of the function g_2 given by (4.17) we can determine $w_1^4 = q^{-1}g_2(1, \tau)$ and, when $g_2 = q = 0$ we have $w_1^6 = p^{-1}g_3(1, \tau)$. When w_1 is found, then w_2 is determined from the formula (4.21). The periods w_1 and w_2 calculated in this way satisfy Eq. (4.20). It is well known[18] that one can uniquely determine a quantity R_0 satisfying the system

$$P(R_0, w_1, w_2) = a_1^2 - a_2 ,$$

$$P'(R_0, w_1, w_2) = 2a_1^3 + a_3 - 3a_1a_2$$

For such R_0 the solution of Eq. (4.13) can be written in the form[20]

$$\phi(R, R_0) = \varepsilon\left[\frac{g(R, R_0)}{\sqrt{2A_6}} - \frac{A_4}{4A_6}\right]^{\frac{1}{2}} \tag{4.22}$$

where the function g is given by the formula

$$g(R, R_0) = \frac{1}{2}\,\frac{P'(R + R_{0/2}, w_1, w_2) + P'(R - R_{0/2}, w_1, w_2)}{P(R + R_{0/2}, w_1, w_2) - P(R - R_{0/2}, w_1, w_2)} - a_1$$

(ii): In the second case we consider separately two cases: (a) when the coefficient $A_0 \geq 0$, and (b) $A_0 < 0$ in the Eq. (4.13).

(a) $A_0 \geq 0$: In this case Eq. (4.13) by a change of variable (4.19) can be reduced to the form

$$\phi = \varepsilon \left[P\left(-\left(\frac{4A_4A_0}{A_6} \right)^{\frac{1}{2}} R, w_1, w_2 \right) - \frac{3A_6}{4A_6A_0} \right]^{\frac{1}{2}} \qquad (4.23)$$

where

$$g_2(w_1, w_2) = \frac{3A_6^2}{A_4^2A_0} \left(9 + \frac{1}{A_0} \right)$$

$$g_3(w_1, w_2) = \frac{-A_6^3}{A_4^3A_0} \left(27 + \frac{27}{2A_0} + \frac{1}{A_0^2} \right)$$

(b) $A_0 < 0$. In this case Eq. (4.13) by a change of variable (4.19) can be reduced to the form

$$\phi = \varepsilon \left[P\left(-\left(\frac{4A_4|A_0|}{A_6} \right)^{\frac{1}{2}} R, w_1, w_2 \right) - \frac{3A_6}{4A_6|A_0|} \right] \qquad (4.24)$$

where

$$g_2(w_1, w_2) = \frac{3A_6^2}{A_4^2A_0} \left(\frac{1}{A_0} - 9 \right)$$

$$g_3(w_1, w_2) = \frac{-A_6^3}{A_4^3A_0} \left(27 + \frac{27}{2A_0} + \frac{1}{A_0^2} \right)$$

Solutions (4.23) and (4.24) have been discussed in[21].

It may happen that one of the periods (say w_2) becomes infinite, $w_2 = \infty$; it takes place when $g_2^3 - 27g_3^2 = 0$. In this case we can express the solutions of Eqs. (4.22) or (4.23) and (4.24) by trigonometric functions using the formula

$$P(R, w_1, w_2 = \infty) = \left(\frac{\pi}{w_1}\right)^2 \left[\sin^{-2}\left(\frac{\pi}{w_1} R\right) - \frac{1}{3}\right] \text{ where } w_1 = \pi\left(\frac{2g_2}{9g_3}\right)^{\frac{1}{2}}$$

In particular when the periods w_1 and w_2 become infinite, i.e. $w_1 = w_2 = \infty$ which takes place when $g_2 = g_3 = 0$ we can express the solutions of Eqs. (4.22) or (4.23) and (4.24) by algebraic functions using the relation $P(R, w_1 = \infty, w_2 = \infty) = R^{-2}$.

We are looking now for a class of solutions of the Eqs. (4.2) being the nonlinear superposition of the two obtained above solutions which can be written in Riemann invariants. We solve this problem using the method presented in section 3. Using the simple integral elements we can write the Eq. (4.2) in the form

$$g^{\mu\nu}\gamma^{\mu}\lambda_{\nu} + b(\phi) = 0, \quad \gamma^{\mu}\lambda_{\nu} - \gamma^{\nu}\lambda_{\mu} = 0 \;, \quad \gamma^{4}\lambda_{\mu} = \phi_{\mu}, \tag{4.25}$$

$$(\mu, \nu = 0, 1, 2, 3)$$

where $\frac{d\phi_{\mu}}{dR} = \gamma^{\mu}$ and $\frac{d\phi}{dR} = \gamma^{4}$. Thus the simple integral nonhomogeneous elements have the form

$$\gamma = \left(-\frac{b(\phi)}{g^{\mu\nu}\lambda_{\mu}\lambda_{\nu}}(\lambda_0, \bar{\lambda}), \frac{\phi_0}{\lambda_0}\right), \quad \lambda = (\lambda_0, \bar{\lambda}), \tag{4.26}$$

$$\bar{\lambda} = (\lambda_1, \lambda_2, \lambda_3)$$

where $\frac{\phi_{\mu}}{\lambda_{\mu}} = \frac{\phi_{\nu}}{\lambda_{\nu}}$ for $\mu,\nu = 0, 1, 2, 3$. According to the assumption of Theorem 3 the conditions (3.2) and (3.3) should be satisfied, i.e.

$$\frac{\partial\phi_{\mu}}{\partial R^{i}} = \frac{b(\phi)}{2\,\xi^{i} g^{\mu\nu}\lambda^{i}_{\mu}\lambda^{i}_{\nu}}\lambda^{i}_{\mu} \quad \text{for } i = 1, 2.$$

(4.26)

$$\frac{\partial\phi}{\partial R^{i}} = \frac{\phi_0}{2\,\xi^{i}\lambda^{i}_{0}} \quad \text{where} \quad \frac{\phi_0}{\lambda^{i}_{0}} = \frac{\phi_1}{\lambda^{i}_{2}} = \frac{\phi_2}{\lambda^{i}_{2}} = \frac{\phi_3}{\lambda^{i}_{3}}$$

and

$$\frac{dR^{i}}{d(\lambda^{i}_{\mu} x^{\mu})} = \xi^{i}(R^{1}, R^{2})$$

where λ^{i} are constant vectors. We choose a normalization of the length of vectors $\lambda^{i} \in E$ such that $\lambda^{i} = (1, \bar{\lambda}^{i})$ for $i = 1, 2$. In particular, when $A_0 = A_1 = 0$ the compatibility conditions of Eq. (4.26) are automatically satisfied. In this case there are essentially three different solutions of Eq. (4.26) and (4.27) given by the Jacobi elliptic functions[18].

(i) $$\phi = \pm \psi_0[\mathrm{cn}(R^1, k_1) + i\,\mathrm{sn}(R^2, k_1) - \frac{A_4}{4A_6\psi_0^2}]^{\frac{1}{2}}$$

where

$$R^1 = \frac{\lambda^1_\mu x^\mu}{\sqrt{\psi_0^2 A_6 - A_4}} + R^1_0 , \quad k_1 = \sqrt{\frac{2\psi_0^2 A_6}{\psi_0^2 A_6 - A_4}} , \quad \bar{\lambda}^1 \cdot \bar{\lambda}_2 = 0$$

$$R^2 = \frac{\lambda^2_\mu x^\mu}{\sqrt{\psi_0^2 A_6 - A_4}} + R^2_0$$

(ii) $$\phi = \pm \psi_0 [\mathrm{dn}(R^1, k_2) - ik\,\mathrm{sn}(R^2, k^2) - \frac{A_4}{4A_6\psi_0^2}]^{\frac{1}{2}}$$

where

$$R^1 = \frac{\lambda^1_\mu x^\mu}{\psi_0\sqrt{2A_6}} + R^1_0 , \quad k_2 = \sqrt{\frac{\psi_0^2 A_6 - A_4}{2\psi_0^2 A_6}} , \quad \bar{\lambda}^1 \cdot \bar{\lambda}^2 = 0$$

$$R^2 = \frac{-\lambda^2_\mu x^\mu}{\psi_0\sqrt{2A_6}} + R^2_0$$

(iii) $$\phi = \pm \psi_0 [\mathrm{dn}(R^1, k_3) \pm k\,\mathrm{cn}(R^2, k_3) - \frac{A_4}{4A_6\psi_0^2}]^{\frac{1}{2}}$$

where

$$R^1 = \frac{\lambda^1_\mu x^\mu}{\sqrt{\psi_0^2 A_6 + A_4}} + R^1_0 , \quad k_3 = \sqrt{\frac{A_4 - \psi_0^2 A_6}{A_4 + \psi_0^2 A_6}} , \quad \bar{\lambda}^1 \cdot \bar{\lambda}^2 = 0$$

$$R^2 = \frac{\lambda^2_\mu x^\mu}{\sqrt{\psi_0^2 A_6 + A_4}} + R^2_0$$

where R^1_0 , R^2_0 and ψ_0 are constants of integration. This class of solutions corresponds to nonlinear superpositions of elliptic functions given by (4.22) and can be interpreted as the interaction of "periodic waves" (periodic potential).

References

1. A. M. Grundland, R. Zelazny, Simple Waves in Quasilinear Hyperbolic Systems. Part 1. Theory of Simple Waves and Simple States. Examples of Applications. Part 2. Riemann Invariants for the Problem of Simple Wave Interactions. J. Math. Phys. 24, (9), 2305-2328. (1983).
2. A. M. Grundland, R. Zelazny, Simple Waves and Their Interactions in Quasilinear Hyperbolic Systems (Polish Academy of Sciences, Publications of the Institute of Geophysics, Polish Scientific Publishers P.W.N., Warsaw, 1-109) (1982).
3. A. M. Grundland, Riemann Invariants for Nonhomogeneous Systems of First Order Partial Quasilinear Differential Equations - Algebraic Aspects. Examples from Gasdynamics, Arch. Mech. Stos. 26, (2), 274-296, (1974).
4. A. M. Grundland, Riemann Invariants for Nonhomogeneous System of Quasilinear Partial Differential Equations. Conditions of Involution. Bull. Acad. Pol. Sci. Ser. Sci. Tech. 22, (4), 177-185, (1974).
5. Z. Peradzynski, On Algebraic Aspects of the Generalized Riemann Invariants Method, Bull. Acad. Pol. Sci, Ser. Sci. Tech. 18, (9), 341-346 , (1970).
6. M. Burnat, Theory of Simple Waves for Nonlinear Systems of Partial Differential Equations and Applications to Gasdynamics. Arc. Mech. Stos. 18, (4), (1966).
7. J. Dieudonne, Foundations of Modern Analysis (Academic Press, New York, 1960).
8. W. I. Arnold, Ordinary Differential Equations, (The MIT Press, Cambridge, 1973).
9. B. L. Rozdestvenskii, N. N. Jamenko, Systems of Quasilinear Equations and Their Applications to Gasdynamics, (Nauka, Physics-Mathematical Literature, Moscow 1978. Translations of Mathematical Monographs A.M.S. Vol. 55, Providence, Rhode Island 1980).
10. E. Goursat, Cours d' Analyse Mathematique (reprint Dover, New York, 1953).
11. M. Burnat, The hyperbolic double waves, Bull. Acad. Pol. Sci. Ser. Sci. 17, (10), 97-106, (1969).
12. A. Jeffrey, Quasilinear Hyperbolic Systems and Waves (Pitman, London 1976).
13. Z. Peradzynski, Geometry of Nonlinear Interactions in Partial Differential Equations Habilitation Desertation. Reports of Inst. of Fund. Tech. Research Warsaw 1981. (in Polish).
14. A. M. Grundland, Riemann Invariants. Wave Phenomena. Modern Theory and Applications, Edited by C. Rogers, T. Moodie (North-Holland, Amsterdam, 1984).
15. E. Cartan, Les Systemes Differentiels en Involution, (Gauthier-Villars, Paris, 1953).
16. W. Slebodzinski, Exterior Differential Forms, (Polish Scientific Publishers, Warsaw, 1957).

17. J. F. Pommaret, Systems of Partial Differential Equations and Lie Pseudogroups (Gordon and Breach Science Publishers Vol. 14 in the Mathematics and its Applications series, New York, 1978).
18. P. F. Byrd, M. D. Friedman, Handbook of Elliptic Integrals for Engineers and Scientists (Springer-Verlag, New York, 1971).
19. S. Saks and A. Zygmund, Analytic Functions, (Polish Scientific, Warsaw, 1971).
20. A. M. Grundland, J. Tuszynski, P. Winternitz, Exact Solutions of the (n + 1)-dimentional classical ϕ^6 -field equations. (to be published).
21. S. N. Behera, A. Khare, ϕ^6- field theory in (1 + 1) dimensions. A model for structural phase transitions, Pramana, (printed in India). 15, pp. 245-269.

UNBOUNDED PERTURBATIONS OF FORCED HARMONIC OSCILLATIONS AT RESONANCE

R. Iannacci and M. N. Nkashama
Universita degli Studi della Calabria
Dipartimento di Matematica
87036 Arcavacata di Rende (CS), ITALIA

1. INTRODUCTION

Initiated by A.C. LAZER and D. E. LEACH [9], much work has been devoted to the study of existence results for the differential equation

$$\begin{aligned} & x''(t) + m^2 x(t) + g(t,x(t)) = e(t) \\ & x(0) - x(2\pi) = x'(0) - x'(2\pi) = 0 \end{aligned} \tag{1.1}$$

where $m \geq 0$ is an integer, $e \in L^1(0,2\pi)$ and g satisfies Caratheodory conditions. Results have been carried out by several authors, the reader is referred to papers [1-6, 8-13].

The aim of this note is to provide an existence result (cfr. [7]) for equation (1.1) in the case when there is resonance at the eigenvalue m^2 of the linear second order differential equation

$$\begin{aligned} & x''(t) + \lambda x(t) = 0, \qquad \lambda \in \mathbb{R} \\ & x(0) - x(2\pi) = x'(0) - x'(2\pi) = 0. \end{aligned} \tag{1.2}$$

The nonlinear function g may be unbounded and "touching" of the eigenvalue $(m + 1)^2$ (respectively $(m - 1)^2$) on a subset of $[0,2\pi]$ of positive measure is allowed. Our results complement those given in [10,11] and generalize those contained in the papers [1-6,8-9, 12-13].

On the other hand, we give a counterexample (Section 3) in the case of "vanishing nonlinearities" which shows that: if $m > 0$, the orthogonality of the forcing term e with respect to the nullspace of the linear part is not sufficient in order to ensure the existence of at least one solution for equation (1.1). This example, being given for a bounded nonlinearity, is valid for papers [1-6,8-13]. This explains the reason for which many authors have been obliged to put some additional conditions on the behaviour of the nonlinearity g when the "Landesman-Lazer's condition" is not fulfilled (cfr. [4, 6, 12] and the bibliography contained in that papers).

S. P. Singh (ed.), Nonlinear Functional Analysis and Its Applications, 277–289.

2. MAIN RESULTS

Let $m \geq 0$ be an integer, $e \in L^1(0,2\pi)$ and $g\colon [0,2\pi] \times \mathbb{R} \to \mathbb{R}$ be a Caratheodory function i.e. $g(., x)$ is measurable on $[0,2\pi]$ for each $x \in R$, $g(t, .)$ is continuous on $\mathbb{R}$ for a.e. $t \in [0,2\pi]$, for each constant $\bar{r} > 0$ there exists a real valued function $\gamma_{\bar{r}} \in L^1(0,2\pi)$ such that

$$|g(t, x)| \leq \gamma_{\bar{r}}(t) \tag{2.1}$$

for a.e. $t \in [0,2\pi]$ and all $x \in \mathbb{R}$ with $|x| \leq \bar{r}$. We have the following

Theorem 1. Assume that for all $\varepsilon > 0$ there exit a constant $B = B(\varepsilon) > 0$ and a real valued function $b_\varepsilon \in L^\infty(0,2\pi)$ such that

$$|g(t, x)| \leq (\Gamma(t) + \varepsilon)\ |x| + b_\varepsilon(t) \tag{2.2}$$

for a.e $t \in [0,2\pi]$ and all $x \in \mathbb{R}$ with $|x| \geq B$, where $\Gamma \in L^1(0,2\pi)$ is such that, for a.e. $t \in [0,2\pi]$,

$$\Gamma(t) \leq (2m + 1) \tag{2.3}$$

with strict inequality on a subset of $[0,2\pi]$ of positive measure. Moreover, suppose that there exit functions $a, A \in L^1(0,2\pi)$ and constants $r, R \in \mathbb{R}$ with $r < 0 < R$ such that

$$g(t,x) \geq A(t) \tag{2.4}$$

for a.e. $t \in [0,2\pi]$ and all $x \geq R$,

$$g(t,x) \leq a(t) \tag{2.5}$$

for a.e. $t \in [0,2\pi]$ and all $x \leq r$.

Then equation (1.1) has at least one solution for each $e \in L^1(0,2\pi)$ provided that

$$\int_0^{2\pi} e(t)v(t)dt < \int_{v>0} g_+(t)v(t)dt + \int_{v<0} g_-(t)v(t)dt \tag{2.6}$$

for all $v \in \mathrm{Span}\{\cos mt, \sin mt\} \setminus \{0\}$, where

$$g_+(t) = \liminf_{x \to +\infty} g(t,x) \text{ and } g_-(t) = \limsup_{x \to -\infty} g(t,x) \tag{2.7}$$

Sketch of the proof:

Step 1: Using some devices due to De FIGUEIREDO [2], MAWHIN and WARD [11], we write equation (1.1) in the equivalent form

$$x''(t) + m^2 x(t) + \tilde{\gamma}(t,x(t))x(t) + h(t,x(t)) = e(t)$$
$$x(0) - x(2\pi) = x'(0) - x'(2\pi) = 0 \tag{2.8}$$

where $\tilde{\gamma}$ and h are Caratheodory functions such that for a.e. $t \in [0,2\pi]$ and all $x \in \mathbb{R}$,

$$0 \le \tilde{\gamma}(t,x) \le \Gamma(t) + (\delta/2), \tag{2.9}$$

$$|h(t,x)| \le \sigma(t) \tag{2.10}$$

where $\delta > 0$ is a constant associated to the function Γ; $\sigma \in L^1(0,2\pi)$ depends only on Γ and γ_B given by (2.1).

In order to apply Mawhin's continuation Theorem [10], we consider (cfr. [7, 10] for details).

$$X = C([0,2\pi),\ Z = L^1(0,2\pi),\ \text{dom } L = W^{2.1}_{2\pi}(0,2\pi)$$

and

$$L\colon \text{dom } L \subset X \to Z,\ x \to x'' + m^2 x$$

$$G\colon X \to Z,\ x \to \tilde{\gamma}\,(\cdot,x(\cdot))x(\cdot),$$

$$H\colon X \to Z,\ x \to h(\cdot,x(\cdot)) - e(\cdot)$$

$$A\colon X \to Z,\ x \to (\delta/2)\ x\,(\cdot).$$

It is routine to check that G, H and A are well defined and L - compact on bounded subsets of X, and that L is a linear Fredholm mapping of index zero. Moreover, problem (2.8) is equivalent to solving the equation.

$$Lx + Gx + Hx = 0 \tag{2.11}$$

in dom L. By Theorem 1.2 in [10] with $\Omega = B(K) = \{x \in C([0,2\pi]) : |x|_C < K\}$, equation (2.11) will have a solution if we can show that for each $\lambda \in [0,1)$ and $x \in \text{dom } L$ such that

$$Lx + (1 - \lambda)Ax + \lambda Gx + \lambda Hx = 0, \tag{2.12}$$

one has $|x|_C < K$.

If $x \in \text{dom } L$ satisfies (2.12) for some $\lambda \in [0,1)$, then

$$x''(t) + m^2 x(t) + [(1-\lambda)(\delta/2) + \lambda\tilde{\gamma}(t,x(t)]x(t) + \lambda h(t,x(t)) - \lambda e(t) = 0, \tag{2.13}$$

and, by inequality (2.9), one has

$$0 \leq (1-\lambda)(\delta/2) + \lambda\tilde{\gamma}(t,x(t)) \leq \Gamma(t) + (\delta/2) \tag{2.14}$$

for a.e. $t \in [0,2\pi]$.

It is clear that for $\lambda = 0$, equation (2.13) has only the trivial solution in dom L since $\delta < (2m+1)$.

Now, let $x \in \text{dom } L$ be a solution of (2.13) for some $\lambda \in (0,1)$, then $x(t) = a_0 + \sum_{n=1}^{\infty} a_n \cos nt + b_n \sin nt$.

Let us consider

$$\bar{x}(t) = a_0 + \sum_{n=1}^{m-1} a_n \cos nt + b_n \sin nt, \quad \bar{x} = 0 \text{ if } m = 0;$$

$$x^0(t) = a_m \cos mt + b_m \sin mt, \quad x^0 = a_0 \text{ if } m = 0;$$

$$\tilde{x}(t) = \sum_{n=m+1}^{\infty} a_n \cos nt + b_n \sin nt, \quad x^{\perp} = \bar{x} + \tilde{x}.$$

Therefore,

$$0 = (2\pi)^{-1}\int_0^{2\pi} (\bar{x}(t) + x^0(t) - \tilde{x}(t))\,\{x''(t) + m^2c(t) + [(1-\lambda)(\delta/2) + \lambda\tilde{\gamma}(t,x(t))]\,x(t)\,\}\,dt + (2\pi)^{-1}\int_0^{2\pi} (\bar{x}(t) + x^0(t) - \tilde{x}(t))(\lambda h(t,x(t)) - \lambda\, e(t))\,dt$$

$$\geq (\delta/2)\,|x^{\perp}|^2_{H^1} - (2\pi)^{-1}(|\bar{x}|_C + |x^0|_C + |\tilde{x}|_C)(|h|_{L^1} + |e|_{L^1}),$$

(cfr. [7] for details). So that by the compact imbedding of $H^1(0,2\pi)$ in $C([0,2\pi])$, one gets

$$0 \geq (\delta/2) \quad |x^{\perp}|^2_{H^1} - \beta(|x^{\perp}|_{H^1} + |x^0|_{H^1})$$

where β depends only on σ and e (but otherwise not on x or λ). Hence, taking $\alpha = \beta(\delta)^{-1}$, one has

$$|x^{\perp}|_{H^1} \leq \alpha + (\alpha^2 + 2\alpha|x^0|_{H^1})^{\frac{1}{2}}. \tag{2.15}$$

Step 2: We claim that there exists a constant $\omega > 0$ such that

$$|x|_{H^1} < \omega \tag{2.16}$$

for any solution $x \in \text{dom } L$ of (2.13) (ω independent of x and λ).

The proof of the claim shall be done by a contradiction argument. Assume that the claim does not hold. Then, there will be a sequence (λ_n) in (0,1) and a sequence (x_n) in dom L with $|x_n|_{H^1} \to +\infty$ such that

$$\begin{aligned} &x_n''(t) + m^2 x_n(t) + (1 - \lambda_n)(\delta/2)x_n(t) + \lambda_n g(t,x_n(t)) \\ &= \lambda_n e(t). \end{aligned} \tag{2.17}$$

It follows immediately from (2.15) that

$$|x_n^0|_{H^1} \to +\infty, \quad |x_n^{\perp}|_{H^1}(|x_n^0|_{H^1})^{-1} \to 0. \tag{2.18}$$

So that the sequence $(x_n(|x_n^0|_{H^1})^{-1})$ is bounded in $H^1(0,2\pi)$.

Using the compact imbedding of $H^1(0,2\pi)$, one can assume, taking a subsequence if it is necessary, that there exits $v \in \text{Span}\{\cos mt, \sin mt\} \setminus \{0\}$ such that

$$\begin{aligned} &x_n(|x_n^0|_{H^1})^{-1} \to v \text{ in } C([0,2\pi]), \\ &x_n(|x_n^0|_{H^1})^{-1} \rightharpoonup v \text{ in } H^1(0,2\pi), \\ &x_n^0(|x_n^0|_{H^1})^{-1} \to v \text{ in } C([0,2\pi]). \end{aligned} \tag{2.19}$$

Let us set

$$v_n = x_n^0(|x_n^0|_{H^1})^{-1}, \tag{2.20}$$

multiplying (2.17) by v_n and using integration by parts, we obtain

$$\lambda_n(2\pi)^{-1}\int_0^{2\pi}(e(t) - g(t,x_n(t))\ v_n(t)\ dt$$

$$= (1-\lambda_n)(2\pi|x_n^0|_{H^1})^{-1}(\delta/2)\int_0^{2\pi}(x_n^0(t))^2\ dt.$$

Since $\int_0^{2\pi}(x_n^0(t))^2\ dt = m^{-2}\int_0^{2\pi}[(x_n^0)'(t)]^2\ dt = m^{-2}\ 2\pi|x_n^0|_{H^1}^2$,
we have

$$(1-\lambda_n)m^{-2}\ (\delta/2)\ |x_n^0|_{H^1} = \lambda_n(2\pi)^{-1}\int_0^{2\pi}[e(t)$$

$$- g(t,x_n(t))]v_n(t)\ dt.$$

Dividing this equality by λ_n (since $0 < \lambda_n < 1$) we get

$$0 \le (2\pi)^{-1}\int_0^{2\pi}[e(t) - g(t,x_n(t))]\ v_n(t)\ dt, \text{ so that}$$

$$0 \le \liminf_{n\to+\infty}\int_0^{2\pi}[e(t) - g(t,x_n(t))]\ v_n(t)\ dt$$

$$\le \liminf_{n\to+\infty}\int_0^{2\pi}e(t)\ v_n(t)\ dt + \limsup_{n\to+\infty}\left(-\int_0^{2\pi}g(t,x_n(t))v_n(t)dt\right)$$

$$= \int_0^{2\pi}e(t)\ v(t)\ dt - \liminf_{n\to+\infty}\int_0^{2\pi}g(t,x_n(t))\ v_n(t)\ dt.$$

Therefore, we have

$$\int_0^{2\pi}e(t)\ v(t)\ dt \ge \liminf_{n\to+\infty}\int_{v>0}g(t,x_n(t))\ v_n(t)\ dt$$

$$+ \liminf_{n \to +\infty} \int_{v<0} g(t,x_n(t))\, v_n(t)dt.$$

Let $I^+ = \{t \in [0,2\pi] : v(t) > 0\}$ and $I^- = \{t \in [0,2\pi] : v(t) < 0\}$. It follows from (2.18) and (2.19) that

$$\begin{aligned} x_n &\to +\infty \quad \text{on } I^+, \\ x_n &\to -\infty \quad \text{on } I^-. \end{aligned} \tag{2.21}$$

Moreover, using assumptions (2.4) and (2.5), one gets that there exists $n_0 \in N$ such that for $n \geq n_0$,

$$g(t,x_n(t))\, v_n(t) \geq -\, 2\alpha\beta_1(\Gamma(t) + (\delta/2)) - \sigma(t)K_1 \text{ a.e. on } I^+,$$

$$g(t,x_n(t))\, v_n(t) \geq -\, 2\alpha\beta_1(\Gamma(t) + (\delta/2)) - \sigma(t)K_1 \text{ a.e. on } I^-$$

for some constant $K_1 > 0$ such that $\sup_{[0,2\pi]} |v_n(t)| < K_1$.

So that by Fatou Lemma and properties of liminf and limsup, one has

$$\int_0^{2\pi} (e(t)\, v(t)dt \geq \int_{v>0} [\liminf_{n \to +\infty} g(t,x_n(t))]\, v(t)dt$$

$$+ \int_{v<0} [\limsup_{n \to -\infty} g(t,x_n(t))]\, v(t)dt.$$

Hence

$$\int_0^{2\pi} e(t)v(t)dt \geq \int_{v>0} g_+(t)v(t)dt + \int_{v<0} g_-(t)v(t)dt,$$

a contradiction to the "Landesman-Lazer's" condition (2.6). Thus the claim holds and by the compact imbedding of $H^1(0,2\pi)$ into $C([0,2\pi])$ one has that there exists a constant $K > 0$ such that $|x|_C < K$ for any solution of (2.13) in dom L and the proof is complete.

The (in some sense) dual version of Theorem 1 has the following form (when $m > 0$):

Theorem 2. *Assume that for all* $\varepsilon > 0$ *there exist a constant* $B = B(\varepsilon) > 0$ *and a function* $b_\varepsilon \in L^\infty(0,2\pi)$ *such that*

$$|g(t,x)| \leq (\Gamma(t) + \varepsilon)\ |x| + b_\varepsilon(t) \tag{2.22}$$

for a.e. $t \in [0,2\pi]$ *and all* $x \in \mathbb{R}$ *with* $|x| \geq B$, *where* $\Gamma \in L^1(0,2\pi)$ *is such that for a.e.* $t \in [0,2\pi]$

$$\Gamma(t) \leq (2m - 1) \tag{2.23}$$

with strict inequality on a subset of $[0,2\pi]$ *of positive measure.*

Moreover, suppose that there exist functions a, $A \in L^1(0,2\pi)$ *and constants* r, $R \in \mathbb{R}$ *with* $r < 0 < R$ *such that*

$$g(t,x) \leq a(t) \tag{2.24}$$

for a.e. $t \in [0,2\pi]$ *and all* $x \geq R$,

$$g(t,x) \geq A(t) \tag{2.25}$$

for a.e. $t \in [0,2\pi]$ *and all* $x \leq r$.

Then, equation (1.1) *has at least one solution for each* $e \in L^1(0,2\pi)$ *provided that*

$$\int_0^{2\pi} e(t)v(t)dt > \int_{v>0} g_+(t)v(t)dt + \int_{v<0} g_-(t)v(t)dt \tag{2.26}$$

for all $v \in \text{Span}\{\cos mt, \sin mt\} \setminus \{0\}$, *where*

$$g_+(t) = \limsup_{x \to +\infty} g(t,x) \text{ and } g_-(t) = \liminf_{x \to -\infty} g(t,x) \tag{2.27}$$

Remark 1. All results of this section hold true for the second order differential equation with dissipation

$$\begin{aligned} &x''(t) + cx'(t) + m^2x(t) + g(t,x(t)) = e(t) \\ &x(0) - x(2\pi) = x'(0) - x'(2\pi) = 0 \end{aligned} \tag{2.28}$$

where $c \in \mathbb{R}$ is an arbitrary constant.

One may consider also the vector second order differential equation

$$\begin{aligned} &x_i''(t) + c_i x_i'(t) + m_i^2 x_i(t) + g_i(t,x(t)) = e_i(t) \\ &x_i(0) - x_i(2\pi) = x_i'(0) - x_i'(2\pi) = 0 \end{aligned} \tag{2.29}$$

$$i = 1, \ldots, n; \; x = (x_1, \ldots, x_n)$$

together with suitable component-wise conditions of the type (2.2)-(2.7) (respectively (2.22)-(2.27)).

3. A COUNTEREXAMPLE.

It is well-known (cfr. e.g. [10,11,13]) that if $m = 0$ in equation (1.1), then the "growth" conditions (2.2)-(2.3), the "sign" conditions (2.4)-(2.5) are sufficient for the existence of at least one solution for equation (1.1) provided that the forcing term e has mean value zero (i.e. e is orthogonal to the nullspace of the linear part in (1.1)).

In contrast to the above situation, we give an example that shows that if $m > 0$ and e is orthogonal to Span $\{\cos mt, \sin mt\}$, then the "growth" conditions (2.2)-(2.3) and the "sign" conditions (2.4)-(2.5) are not sufficient in order to ensure the existence of at least one solution for equation (1.1) when the function g is not linear. This example, being given for a bounded nonlinearity, is valid for papers [1-6, 8-13].

Let us consider the following equation:

$$\begin{aligned} &x''(t) + x(t) + g(t,x(t)) = d \\ &x(0) - x(2\pi) = x'(0) - x'(2\pi) = 0 \end{aligned} \tag{3.1}$$

where $d \in \mathbb{R}$, $g : [0,2\pi] \times \mathbb{R} \to \mathbb{R}$ is defined by

$$g(t,x) = \begin{cases} \sin t \;\; \sin^2 x & \text{for } 0 \leq t \leq \pi \text{ and } x \geq 0 \\ 0 & \text{for } \pi \leq t \leq 2\pi \text{ and } x \geq 0 \\ 0 & \text{for } 0 \leq t \leq 2\pi \text{ and } x \leq 0 \end{cases} \tag{3.2}$$

Observe that g is continuous, bounded and the forcing term $d \in \mathbb{R}$ is orthogonal to cos t and sin t. We claim that:

If $0 < d \neq K\pi$, $K = 1, 2, 3, \ldots$, then equation (3.1) has no solution.

Indeed, assume that $x \in W^{2,1}_{2\pi}(0,2\pi)$ is a solution of (3.1), then by the theory of linear differential equation $d - g(t,x(t))$ has to be orthogonal to sin t and cos t. Therefore

$$0 = \int_0^{2\pi} [d - g(t,x(t))] \sin t \, dt = - \int_0^{2\pi} g(t,x(t)) \sin t \, dt.$$

Now

$$0 = \int_0^{2\pi} g(t,x(t)) \sin t \, dt = \int_{x>0} g(t,x(t)) \sin t \, dt$$

$$+ \int_{x<0} g(t,x(t)) \sin t \, dt$$

The last term being zero by definition of g, one has

$$\int_{x>0} g(t,x(t)) \sin t \, dt = 0$$

Let us set

$$J_1 = \{ t \in [0,\pi] : x(t) > 0 \} , \; J_2 = \{ t \in [\pi,2\pi] : x(t) > 0 \},$$

then

$$\int_{x>0} g(t,x(t)) \sin t \, dt = \int_{J_1} g(t,x(t)) \sin t \, dt$$

$$+ \int_{J_2} g(t,x(t)) \sin t \, dt$$

The last term being zero by the construction of g, one has

$$\int_{J_1} \sin^2 t \, \sin^2 x(t) \, dt = 0 \tag{3.3}$$

Since $0 < d \neq K\pi$ for any $K \in \mathbb{N}$, it follows from (3.3), by means of easy computations and the construction of g, that $x(t) \leq 0$ on $[0,\pi]$.

$$x(t) \leq 0 \text{ on } [0,\pi] \text{ so that } g(t,x(t)) = 0 \text{ on } [0,2\pi] \tag{3.4}$$

by construction.

Therefore, x is a solution of the linear equation

$$x''(t) + x(t) = d \text{ on } [0,2\pi] \tag{3.5}$$

$$x(0) - x(2\pi) = x'(0) - x'(2\pi) = 0$$

i.e.

$$x(t) = d + a \cos t + b \sin t, \quad a, b \in \mathbb{R} \text{ with}$$

$x(t) \leq 0$ on $[0,\pi]$ (cfr. (3.4)).

The last inequality implies that

$0 \geq x(0) = d + a$ and $0 \geq x(\pi) = d - a$, so that

$a \leq - d < 0$ and $0 < d \leq a$ a contradiction.

Thus, equation (3.1) has no solution.

The above counterexample explains the reason for which many authors have been obliged to put some additional conditions on the (asymptotic) behaviour of the nonlinearity g (cfr. e.g. [4,6]) or on the size of the norm of the forcing term (cfr. e.g. [12]). Let us mention that it seems that there was no such an example in the literature, at least for equation (1.1).

REFERENCES

1. H. Amann and G. Mancini, *Some applications of monotone operator theory to resonance problems*, Nonlinear An., TMA, 3(6) (1979), 815 - 830.

2. D. G. De Figueiredo, *Semilinear elliptic equations at resonance: higher eigenvalues and unbounded nonlinearities*, in Recent Advances in Differential Equations (R. Conti, Ed.), pp. 89-99, Academic Press, London, 1981.

3. T. R. Ding, *Unbounded perturbation of forced harmonic oscillations at resonance*, Proc. of the AMS (1) 88 (1983), 59-66.

4. P. Drabek, *Solvability of nonlinear problems at resonance*, Commentationes Math. Univ. Carolinea 23(2) (1982), 359-368.

5. C. Fabry and C. Franchetti, *Nonlinear equations with growth restrictions on the nonlinear term*, J. of Diff. Eq. 20(1976), 283-291.

6. S. Fucik and P. Hess, *Nonlinear perturbations of linear operators having nullspace with strong unique continuating property*, Nonlinear An., TMA, 3(2) (1979), 271-277.

7. R. Iannacci and M. N. Nkashama, *Unbounded perturbations of forced second order ordinary differential equations at resonance*, (To appear).

8. R. Kannan, *Perturbation methods for nonlinear problems at resonance*, in Nonlinear Functional Analysis and Differential Equations, (L. Cesari, R. Kannan and J. Schuur, Eds.), pp. 209-225, Marcel Dekker Inc., New York, 1976.

9. A. C. Lazer and D. E. Leach, *Bounded perturbations of forced harmonic oscillators at resonance*, Ann. Mat. Pura ed Appl., (4) 82 (1969), 49-68.

10. J. Mawhin, *Compacité, monotonie et convexité dans l'étude des problèmes aux limites semi-linéaires*, Sém. Anal. Moderne n° 19, Université de Sherbrooke, Québec, 1981.

11. J. Mawhin and J. R. Ward, *Periodic solutions of some forced Liénard differential equations at resonance*, Arch. Math., 41 (1983), 337-351.

12. P. Omari and F. Zanolin, *Existence results for forced nonlinear periodic BVP's at resonance*, Ann. Mat. Pura ed Appl. (to appear).

13. R. Reissig, *Continua of periodic solutions of the Liénard equations*, in Constructive Methods for Nonlinear Boundary Value Problems and Nonlinear Oscillations ISNM, pp. 126-133, Birkhaüser, Basel, 1979.

MINIMIZING THE NUMBER OF FIXED POINTS

Michael Kelly
Mathematics Department
SUNY-Binghamton
Binghamton, NY 13901
USA

ABSTRACT. Let P denote the topological space obtained by taking a closed neighborhood of the figure-eight in the plane. Let MF(f) denote the minimum number of fixed points achievable among maps homotopic to a given self-map f of P. We present here a formula for the value of MF(f). Note that MF(f) depends on the induced homomorphism, $f_{\#}$, on fundamental group, so our formula concerns the two relevant words in the free group on the letters a and b corresponding to the loops which comprise the figure eight. <u>Special case</u>: Let $g_m : P \to P$, $m \geq 0$, be given such that $(g_m)_{\#}(a) = (bab^{-1}a^{-1})^m ba$ and $(g_m)_{\#}(b) = 1$. It is easy to show that the <u>Nielsen number</u> of g_m, $N(g_m)$, is equal to zero. On the other hand, our formula shows that $MF(g_m) = 2m$. Hence the difference between N(f) and MF(f) can be made arbitrarily large.

1. INTRODUCTION

Let X^n be a compact n-dimensional manifold (with or without boundary) and f a self-map of X. Let MF(f) denote the minimum number of fixed points occurring among maps homotopic to f, and let N(f) be the Nielsen number of f [1]. These are related by the following classical result.

<u>Theorem 0</u>. If $n > 2$ or if X is a surface with non-negative Euler characteristic then $MF(f) = N(f)$.

For the proof when X is a surface with non-negative Euler characteristic the reader is referred to the work of Nielsen [6] using the Torus. The case $n > 2$ was first proved by Wecken [7]. A modern treatment, for smooth manifolds, is given in [4]. The key to the proof is the 2n-dimensional Whitney Lemma. Fixed points of f correspond to intersection points of $\Delta = \{(x,x)|x \in X\}$ and $\Gamma = \{(x,f(x))|x \in X\}$ as

S. P. Singh (ed.), Nonlinear Functional Analysis and Its Applications, 291–297.

subsets of $X \times X$. The Whitney Lemma enables us to deform Γ so as to minimize the number of points in $\Gamma \cap \Delta$. Consequently, f can be deformed to a map having $N(f)$ fixed points.

As the Whitney Lemma fails in dimension 4 it should not be surprising that, in general, Theorem 0 does not hold when $n = 2$. In fact, Jiang [2,3] has recently produced examples for which $N(f) = 0$ but $MF(f) > 0$. The purpose of this talk is to announce a formula for the value of $M(f)$ for any self-map, f , of the disk with two open holes removed. This is stated as Theorems 1 and 3. As a consequence we state Corollary 4 which shows that the difference between $N(f)$ and $MF(f)$ can be arbitrarily large.

2. PRELIMINARIES

Let P denote the regular neighborhood of the loops S_1 , S_2 as indicated in Figure 1. Identify $\pi_1(P,x_0)$ with the free group, G , on the letters a , b by associating the homotopy classes of the oriented loops S_1 and S_2 with a and b respectively. Given words $X,Y \in G$ we define $f_{(X,Y)} : S_1 \vee S_2 \to S_1 \vee S_2$ by sending S_1 and S_2 to the obvious loops corresponding to X and Y respectively. Define $F_{(X,Y)} : P \to P$ by applying $f_{(X,Y)}$ after the obvious retraction of P onto $S_1 \vee S_2$.

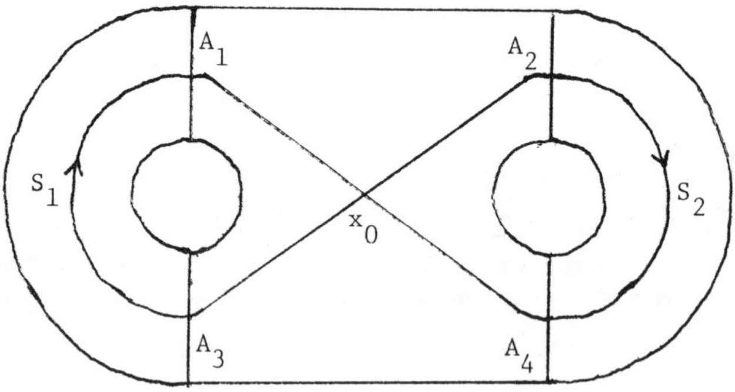

Figure 1: $P \equiv$ disk with two holes, indicating S_1 , S_2 and $A = \cup A_i$.

Notation: (1) Given the reduced word $W \in G$ let $|W|$ denote the length of W. If $W \neq 1$ let W_b denote the prefix and W_e the suffix of W with $|W_b| = |W_e| = 1$. Also, $1_b = 1$, $1_e = 1$, $W_b^{-1} = (W_b)^{-1}$, and $W_e^{-1} = (W_e)^{-1}$. (2) If $\mathcal{W}$ is a set of words and X is a single word in G define $\Phi_{\mathcal{W}}(X)$ = number of appearances in X of any member of $\mathcal{W} \cup \mathcal{W}^{-1}$ where $\mathcal{W}^{-1} = \{R^{-1} | R \in \mathcal{W}\}$. (3) If ℓ is a letter and W, X, Y are reduced words in G define

$$\rho_{\ell}(W) = \begin{cases} 2 & \text{if } W_b = \ell = W_e \text{ and } |W| > 1 \\ 0 & \text{if } W_b \neq \ell \text{ and } W_e \neq \ell \\ 1 & \text{otherwise} \end{cases}$$

and $\sigma(X,Y) = \begin{cases} -\rho & \text{if } \rho = 1 \\ 1 - \rho & \text{otherwise} \end{cases}$ where $\rho = \rho_a(X) + \rho_b(Y)$.

Let $X,Y \in G$ and consider the following conditions on (X,Y). Unless stated otherwise all elements of G and their products are reduced words.

Condition (T_1): $Y = 1$ and X is cyclically reduced.

Condition (T_2): $Y \neq 1$ is cyclically reduced and $X = U\overline{X}U^{-1}$ where $\overline{X} \neq 1$ is cyclically reduced. Also, if $U = 1$ then $X_b \neq Y_b$, $X_b \neq Y_e^{-1}$, $X_e \neq Y_e$, and $X_e \neq Y_b^{-1}$.

Condition (T_3): $X = W\overline{X}$, $Y = W\overline{Y}$ where $W \neq 1$ is the maximal common prefix of X and Y. Also, $\overline{X} = 1$ implies that $\overline{Y} = 1$, and $X_e = Y_e$ implies that $\overline{Y} = 1$ and there exists V such that $X = V^r$, $Y = V^s$, $r \geq s > 0$.

Condition (T_4): $X = W\overline{X}$, $Y = \overline{Y}W^{-1}$ where $W \neq 1$ is the maximal common prefix of X and Y^{-1}. Also, $\overline{X} \neq 1$, $\overline{Y} \neq 1$, and $X_e \neq Y_b^{-1}$.

Condition (T_5): $X = \ell_1\ell_2\overline{X}$, $Y = \ell_1^{-1}$ where $\ell_1\ell_2 \in \{ab, ba, a^{-1}b^{-1}, b^{-1}a^{-1}\}$ and $\overline{X} \neq 1$ implies that $\overline{X}_e = \ell_2$.

The ordered pair (X,Y) <u>is of type</u> T_K, $1 \leq K \leq 5$, iff (X,Y) satisfies condition (T_k) but does not satisfy condition $(T_{K'})$ when $K' \neq K$.

We define a function M from $\{(X,Y) \mid (X,Y) \text{ has a type}\}$ into the non-negative integers as follows. If (X,Y) is of type T_K then

For $K = 1$; $M(X,Y) = \Phi_a(X) + \lambda_1$ where $\lambda_1 = \begin{cases} -1 & \text{if } a \text{ appears in } X \\ +1 & \text{if } a \text{ does not appear in } X\,. \end{cases}$

For $K = 2$; If $U_e\overline{X}_b = aa$ or $\overline{X}_e U_e^{-1} = aa$ or $\rho_a(\overline{X}) = 2$ or $\Phi_{aa}(U) > 0$ then $M(X,Y) = \Phi_\alpha(\overline{X}) + \Phi_b(Y) + \Phi_{\{ab,ab^{-1},a^{-1}b,a^{-1}b^{-1}\}}(U) + \Phi_a(U_b) + \sigma(X,Y) + \lambda_2$ where

$$\lambda_2 = \begin{cases} -1 & \text{if } U_e\overline{X}_b = aa \text{ or } \overline{X}_e U_e^{-1} = aa \text{ or } \rho_a(\overline{X}) = 2 \\ 0 & \text{if } U_e = b^{\pm 1} \text{ and } \rho_a(\overline{X}) < 2 \\ +1 & \text{otherwise} \end{cases}$$

otherwise $M(X,Y) = \Phi_a(X) + \Phi_b(Y) + \sigma(X,Y)$.

For $K = 3$; If $\overline{Y} = 1$ write $\overline{X} = W^{N-1}\overline{W}\,\overline{\overline{X}}$ where N is chosen maximal and $\overline{W}$ is the maximal common prefix of W and the reduced form of $W^{1-N}\overline{X}$. Then $M(X,Y) = \Phi_a(X) + \Phi_b(Y) + \min\{\lambda_3, \sigma(X,Y)\}$ where

$$\lambda_3 = \begin{cases} -2 & \text{if either } \overline{X}_b\overline{Y}_b = ab \text{ or } (\overline{Y} = 1 \text{ and } \overline{\overline{X}}_b(\overline{W}^{-1}W)_b = ab)\,. \\ -1 & \text{if } \lambda_3 = -2 \text{ is not satisfied and if either } W_e\overline{X}_b = b^{-1}a \text{ or} \\ & W_e\overline{Y}_b = a^{-1}b \text{ or } \Phi_{a^{-1}b}(W) > 0 \text{ or } (\overline{Y} = 1 \text{ and } (W^{N-1}\overline{W})_e\overline{\overline{X}}_b = b^{-1}a)\,. \\ +1 & \text{otherwise.} \end{cases}$$

For $K = 4$; $M(X,Y) = \Phi_a(X) + \Phi_b(Y) + \tau$ where the value of τ is obtained from the following table.

CONDITION	VALUE OF τ
(I) $\overline{X} = aX'$, $\overline{Y} = Y'b$ or $\overline{X} = aX'$, $W_e = b$ (set $Y' = Y$) or $\overline{Y} = Y'b$, $W_e = a^{-1}$ (set $X' = X$)	$\lambda_4 - \Phi_{\{ab,ba\}}(W) - 2$
(II) not (I) but $W = W_1(ba)^{\delta}W_2$ where $\delta = \pm 1$ and $\Phi_{ba}(W_2) = 0$ (set $X' = X$, $Y' = Y$)	$\lambda_4 - \Phi_{\{ab,ba\}}(W_1(ba)^{\delta}) - 1$
(III) neither (I) nor (II)	$\sigma(X,Y)$

Here $\lambda_4 = \begin{cases} +1 & \text{if } X'_e \neq a \text{ , } Y'_b \neq b \text{ , and } W_b = a^{-1} \text{ or } b \text{ .} \\ 0 & \text{if } W_b = a \text{ or } b^{-1} \text{ and } \rho_a(X'_e) + \rho_b(Y'_b) < 2 \text{ .} \\ -1 & \text{otherwise.} \end{cases}$

For $K = 5$; $M(X,Y) = \Phi_a(X) + \Phi_b(Y) + \sigma(X,Y) - 2\rho_a(\overline{X}_e)$.

3. STATEMENT OF RESULTS AND INDICATION OF THEIR PROOFS

Theorem 1. Let (X,Y) have a type. Then $MF(F_{(X,Y)}) = M(X,Y)$.

Procedure 2. Let ϕ and ψ be homeomorphisms of P such that $\phi_{\#}(a) = b$, $\phi_{\#}(b) = a$, $\psi_{\#}(a) = ab$, $\psi_{\#}(b) = a^{-1}$. Given $f : P \to P$ define three moves: (I) $f \Longrightarrow f'$ ($\simeq f$ satisfying $f'_{\#}(\cdot) = Q \cdot f_{\#}(\cdot) \cdot Q^{-1}$ on π_1) ; (II) $f \Longrightarrow \phi f \phi^{-1}$; (III) $f \Longrightarrow \psi f \psi^{-1}$. For $i \geq 1$, f^{i+1} is obtained from f^i ($f^1 \equiv f$) by applying Step (i) as follows, where $(R^i,S^i) \equiv (f^i_{\#}(a),f^i_{\#}(b))$ [convention: if (R^i,S^i) has a type then $f^{i+1} = f^i$] .

STEP (1): apply (I) so that (a) $|R^2| + |S^2| \leq |VR^2V^{-1}| + |VS^2V^{-1}|$ for each $V \in G$, (b) R^2 or S^2 is cyclically reduced, and (c) $R^2_e = S^2_e$ or $R^2_e = (S^2_b)^{-1}$ implies that $R^2 = W^n$ and $S^2 = W^m$ for some $W \in G$.

STEP (2): apply (II) iff either (R^3,S^3) has a type or

$(R^2,S^2) = (V^{-1},VS)$. STEP (3): apply (III). STEP (4): apply (I) as in Step (1). STEP (5): apply (II). We can prove that (R^6,S^6) has a type and thus,

<u>Theorem 3</u>. Given $f : P \to P$, Procedure 2 yields a pair (R,S) having a type, such that $MF(f) = MF(F_{(R,S)})$. Hence $MF(f) = M(R,S)$.

<u>Corollary 4</u>. For each integer $m \geq 0$, any map $g_m : P \to P$ satisfying $(g_m)_{\#}(a) = (bab^{-1}a^{-1})^m ba$ and $(g_m)_{\#}(b) = 1$ has $N(g_m) = 0$ and $MF(g_m) = 2m$.

We now give an outline of the proof of Theorem 1. The details can be found in [5]. Let (X,Y) be of type T_K . The approach is to find representative maps in the homotopy class of $F_{(X,Y)}$ rel ∂P which achieve $MF(F_{(X,Y)})$ fixed points. Let $f \simeq F_{(X,Y)}$ rel ∂P . The first step is a "clean up" procedure on f (within its homotopy class) to improve $f^{-1}(A)$ where $A = \bigcup_{i=1}^{4} A_i$ (see Figure 1). This includes; (i) general positioning so that $f^{-1}(A)$ is a 1-dimensional proper submanifold of P transverse to A and (ii) various lemmas which reduce the geometric intersection of $f^{-1}(A)$ with A : all without increasing the number of fixed points. Next, we explicitly describe a set $\mathcal{P}(X,Y)$ of representatives from certain isotopy classes of 1-dimensional proper submanifolds of P . Precisely: $\Gamma \in \mathcal{P}(X,Y)$ iff (1) there is a map g homotopic to $F(X,Y)$ rel ∂P such that Γ is ambient isotopic (rel ∂P) to $g^{-1}(A)$, (2) whenever Δ is isotopic (rel ∂P) to Γ then $\#(\Gamma \cap A) \leq \#(\Delta \cap A)$, and (3) if $\Gamma' \in \mathcal{P}(X,Y)$ is isotopic (rel ∂P) to Γ then $\Gamma' = \Gamma$.

We are interested in the subset $\mathcal{P}'$ of $\mathcal{P}(X,Y)$ given by $\Gamma \in \mathcal{P}'$ iff $\Gamma \in \mathcal{P}(X,Y)$ and there is a "cleaned up" map g satisfying (1). We can prove that $\mathcal{P}'$ is non-empty. Pick $\Gamma \in \mathcal{P}'$ and a corresponding "cleaned up" g . Since Γ is known in terms of the explicit description of $\mathcal{P}$, the fact that $g^{-1}(A)$ is "cleaned up" yields enough information to explicitly construct a preferred ambient isotopy carrying Γ onto $g^{-1}(A)$. If h is the end of this isotopy $g^{-1}(A) = h(\Gamma)$; thus $g^{-1}(A)$ is explicitly known. By using a Brouwer-Lefschetz type fixed point theorem we are able to find a lower bound, $m(\Gamma,g)$, for the number of fixed points of g . We can prove that $MF(F_{(X,Y)}) = \min_{\mathcal{P}'}\{m(\Gamma,g)\}$ and finally that $\min_{\mathcal{P}'}\{m(\Gamma,g)\} = M(X,Y)$.

REFERENCES

[1] Brown, R.F., 'The Lefschetz Fixed Point Theorem,' Scott-Foresman Chicago, 1971.
[2] Jiang, B., 'Fixed Points and Braids,' Invent. Math. 75 (1984), 69-74.
[3] Jiang, B., 'Fixed Points and Braids, II,' preprint.
[4] Jiang, B., 'Fixed Point Classes from a Differential Viewpoint,' Lecture Notes in Mathematics, Vol. 866, 163-170, Springer-Verlag, 1981.
[5] Kelly, M., 'Minimizing the Number of Fixed Points for Self-Maps of Compact Surfaces,' Thesis, State University of New York, Binghamton, N.Y., 1985.
[6] Nielsen, J., 'Über die Minimalzahl der Fixpunkte bei Abbildungstypen der Ringflächen,' Math. Ann. 82 (1921), 83-93.
[7] Wecken, F., 'Fixpunktklassen, I, II, III,' Math. Ann., 117 (1941) 549-671; 118 (1942) 216-234, 544-577.

APPROXIMATE FIXED POINTS FOR MAPPINGS IN BANACH SPACES

W. A. Kirk
Department of Mathematics
University of Iowa
Iowa City, Iowa, 52242

ABSTRACT. Let K be a fixed bounded closed convex subset of a Banach space, let $\mathfrak{F}_\epsilon(k)$ denote the collection of all mappings of K into an ϵ-neighborhood of K having Lipschitz constant k, and let

$$\rho_k(\epsilon) = \sup_{T\in\mathfrak{F}_\epsilon(k)} \{\inf\{\|x-T(x)\| \;\big|\; x \in K\}\}.$$

It is shown that $\rho_k(\epsilon) \leq [2(1-k)^{-1}+1]\epsilon$ if $k \in (0,1)$; $\rho_1(\epsilon) \to 0$ as $\epsilon \to 0^+$, and if K has nonempty interior,

$$\rho_k(\epsilon) \leq \left(\frac{\mathrm{diam}(k)-\bar{r}+\epsilon}{\bar{r}+\epsilon}\right)\epsilon$$

where $\bar{r} = \sup\{r > 0 \mid B(x;r) \subset K \text{ for some } x \in K\}$.

1. INTRODUCTION

Let K be a subset of a Banach (or normed) space X, and for $\epsilon > 0$ let

$$N_\epsilon(K) = \{x \in X \mid \inf\{\|x-y\| \;\big|\; y \in K\} \leq \epsilon\}.$$

In this paper we take up the question of how near mappings of the form $f : K \to N_\epsilon(K)$ come to have fixed points. Our basic results, which are geometric in nature, are motivated by the following observation.

<u>Theorem 1</u>. Let X be a Banach space, K a closed convex subset of X, and $T : K \to N_\epsilon(K)$ a continuous mapping which has pre-compact range. Then $\inf\{\|x-T(x)\| \;\big|\; x \in K\} \leq \epsilon$.

Theorem 1 is a direct consequence of the following result of Fan [2].

<u>Theorem I</u> ([2]). Let X be a Banach space, H a compact convex subset

S. P. Singh (ed.), Nonlinear Functional Analysis and Its Applications, 299–303.

of X, and $f : H \to X$ continuous. Then there exists $y \in H$ such that

$$\|y-f(y)\| = \inf\{\|x-f(y)\| \mid x \in H\}.$$

Proof of Theorem 1. Let $\eta > 0$ be arbitrary, choose $\{x_1, \cdots, x_n\} \subset K$ so that $\{T(x_1), \cdots, T(x_n)\}$ is an $\eta/2$-net in $T(K)$, and for $i = 1, \cdots, n$ select $w_i \in K$ such that $\|w_i - T(x_i)\| \le \epsilon + \eta/2$. Now let

$$H = \overline{\text{conv}}\{x_1, \cdots, x_n, w_1, \cdots, w_n\}.$$

Note that if $x \in H$ then $\|T(x)-T(x_i)\| \le \eta/2$ for some i; hence $\|T(x)-w_i\| \le \epsilon+\eta$. In view of Theorem 1 there exists $y \in H$ such that

$$\|y-T(y)\| = \inf\{\|x-T(y)\| \mid x \in H\}$$
$$\le \epsilon + \eta.$$

Since η was arbitrary, the conclusion follows.

2. MAIN RESULTS

We now turn to our basic geometric results. Let K be a fixed bounded closed convex subset of a Banach space X, let $\epsilon > 0$, and for $k > 0$ set

$$\mathfrak{F}_\epsilon(k) = \{T : K \to N_\epsilon(K) \mid \|T(x)-T(y)\| \le k\|x-y\|, x,y \in K\};$$

$$\rho_k(\epsilon) = \sup_{T \in \mathfrak{F}_\epsilon(k)} \{\inf\{\|x-T(x)\| \mid x \in K\}\}.$$

Our principal results are for nonexpansive mappings (the case $k = 1$), and the first is quite simple.

Theorem 2. If K is a bounded closed convex subset of a Hilbert space X, then $\rho_1(\epsilon) = \epsilon$.

Proof. Obviously $\rho_1(\epsilon) \ge \epsilon$. Let $P : X \to K$ be defined by $\|x-P(x)\| = \inf\{\|x-y\| \mid y \in K\}$. It is known ([1]) that P is nonexpansive; thus $P \circ T : K \to K$ is nonexpansive and it follows that $P \circ T(x) = x$ for some $x \in K$. Hence

$$\|x-T(x)\| = \|P \circ T(x)-T(x)\|$$
$$= \inf\{\|T(x)-y\| \mid y \in K\}$$
$$\le \epsilon.$$

The fact that $\rho_1(\epsilon) = \epsilon$ if K is a ball is a corollary of the following. (We use $B(x;r)$ to denote a closed ball with center x and radius r.)

Theorem 3. If K is a bounded closed convex subset of a Banach space X, and if $\mathrm{int}(K) \neq \emptyset$, then

$$\rho_1(\varepsilon) \leq \left(\frac{\delta(K)-\bar{r}+\varepsilon}{\bar{r}+\varepsilon}\right)\varepsilon,$$

where $\delta(K)$ denotes the diameter of K and

$$\bar{r} = \sup\{r > 0 \mid B(x;r) \subset K \text{ for some } x \in K\}.$$

For arbitrary K we have only the following estimates.

Theorem 4. If K is any bounded closed convex subset of a normed linear space X,

a) $\rho_k(\varepsilon) \leq [2(1-k)^{-1}+1]\varepsilon$ for all $k \in (0,1)$;

b) $\lim_{\varepsilon\to 0^+} \rho_1(\varepsilon) = 0.$

For the proof of Theorem 3 we require the following well-known fact (cf., Proposition 1 of Petryshyn [3]).

Proposition. Suppose G is a bounded open convex subset of a Banach space X with $0 \in G$, and suppose $T: \bar{G} \to X$ is nonexpansive and satisfies

$$\text{(L-S)} \qquad T(x) \neq \lambda x \quad \text{for } x \in \partial G \text{ and } \lambda > 1.$$

Then $\inf\{\|x-T(x)\| \mid x \in G\} = 0$.

Proof of Theorem 3. By assumption there exists $x_0 \in K$ and $r > 0$ such that $B(x_0;r) \subset K$, and we may suppose $x_0 = 0$. Let $\varepsilon > 0$, choose $\mu \in (0,\rho_1(\varepsilon))$, and select $T \in \mathfrak{F}_\varepsilon(1)$ so that $\inf\{\|x-T(x)\| \mid x \in K\} \geq \rho_1(\varepsilon)-\mu$. Since T has no fixed points, (L-S) must fail for some $y \in \partial K$, i.e., $T(y) = \lambda y$ for some $\lambda > 1$. Choose $z \in \partial K$ so that

$$\begin{aligned}\|z-T(y)\| &\leq \mathrm{dist}(T(y),K)+\mu \\ &\leq \varepsilon+\mu.\end{aligned}$$

If $z = y$, take $w = y$. Otherwise, let w be the point of X defined by

$$(1) \qquad y = (1-\lambda^{-1})w+\lambda^{-1}z.$$

Then, since $y = \lambda^{-1}T(y)$, (1) implies

$$(2) \qquad (1-\lambda^{-1})w = \lambda^{-1}(T(y)-z).$$

From (2),

$$(1-\lambda^{-1})\|w\| = \lambda^{-1}\|T(y)-z\| \leq \lambda^{-1}(\epsilon+\mu),$$

so

$$(3) \qquad (1-\lambda^{-1})[\|w\| + (\epsilon+\mu)] \leq \epsilon+\mu.$$

Since $(1-\lambda^{-1})\|T(y)\| = \|y-T(y)\| \geq \rho_1(\epsilon)-\mu$, by (3)

$$\rho_1(\epsilon)-\mu \leq \frac{(\epsilon+\mu)\ \|T(y)\|}{\|w\|+\epsilon+\mu}.$$

On the other hand, $\|T(y)\| \leq \delta(K)-r+\epsilon$, and since both y and z lie on ∂K, $w \notin \text{int}(K)$, so in particular $\|w\| \geq r$. Therefore

$$\rho_1(\epsilon)-\mu \leq \frac{(\epsilon+\mu)(\delta(k)-r+\epsilon)}{r+\epsilon+\mu}.$$

Since $\mu > 0$ was arbitrary and $B(x_0;r)$ an arbitrary ball in K, the conclusion follows.

Proof of Theorem 4. (a) Let $T \in \mathfrak{F}_\epsilon(k)$ and select: $p \in (0,1-k)$, $s \in (0,1-(k+p))$, $r \in (0,s\epsilon/p)$. For each $x \in N_\epsilon(k)$ let $P(x)$ denote a point of K for which $\|x-P(x)\| \leq \epsilon+r$. Then $f = P\circ T: K \to K$, and for $x,y \in K$,

$$\|f(x)-f(y)\| \leq \|P\circ T(x)-T(x)\| + \|T(x)-T(y)\| + \|T(y)-P\circ T(y)\| \leq k\|x-y\| + 2\epsilon + 2r.$$

In particular, if $\|x-y\| \geq 2\epsilon/p$ then $\|x-y\| \geq 2r/s$ and

$$\|f(x)-f(y)\| \leq (k+p+s)\|x-y\| = k'\|x-y\|$$

where $k' < 1$. Therefore, given $n \in \mathbb{N}$ and $x \in K$, either $\|f^n(x)-f^{n+1}(x)\| \leq (k')^n\|x-f(x)\|$ or for some $i < n$, $\|f^i(x)-f^{i+1}(x)\| < 2\epsilon/p$. Since $(k')^n \to 0$ as $n \to \infty$ it follows that for some $x \in K$, $\|x-f(x)\| < 2\epsilon/p$ from which

$$\|x-T(x)\| \leq \|x-f(x)\| + \|P\circ T(x)-T(x)\| < 2\epsilon/p+\epsilon+r.$$

Since r can be chosen arbitrarily small and p arbitrarily near $1-k$, it follows that $\rho_k(\epsilon) \leq 2\epsilon(1-k)^{-1}+\epsilon$.

We prove (b) by contradiction. Suppose there exists $\rho > 0$ such that $\rho_1(\epsilon) > \rho$ for all $\epsilon > 0$. We assume (without loss of generality) that $0 \in K$ and choose $k \in (0,1)$ so that $(1-k)\|x\| < \rho/4$ for all

$x \in K$. With k fixed, choose $\epsilon > 0$ so that $\rho_k(\epsilon) < \rho/2$ (using (a)). Now choose $T \in \mathfrak{F}_\epsilon(1)$ so that $\inf\{\|x-T(x)\| \mid x \in K\} \geq \rho$. Since $kT \in \mathfrak{F}_{k\epsilon}(k) \subset \mathfrak{F}_\epsilon(k)$, there exists $x \in K$ such that $\|x-kT(x)\| < \rho_k(\epsilon) + \rho/4$. Thus we have the contradiction

$$\begin{aligned} \rho \leq \|x-T(x)\| &\leq \|x-kT(x)\| + (1-k)\|T(x)\| \\ &< \rho_k(\epsilon) + \rho/4 + \rho/4 \\ &< \rho. \end{aligned}$$

We should remark that the above observations are preliminary in nature; indeed, we do not even have an example of a case in which $\rho_1(\epsilon) \neq \epsilon$.

REFERENCES

1. W. Chaney and A. Goldstein, 'Proximity maps for convex sets', Proc. Amer. Math. Soc. 10(1959), 448-450.

2. K. Fan, 'Extensions of two fixed point theorems of Browder', Math. Z. 112(1969), 234-240.

3. W.V. Petryshyn, 'Structure of the fixed point sets of k-set contractions', Arch. Rational Mech. Anal. 40(1970/71), 312-328.

INVARIANTLY COMPLEMENTED SUBSPACES AND GROUPS WITH FIXED POINT PROPERTY

Anthony To-Ming Lau[(1)]
Department of Mathematics
University of Alberta
Edmonton, Alberta
Canada T6G-2G1

Abstract A locally compact group G is said to have the fixed point property if whenever G acts affinely on a compact convex subset K of a separated locally convex space, K contains a common fixed point for G. In this note, we characterize locally compact groups G with fixed point property in terms of existence of invariant complement for certain weak*-closed invariant subspaces when G acts on a dual Banach space.

1. Introduction

Let $T = \{\lambda \in C , |\lambda| = 1\}$ be the circle group. D.J. Neuman proved in [8] that the Hardy space H^1 consisting of all $f \in L_1(T)$ such that $\hat{f}(n) = 0$ for all $n < 0$ does not have a closed complement in $L_1(T)$ i.e. H^1 is not the range of a continuous projection on $L_1(T)$. In order to characterize closed translation invariant complemented subspaces of $L_1(T)$ (or more generally $L_1(G)$ of a compact abelian group), Rudin proved in [12] that if G is a compact group, then G has the G-invariant complemented subspace property: Whenever G acts continuously on a Banach space X and L is a closed complemented G-invariant subspace of X, then L is the range of a continuous projection which commutes with the action of G.

In fact, by considering the action of G on $L_1(G)$ by left translation, it is not difficult to see ([6, Proposition 5.1]) that if G is a locally compact group with the G-invariant complemented subspace property then G is compact.

Footnote(1): This research is supported by NSERC grant A7679.

S. P. Singh (ed.), Nonlinear Functional Analysis and Its Applications, 305–311.

In this report, we shall consider some variants of the G-invariant complemented subspace property for locally compact groups with fixed point property. Such groups include all solvable groups and all compact groups.

This report contains details of lecture presented at the NATO Advanced Study Institute on nonlinear functional analysis and fixed point theory held in Maratea, Italy. We thank the organisers for their kind invitation to speak and warm hospitality during the conference.

2. Preliminaries

Let G be a locally compact group and X be a Banach space. We say that X is a left Banach G-module if there exists a map $G \times X \to X$, denoted by $(g,x) \to g \cdot x$, such that the following holds:

(i) $(g_1 g_2) \cdot x = g_1(g_2 \cdot x)$ for $g_1, g_2 \in G$, $x \in X$;

(ii) $\|g \cdot x\| \leq \|x\|$ for all $x \in X$, $g \in G$;

(iii) for each $x \in X$, the map $g \to g \cdot x$ from G into X is continuous.

In this case we can define for each $x \in X$, $g \in G$, $f \in X^*$,

$$\langle f \cdot g, x \rangle = \langle f, g \cdot x \rangle$$

A closed subspace L of X^* is G-variant if $L \cdot g \subset L$ for all $g \in L$. We say that L is invariantly complemented if there exists a closed G-invariant subspace N of X^* such that $X^* = L + N$; or equivalently, there exists a continuous projection P of X^* onto L such that $P(f \cdot g) = P(f) \cdot g$ for all $g \in G$, $f \in X^*$.

A left Banach G-module X is called non-degenerate if the closed linear span of $\{g \cdot x ; g \in G, x \in G, x \in X\}$ is X.

We say that a locally compact group G has the fixed point property if whenever G acts affinely on a compact convex set K of a separately locally convex space and the map $G \times G \to K$ is continuous, then there exists $x_0 \in K$ such that $g \cdot x_0 = x_0$ for all $g \in G$. It follows from the Markov-Kakatani fixed point theorem and the Kakutani fixed point theorem (see [2, pp. 456-457]) that any abelian group and any compact group has the fixed point property. In [3], Furstenburg defined this property and proved that a connected semi-simple Lie group does not have the fixed point property unless it is compact. Finally in [10] Rickert proved that G has the fixed point property if and only if G is amenable i.e. the space $UB_r(G)$ of bounded right uniformly continuous functions on G has a left invariant mean m, i.e. m is a positive continuous linear functional on $UB_r(G)$ with norm one, and $m(\ell_g f) = m(f)$ for all $f \in UB_r(G)$, $g \in G$, where $(\ell_g f)(x) = f(g^{-1}x)$ for all $x, g \in G$. As shown in [4], this is equivalent to the space $L_\infty(G)$ of essentially bounded measurable functions on G has a left invariant mean. (See [4] and [9] for excellent expositions of such groups).

3. G-invariant complemented subspace property on dual Banach spaces

A locally compact group G is said to have the G-invariant complemented subspace property on dual Banach spaces if whenever X is a non-degenerate left Banach G-module and L is a complemented weak*-closed G-invariant subspace of X^*, then L is invariantly comlemented.

It follows from Rosenthal [11, Theorem 1.1] and its proof that any locally compact group G which has the fixed point property as a discrete group (e.g. when G is abelian) has the G-invariant complemented subspace property on dual Banach space when G acts on $L_\infty(G)$. However, as pointed out in [Zentralblatt fur Mathematik 1982:483.43002], his proof, based on Rosenthal's idea in [11, Theorem 1.1] has a gap. Note that, as well known, the compact groups $SO(n,\mathbf{R})$, $n \geq 3$, do not have the fixed point property as discrete groups. (See [9]).

Recently Lau and Losert [7]) establishes the following definite link between locally compact groups with fixed point property and invariant complementation of weak*-closed invariant subspaces of the dual of a Banach G-module:

Theorem 1 (Lau and Losert [7]) Let G be a localy compact group. Then G has the fixed point property if and only if G has the G-invariant complemented subspace property on dual Banach spaces.

A left G-module on a dual Banach space E is weak*-continuous if the map $G \times E \to E$ is continuous when E has the weak*-topology. We proved also in [6]:

Theorem 2 Let G be a locally compact group. Then G has the fixed point property if and only if G has the following property:

(C) Whenever G is a weak*-continuous left Banach G-module on a dual Banach space E and A is a norm closed G-invariant subspace of E such that the map $G \times A \to A$ is continuous when A has the norm topology, then any weak*-closed G-invariant subspace contained in A has a G-invariant closed complement in A.

Theorem 3 Let G be a locally compact group. Then G has the fixed point property if and only if G has the following property:

(D) Whenever X is a left Banach G-module and M is a weak*-closed subspace of X^* of the form $\{\varphi \in X^* ; \varphi \cdot g = \varphi \text{ for } g \in H\}$ where H is a closed subgroup of G, then M has a closed complement which is invariant under any weak*-weak* continuous linear operator from X^* into X^* which commutes with the action of G.

Proof: If G has the fixed point property, then any closed subgroup H has the fixed point property [4, Theorem 2.3.3]. For each $\varphi \in X^*$, let K_φ denote the weak*-closure of the convex hull of $\{\varphi \cdot g ; g \in H\}$. Consider the action of H on K_φ defined by $(g,\psi) \to \psi \cdot g$. Then there exists $\psi \in K_\varphi$ such that $\psi \cdot g = \psi$ for all $g \in G$. In particular, $K_\varphi \cap M$ is non-empty for each $\varphi \in X^*$. By a result of Yeadon [15] (see also [5]), there exists a projection Q from X^* onto M such that Q commutes with any weak*-weak* continuous linear operator from X^* into X^* which commutes with the action of G on X^*.

The converse can be proved by an argument similar to the proof of Theorem 3.3 in [6].

4. Subspaces of $L_p(G)$

When M is a weak*closed subalgebra of $L_\infty(G)$ invariant under left translation and closed under conjugation, then there exists a closed subgroup H of G such that $M = \{f \in L_\infty(G)$ and $r_g f = f$ for all $g \in H\}$, where $r_g f(t) = f(tg)$, $t \in G$ ([13, Theorem 2]). In particular, if G is has the fixed point property, then M is invariantly complemented by Theorem 3 (see [6, Theorem 3.3]).

If $1 < p < \infty$, and G is amenable, then any closed complemented left translation invariant subspace of $L_p(G)$ is invariantly complement ([11, Lemma 3.1]).

Problem 1: Let $1 < p < \infty$ and M is a closed complemented left translation invariant subspace of $L_p(G)$. Is M necessarily invariantly complemented?

Note that this is the case for $p = 2$, since the orthogonal complement of M is also left translation invariant.

More generally:

Problem 2 Let G a locally compact group. Does G have the G-invariant complemented subspace property for reflexive Banach spaces?

5. Semigroups

It follows from Day's fixed point theorem [1] and an argument similar to that of the proof for Theorem 3.3 in [6] that if G is a discrete semigroup and G has the G-invariant complemented subspace property for dual Banach space, then G has the fixed point property (i.e. G is left amenable).

Problem 3 Let G be a discrete semigroup with the fixed point property, does G has the G-invariant complemented subspace property for dual Banach space?

Problem 4 Let X be a weak*-closed complemented left translation invariant subspace of $\ell_\infty(G)$ when G is a semigroup with fixed point property. Is X invariantly complemented?

We do not know the answer to Problems 3 and 4 even when G is abelian.

References

1. M.M. Day, 'Fixed point theorems for compact convex sets', Illinois J. Math 5 (1961) pp. 585-589.

2. N. Dunford and J. Schwartz, Linear Operators I, John Wiley and Sons (1957).

3. H. Furstenberg, 'A Poisson formula for semi-simple Lie groups,' Annals of Math. 77 (1963), pp. 335-386.

4. F.P. Greenleaf, Invariant means on topological groups and their applications, Van Nostrand 1969.

5. A.T. Lau, 'Semigroup of operators on dual Banach space', Proc. A.M.S. 54 (1976). pp. 393-396.

6. A.T. Lau, 'Invariantly complemented subspaces of $L_\infty(G)$ and amenable locally compact groups' Illinois J. Math 26 (1982) pp. 226-235.

7. A.T. Lau and V. Losert, 'Weak*-closed complemented invariant subspace of $L_\infty(G)$ and amenable locally compact groups', Pacific Journal of Math (to appear).

8. D.J. Neuman, 'The nonexistence of projections from L^1 to H^1 ', Proc. A.M.S. (1961), pp. 98-99.

9. J.P. Pier, Amenable locally compact groups, John Wiley and Sons, 1984.

10. N. Rickert, 'Amenable groups and groups with fixed point property', Trans. A.M.S. 127 (1967), pp. 221-232.

11. H.P. Rosenthal, 'Projections onto translation invariant subspace of $L^p(G)$', Memoirs A.M.S. 63 (1966).

12. W. Rudin, 'Invariant means on L^∞', Studia Mathematica 44 (1972), pp. 219-227.

13. M. Takesaki and N. Tatsuuma, 'Duality and subgroups', Ann. of Math 93 (1971), pp. 344-364.

14. Y. Takahashi, 'A characterization of certain weak*-closed subalgebras of $L^\infty(G)$ ', Hokhaido Math. Journal 11 (1982), pp. 116-124.

15. F.J. Yeadon, 'Fixed points and amenability: a counterexample', J. Math. Anal. Appl. 45 (1974), pp. 718-720.

ON A CERTAIN DIFFERENCE-DIFFERENTIAL EQUATION

B. Lawruk
Department of Mathematics
McGill University
Montreal, Quebec, Canada H3A 2K6

ABSTRACT. A general difference-differential equation (1) is considered. Applying the method of distributions, it is proved that (1) is equivalent to an infinite system of linear partial differential equations with constant coefficients (3). In the particular case when $g(x,t)\equiv 0$, this system reduces to a finite system.

An equation of the form

$$\sum_{k=1}^{\infty} a_k(t,D)f(x+\varphi_k(t)) = g(x,t) \tag{1}$$

will be studied, where $x\in R^n$, $t\in\Delta = (-\varepsilon,\varepsilon)\subset R$, $\varphi_k:\Delta \to R^n$, $f:R^n \to \mathbb{C}$, $g:R^n\times\Delta \to \mathbb{C}$, the a_k are given linear differential operators in x, $D = \frac{1}{i}\frac{\partial}{\partial x} = \frac{1}{i}(\frac{\partial}{\partial x_1},\ldots,\frac{\partial}{\partial x_n})$, of order p_k with coefficients depending on t, and the φ_k are given functions. It is natural to assume that $\varphi_k(0) = 0$ for k=1,2,..., and hence, that $\sum_{k=1}^{\infty} a_k(0,D) = g(x,0) = 0$. The equation (1) is considered in the class of distributions $\mathcal{D}'(R^n)$, i.e. for every $t\in\Delta$, $g\in\mathcal{D}'(R^n)$ is given, and $f\in\mathcal{D}'(R^n)$, independent of t is sought.

In the particular case of $a_k(t,D) = a_k(t)$ for k=1,2,..., i.e. when the a_k's are differential operators in x of order zero, the equation (1) becomes

$$\sum_{k=1}^{\infty} a_k(t)f(x+\varphi_k(t)) = g(x,t). \tag{2}$$

S. P. Singh (ed.), Nonlinear Functional Analysis and Its Applications, 313–316.

Theorem 1. If (i) the series $\sum_{k=1}^{\infty} a_k(t,z)$, where $z = x+iy \in \mathbb{C}^n$ is absolutely convergent and there exist constants C, $m \geq 0$ and b such that

$$\sum_{k=1}^{\infty} |a_k(t,z)| \leq C(1+|z|^m)e^{b|y|}$$

for every $t \in \Delta$, (in the case of equation (2), only the absolute convergence of the series $\sum_{k=1}^{\infty} a_k(t)$ for every $t \in \Delta$ is required);
(ii) there exists constant c, independent of k or t and such that $|\varphi_k(t)| \leq c$; and if in addition $a_k, \varphi_k \in C^\infty(\Delta)$ for every k, and for every $\varphi \in \mathcal{D}(R^n)$, $<g(x,t),\varphi(x)> \in C^\infty(\Delta)$, then every solution of the equation (1) satisfies the following infinite system of linear partial differential equations with constant coefficients

$$\sum_{k=1}^{\infty} \sum_{\ell=0}^{m} \frac{m!}{(m-\ell)!\ell!} \sum_{\upsilon=0}^{\ell-1} \sum_{\substack{p_1+\ldots+p_\ell=\ell-\upsilon \\ p_1+\ldots+\ell p_\ell=\ell}} \frac{1}{p!} a_k^{(m-\ell)}(0,D) \cdot$$

$$\cdot\left(\frac{\varphi_k^{(1)}(0)}{1!} \cdot \frac{\partial}{\partial x}\right)^{p_1} \cdot \ldots \cdot \left(\frac{\varphi_k^{(\ell)}(0)}{\ell!} \cdot \frac{\partial}{\partial x}\right)^{p_\ell} f(x) = g^m(x,0) \tag{3}$$

$(m=1,2,\ldots)$.

Proof. The equation (1) means that for every $\varphi \in \mathcal{D}(R^n)$,

$$\sum_{k=1}^{\infty} <f(x),a_k(t,-D)\check{\varphi}(x-\varphi_k(t))> = <g(x,t),\check{\varphi}(x)> \tag{4}$$

is true for every $t \in \Delta$, where $\check{\varphi} = \varphi(-x)$.
The Fourier transform of φ is denoted by $\hat{\varphi}$, i.e.

$$\hat{\varphi}(\xi) = \int_{-\infty}^{\infty} \varphi(x)e^{-i(x\cdot\xi)}dx .$$

The Fourier transform $\hat{f}$ of a distribution $f \in \mathcal{D}'(R^n)$ is defined by the Parseval equality

$$<\hat{f},\hat{\varphi}> = (2\pi)^n<f,\check{\varphi}>$$

(see, e.g. [1] or [3]).
Applying the Fourier transform with respect to x, one obtains from (4)

$$\sum_{k=1}^{\infty} a_k(t,\xi)<\hat{f}(\xi),\hat{\varphi}(\xi)e^{-i(\varphi_k(t)\cdot\xi)}> = <\hat{g}(\xi,t),\hat{\varphi}(\xi)>,$$

which means that

$$\sum_{k=1}^{\infty} a_k(t,\xi)e^{-i(\varphi_k(t)\cdot\xi)}\hat{f}(\xi) = \hat{g}(\xi,t). \tag{5}$$

By conditions (i) and (ii) the series

$$\sum_{k=1}^{\infty} a_k(t,\xi)e^{-i(\varphi_k(t)\cdot\xi)}$$

converges absolutely to a function $a(t,\xi)$ which can be continued in $\mathbb{C}^n$ as en entire function of the exponential type $\le c+b$. Therefore the product on the left hand side of (5) is well defined in the space of Fourier transforms of distributions.

In the case of equation (2) the series

$$\sum_{k=1}^{\infty} a_k(t)e^{-i(\varphi_k(t)\cdot\xi)}$$

converges absolutely to a function which can be continued in $\mathbb{C}^n$ as n entire function of the exponential type $\le c$.

Differentiaion of both sides of (1) with respect to t and subsequent substitution t=0 leads to the system (3).

Remark 1. Observe that a necessary condition for the existence of a solution of the equation (1) is that for every s and t in Δ

$$\sum_{k=1}^{\infty} a_k(t,D)g(x+\varphi_k(t),s) = \sum_{k=1}^{\infty} a_k(s,D)g(x+\varphi_k(s),t).$$

Remark 2. In the case of $g(x,t) \equiv 0$, the system (3) is, in fact, equivalent to a finite system of partial differential equations with constant coefficients. This follows from the Hilbert Nullstellensatz which implies that every polynomial ideal in the ring of polynomials in R^n is Noetherian.

If the functions involved are analytic with respect to t, then the inverse theorem holds. More precisely, there is

Theorem 2. If in addition to the assumptions (i) and (ii) in Theorem 1, functions $a_k(t,D)$, $\varphi_k(t)$ and $\langle g(x,t),\varphi(x)\rangle$ for every $\varphi \in \mathcal{D}(R^n)$, are analytic in Δ, then every solution of the system (3) is a solution of the equation (1).

Proof is an immediate consequence of the statement that two analytic functions are equal if and only if the coefficients of their power series expansions about the same point are equal.

Corollary. In the case of $g(x,t) \equiv 0$ every solution of (2) admits an integral representation in the sense of L. Ehrenpreis (see [2], Chapter VII) and V.P. Palamodov (see [4], Chapter VI), since the system (3) is equivalent to a finite system of partial differential equations with constant coefficients.

The author is grateful to Z. Zielezny from SUNY at Buffalo for reading the paper and making valuable remarks.

REFERENCES

1. L. Ehrenpreis, 'Solutions of some problems of division I', *Amer.J.of Math.* vol.**76** (1954), pp.883-903.
2. L. Ehrenpreis, *Fourier analysis in several complex variables*, Wiley-Interscience Publishers, 1970.
3. I.M. Gelfand and G.E. Shilov, *Generalized functions*, Academic Press, 1968.
4. V.P. Palamodov, *Linear differential operators with constant coefficients*, Springer-Verlag, 1970.

LIMIT CYCLES OF CERTAIN POLYNOMIAL SYSTEMS

N.G. Lloyd
Department of Pure Mathematics
The University College of Wales
Aberystwyth
Dyfed

ABSTRACT. In this paper I shall give a brief description of some of the results on polynomial systems in the plane which have recently been obtained by my students and myself. I shall also discuss some connections with the existing literature. In Section 1, so-called small-amplitude limit cycles will be discussed, while in Sections 2 and 3, the relationship between the two-dimensional system and a certain periodic one-dimensional non-autonomous equation ((2.2)) will be exploited.

1. SOME BACKGROUND

Consider two-dimensional differential systems of the form

$$\dot{x} = P(x,y), \quad \dot{y} = Q(x,y) \tag{1.1}$$

in which P and Q are polynomials. One of the most famous outstanding problem in the theory of nonlinear oscillations is Hilbert's sixteenth problem, which is concerned with the number of limit cycles that systems such as (1.1) can have. Let S_n be the set of systems of this form with P and Q of degree at most n, and let $\pi(P,Q)$ be the number of limit cycles of (1.1). The problem is to estimate

$$H_n = \sup\{\pi(P,Q); (P,Q) \in S_n\}$$

in terms of n and to seek information on the possible configurations of limit cycles.

Very little is known about the numbers H_n - it has not even been established whether or not they are finite. Perhaps more remarkably, it has not been proved that a polynomial system cannot have infinitely many limit cycles - the proof proposed by Dulac in 1923 [10] is now known to be incomplete. In a notable recent development, Bamón [4] has shown that quadratic systems cannot have infinitely many limit cycles. Indeed, most of the work on Hilbert's problem has been concerned with quadratic systems. Although it was thought for some years that $H_2 = 3$, it was shown by Shi [21] and a number of other authors that there are quadratic

S. P. Singh (ed.), Nonlinear Functional Analysis and Its Applications, 317–326.

systems with at least four limit cycles (see [6] for references); there are now claims that $H_2 = 4$.

I has proved useful to consider limit cycles which bifurcate out of a critical point; this is a technique which has been used successfully in several investigations of polynomial systems, and we refer the reader to [18] and [6] for details. To summarize the idea, recall that a critical point is a fine focus for (1.1) if it is a centre for the linearized system

$$\dot{\xi} = \left[\frac{\partial(P,Q)}{\partial(x,y)}(p)\right]\xi \qquad (\xi \in \mathbb{R}^2).$$

We start with a fine focus of as high an order as possible and make a sequence of perturbations of the coefficients of P and Q each of which reverses the stability of the critical point. At each stage, a limit cycle bifurcates; such limit cycles are said to be of small amplitude.

To be precise suppose that the origin is a critical point of (1.1) of focus type; in canonical coordinates, the system is of the form

$$\left.\begin{aligned} \dot{x} &= \lambda x + y + p(x,y) \\ \dot{y} &= -x + \lambda y + q(x,y) \end{aligned}\right\} \qquad (1.2)$$

where p and q are polynomials without linear terms. A Liapunov function V is sought such that $\dot{V}$, the rate of change of V along orbits, is of the form $\eta_2 r^2 + \eta_4 r^4 + \ldots$, where $r^2 = x^2 + y^2$ and the coefficients η_{2i} are polynomial functions of the coefficients of P and Q. It is known that such a choice of V is possible (see [20], for example), and it is easily checked that $\eta_2 = \lambda$. The η_{2i} are termed the focal values, and the order of the fine focus is k if $\eta_2 = \eta_4 = \ldots = \eta_{2k} = 0$ and $\eta_{2k+2} \neq 0$. The ring generated by the η_{2i} has a finite basis; the elements of the basis are called the Liapunov quantities, and are calculated recursively by setting $\eta_2 = \eta_4 = \ldots = \eta_{2i-2}$ in the expression for η_{2i} for i=2,3,... For a given class of systems, the aim is to find the maximum possible number of limit cycles which can be generated by bifurcation out of the origin. For this, we need to know the smallest value of k such that $\eta_{2i} = 0$ for all i if $\eta_2 = \eta_4 = \ldots = \eta_{2k} = 0$; we are then in a position to generate k-1 small amplitude limit cycles.

The necessary calculations (of the η_{2k}) are extremely complicated, involving polynomials in the coefficients occurring in p and q whose degree increases with k. For this reason, an algorithm was developed which can be implemented on a computer using Symbolic Manipulation Techniques. This was described in [6], where it was shown that, if p and q are homogeneous cubic polynomials, then there can be no more than five small-amplitude limit cycles; examples of systems with exactly five were given. Thus $H_3 \geq 5$. The well-known result that quadratic systems can have no more than three small-amplitude limit cycles can be verified using the same technique.

In view of these results, it was natural to consider systems in which p and q contain both quadratic and cubic terms. The calculations in this case are indeed massive, and we have been able to calculate relatively few of the focal values. The computations were performed using the REDUCE symbolic manipulation package, and the results can be found in [11]. For particular classes of systems, the calculations are, of course, more manageable. For example, we have considered systems of the form

$$\left.\begin{aligned}\dot{x} &= \lambda x + y + Ax^2 + Bxy + Cy^2 - G(x^2+y^2)y + \mu(x^2+y^2)x \\ \dot{y} &= -x + \lambda y + Dx^2 + Exy + Fy^2 + G(x^2+y^2)x + \mu(x^2+y^2)y\end{aligned}\right\} \quad (1.3)$$

Such systems have no critical points at infinity, and it proves possible to bifurcate a limit cycle from the line at infinity by taking μ to be sufficiently small. With suitably chosen coefficients, (1.3) has a fine focus of order four at the origin, from which four small-amplitude limit cycles bifurcate under appropriate perturbation. There is in addition another fine focus, which is encircled by a limit cycle. From this description, it is seen that $H_3 \geq 6$. The interesting feature is that limit cycles around the origin have opposite orientation to that 'at infinity'. Thus in these systems, all the nice properties of quadratic systems are violated (recall that for quadratic systems, a limit cycle can contain only one critical point in its interior domain, limit cycles encircling the same critical point have the same orientation, while limit cycles encircling different critical points have opposite orientations). These results can be refined and the estimate given above improved; the details appear in [7]. An example is also given in [11] of a cubic system with six small-amplitude limit cycles about the origin.

Since our aim is to generate as many small-amplitude limit cycles as we can, it would seem sensible to start with systems with several fine foci. It is fairly easy to verify that for quadratic systems, fine foci of order greater than one cannot coexist, and that no more than three small-amplitude limit cycles can be produced in total. This question of the coexistence of fine foci of (1.2) when p and q are homogeneous cubics is considered in [11]. It is shown that if the origin is a fine focus of order 3,4 or 5, then there can be no other fine foci; if the origin is of order 2, there can be a pair of other fine foci, but of order 1 only; if the origin is of order 1, there can be four other fine foci, none of which can have order greater than 1. It follows immediately that no more than five small-amplitude limit cycles can be produced in total.

Finally in this section, we note that results can be proved about small-amplitude limit cycles without embarking on the calculation of the focal values on a computer. This was done in [5], where Liénard type equations

$$\dot{x} = y - F(x), \quad \dot{y} = -g(x)$$

were considered. It was shown that if g is odd and the degree of F is 2n+1 or 2n+2, then there can be n small-amplitude limit cycles and no more. If g is not odd, then this result breaks down (see [13]). It had been conjectured in [15] that this result was true without the restriction to small amplitude, raising the interesting question, relating to all the work described above, whether the local results which have been proved are true 'globally'.

2. THE RELATED NON-AUTONOMOUS EQUATION

We now consider systems of the form (1.2) in which p and q are homogeneous polynomials (of degree n, say). In polar form, the system is of the form

$$\dot{r} = \lambda r + r^n f(\theta), \quad \dot{\theta} = -1 + r^{n-1} g(\theta),$$

where f and g are homogeneous polynomials in cos θ and sin θ of degree n+1. It was shown in [18] that limit cycles of these systems can be investigated by making the transformation

$$\rho = r^{n-1}(1-r^{n-1}g(\theta))^{-1}, \tag{2.1}$$

which was noted for quadratic systems by Lins Neto [14]. We then have

$$\frac{d\rho}{d\theta} \quad \alpha(\theta)\rho^3 + \beta(\theta)\rho^2 - \lambda(n-1)\rho, \tag{2.2}$$

where and

$$\left.\begin{aligned} \alpha(\theta) &= -(n-1)g(\theta)(f(\theta)+\lambda g(\theta)) \\ \beta(\theta) &= -(n-1)f(\theta) + g'(\theta) - 2\lambda(n-1)g(\lambda). \end{aligned}\right\} \tag{2.3}$$

Equation (2.2) is 2π-periodic, α and β being homogeneous polynomials in cos θ and sin θ; α is of degree 2(n+1), while β is of degree n+1. The transformation (2.1) is defined and is invertible in an open set $\mathfrak{D}$ containing the origin whose boundary is the curve $\mathfrak{C} : r^{n-1} = (g(\theta))^{-1}$. The branches of $\mathfrak{C}$ 'join' the critical points at infinity (which are given by the zeros of g); if there are no such points, $\mathfrak{C}$ is a closed curve. For quadratic systems, it is known that all limit cycles encircling the origin are contained in $\mathfrak{D}$. In general, positive 2π-periodic solutions of (2.2) yield limit cycles of (1.2) in $\mathfrak{D}$, and certainly all small-amplitude limit cycles around the origin are contained in $\mathfrak{D}$.

In [1] the complexified form of (2.2) is condidered - for reasons there explained:

$$\dot{z} = \alpha(\theta)z^3 + \beta(\theta)z^2 - \lambda_n z, \tag{2.4}$$

where $\lambda_n = (n-1)\lambda$. Most of [1] is devoted to a detailed investigation of the multiplicity of the origin (as a periodic solution) for various classes of coefficients. It is supposed that α and β are polynomials in θ or in cos θ and sin θ. If (2.2) is derived from (1.2), we say that it is of Hilbert type and the multiplicity of the origin is then one more than the order of the fine focus of (1.2) at x = y = 0.

The study of (2.4) is closely related to the subject matter of [16], where it was shown that, if α,β,γ and δ are periodic functions (of period ω, say) and α is never zero, then

$$\dot{z} = \alpha(t)z^3 + \beta(t)z^2 + \gamma(t)z + \delta(t) \tag{2.5}$$

has exactly three ω-periodic solutions, counting multiplicity. If α has zeros and changes sign, there is no upper bound for the number of periodic solutions without imposing some restrictions on the number of zeros of the coefficients. However, if α is permitted to have zeros but not to change sign, there are at most three periodic solutions (see [ ] for a proof). Our aim here is to consider equations (2.4) of Hilbert type when α does not change sign and to explore some of the consequences for the corresponding system (1.2). We therefore suppose that α and β are as given in (2.3), and restrict consideration to real solutions of (2.4). We are motivated by the paper of Chicone [8], where it is supposed that p and q in (1.2) have a common factor of degree n-1, that of Koditschek and Narendra [12], where it is further supposed that $n = 2$, and the comments of Coppel [9] thereon.

We distinguish between the two cases (i) n even and (ii) n odd. When n is even, g and β are of odd degree; hence (1.2) has critical points at infinity and $\int_0^{2\pi} \beta = 0$. Moreover, if $\phi(t)$ is a real 2π-periodic solution of (2.4), then so is $-\phi(\pi+t)$, and the two are of opposite signs.

The results which follow relate to equation (2.4) in general, not only if it is derived from the two-dimensional system (1.2).

Proposition 2.1 *Suppose that* n *is even and that* α *does not change sign. If* $\lambda \neq 0$, *there are no positive* 2π-*periodic solutions if* $\lambda\alpha \leq 0$ *and at most one if* $\lambda\alpha \geq 0$. *If* $\lambda = 0$, *there are no non-trivial periodic solutions (of either sign).*

Proof Since α does not change sign, there are at most three 2π-periodic solutions. But $z = 0$ is one of them, and by the remarks immediately preceding the statement of the theorem, there can only be one positive periodic solution. Suppose now that $\lambda\alpha \leq 0$; if ϕ is a positive periodic solution, since $\int_0^{2\pi} \beta = 0$, we have

$$0 = \int_0^{2\pi} \phi^{-2}\dot{\phi} = \int_0^{2\pi} \alpha\phi - \lambda_n \int_0^{2\pi} \phi^{-1}.$$

The right hand side is positive or negative according to whether $\lambda < 0$ or $\lambda > 0$, leading to a contraction. Finally, suppose that $\lambda = 0$. In this case, the origin is a periodic solution of multiplicity at least two; since non-trivial periodic solutions occur in pairs, there can be no such solutions (real or complex).

Proposition 2.2 *Suppose that* n *is odd and that* α *does not change sign. If* $\lambda \neq 0$, *there are at most two positive* 2π-*periodic solutions; there can be no more than one if* $\alpha \geq 0$ *and* $\lambda > 0$ *or* $\lambda = 0$.

Proof In general, there are at most two positive 2π-periodic solutions simply because there are no more than three in total, and $z = 0$ is one such solution. The conclusion in the case when $\lambda = 0$ follows as in the proof of Proposition 2.1. When $\alpha \geq 0$ and $\lambda > 0$, we have recourse to the formulae for the derivatives of the Poincaré map given in [17]. Though these were presented in a slightly different setting, they apply perfectly well to the map $q(c) = z(2\pi;0,c) - c$ defined here for $c > 0$. Since $\alpha \geq 0$, reference to [17] tells us that $q'''(c) > 0$. But because $\lambda > 0$, we have that $q(c) < 0$ if c is sufficiently small. It follows that q cannot have more than one zero.

We can make one further comment when n is odd. We allow α to take both signs but require β to be non-zero (this is not possible, of course, when n is even). We use a result proved in [2], using the methods of [16], that when $\beta \neq 0$, (2.5) has at most four periodic solutions. We therefore have the following.

Proposition 2.3 Suppose that $\beta \neq 0$ (so that n must be odd). Then (2.4) has at most three positive periodic solutions.

Although the results noted above apply to equations of the form (2.4) which are not of Hilbert type, we are here mainly interested in their counterparts for system (1.2). The conclusions then relate to limit cycles in the set $\mathfrak{D}$; as we have noted, this is no restriction when $n = 2$ and always encompasses all small-amplitude limit cycles. Our next task is therefore to interpret the condition that α does not change sign in terms of system (1.2). We thus seek hypotheses under which $g(\theta)$ and $h(\theta) = f(\theta) + \lambda g(\theta)$ change sign at the same values of θ.

Let us first follow Chicone [8] and suppose that

$$p(x,y) = (ax+dy)k(x,y), \quad q(x,y) = (cx+dy)k(x,y),$$

where k is a homogeneous polynomial of degree $n-1$. Define matrices

$$A = \begin{pmatrix} \lambda & 1 \\ -1 & \lambda \end{pmatrix}, \quad B = \begin{pmatrix} a & b \\ c & d \end{pmatrix}, \quad J = \begin{pmatrix} 0 & -1 \\ 1 & 0 \end{pmatrix}.$$

Chicone's hypotheses are that $\lambda \geq 0$ and that the symmetric parts of JB and B^*JA are definite and agree in sign. It is an easy matter to verify that the corresponding quadratic forms are $-g(\theta)/k(\theta)$ and $h(\theta)/k(\theta)$, respectively, where $k(\theta) = k(\cos\theta, \sin\theta)$. The hypotheses therefore imply that $\alpha(\theta) \geq 0$ for all θ, so that Propositions 2.1 and 2.2 are applicable; we immediately have that there can only be one limit cycle in $\mathfrak{D}$. The hypotheses of Koditschek and Narendra [12] are the same, except that they have $n = 2$; our restriction to the set $\mathfrak{D}$ is then unnecessary. It should be noted that we have not proved the existence of a periodic solution.

Chicone shows that his hypotheses imply that there are no critical points in the finite plane except for the origin. Coppel [9] emphasises the significance of this. The following is very easily verified.

Lemma There is a critical point in the finite plane other than the origin if and only if there is θ_o such that $g(\theta_o) \neq 0$ and $h(\theta_o) = 0$.

Proof If there is a critical point with $\theta = \theta_0$, $r = r_0$, then

$g(\theta_o) = r_o^{-(n-1)}$ and $f(\theta_o) = -\lambda r_o^{-(n-1)}$; hence $h(\theta_o) = 0$. Conversely, if $h(\theta_o) = 0$ and $g(\theta_o) \neq 0$, there is a critical point at $r = (g(\theta_1))^{-1}$, $\theta = \theta_1$, where $\theta_1 = \theta_o$ if $g(\theta_o) > 0$ and $\theta_1 = \theta_o + \pi$ if $g(\theta_o) < 0$.

We see that only exceptionally is α of one sign when there are finite critical points other than the origin; we therefore adopt this as a hypothesis. In this case, $g(\theta) = 0$ whenever $h(\theta) = 0$. Exploiting this implication, we have the following.

Proposition 2.4 *Suppose that the origin is the only critical point in the finite plane. Then* α *does not change sign if one of the following conditions holds:*
(1) *there are no critical points at infinity (in which case* n *is odd);*
(2) *there is only one pair of critical points at infinity (in which case* n *is even) and at these* g *changes sign;*
(3) *if* g *changes sign at* θ, *then so does* h.

Proof The proof is straightforward. For (2), we note that if g only has one pair of zeros, then these are the only candidates for the zeros of h; but h is of odd degree, and so must change sign.

Remark Proposition 2.4, part (2) is very similar to the result of Coppel [9].

Finally in this section, we comment on systems of the form

$$\dot{x} = \lambda x + y + A(x,y)k(x,y)$$
$$\dot{y} = -x + \lambda y + B(x,y)k(x,y)$$

where k is a homogeneous polynomial of degree at most n-2. As far as the results we have described are concerned, it is only necessary to consider the system with $k \equiv 1$.

$$\dot{x} = \lambda x + y + A(x,y), \quad y = -x + \lambda y + B(x,y).$$

Let the corresponding functions α be α_1 and α_2. Then $\alpha_1 = k^2\alpha_2$.

3. EQUATION (2.4) WHEN λ IS VARIED

We now give another example of how equation (2.4) can throw light on (1.2). We suppose that α and β in (2.4) are fixed and that λ varies. Initially, our results relate to (2.4) without the restriction that it is of Hilbert type; we then deduce some results for the system (1.2), which for quadratic systems were essentially proved by Mieussens [19]. More details are to be found in the dissertation of Alwash [3].

We fix α and β, denote the equation (2.4) by E_λ and write $\phi(t;c,\lambda)$ for the solution of E_λ with initial point c. We say that the origin is a centre for E_λ if every solution in some neighbourhood of 0 is periodic.

Proposition 3.1 (i) Given c_0, there is at most one λ such that $\phi(t;c_0,\lambda)$ is periodic. (ii) Suppose that $x = 0$ is a centre for E_0; then for $\lambda \neq 0$, E_λ has no periodic solutions.

Proof (i) Suppose that $\phi(t;c_0,\nu)$ is periodic and $c_0 > 0$; a similar argument applies when $c_0 < 0$. We show that if $\lambda \neq \nu$, then $\phi(t;c_0,\lambda) \neq \phi(t;c_0,\nu)$ for all t for which the solutions are defined. Let $\phi_\lambda(t) = \phi(t;c_0,\lambda)$. From the equation, $\dot\phi_\lambda(0) > \dot\phi_\nu(0)$ when $\lambda < \nu$ and $\dot\phi_\lambda(0) < \dot\phi_\nu(0)$ when $\lambda > \nu$. Therefore, for sufficiently small t, $\phi_\lambda(t) > \phi_\nu(t)$ when $\lambda < \nu$ and $\phi_\lambda(t) < \phi_\nu(t)$ when $\lambda > \nu$. Suppose that $\lambda < \nu$. We show that $\phi_\lambda(t) > \phi_\nu(t)$ for $0 \leq t \leq 2\pi$. If not, there is s such that $\phi_\lambda(s) = \phi_\nu(s)$ and $\phi_\lambda(t) > \phi_\nu(t)$ for $0 \leq t < s$. Since $\dot\phi_\lambda(s) > \dot\phi_\nu(s)$, then, for t near s, $\phi_\lambda(t) < \phi_\nu(t)$ when $t < s$ and $\phi_\lambda(t) > \phi_\nu(t)$ when $t > s$. Hence, there is $s_1 < s$ with $\phi_\lambda(s_1) = \phi_\nu(s_1)$. This contradicts the choice of s. A similar argument applies when $\lambda > \nu$.
(ii) If E_0 has a centre at $x = 0$, then there is ξ such that $\phi(t;c,0)$ is periodic for all $c \in [0,\xi)$. By (i), $\phi(t;c,\lambda)$ is not periodic for $0 \leq c < \xi$ if $\lambda \neq 0$. If $\xi < \infty$, then $\phi(t;\xi,0)$ is not defined for all $t \in [0,2\pi]$. If $\lambda < 0$ then, by the argument in (i), $\phi(t;\xi,\lambda)$ is also not defined for all $t \in [0,2\pi]$; it follows that, for $c \geq \xi$, $\phi(t;c,\lambda)$ cannot be periodic. A similar argument applies for $c < 0$. If $\lambda > 0$, we apply the same argument under reversed time.

Proposition 3.2 Given c_0, suppose that there is ν such that $\phi(t;c_0,\nu)$ is defined for $0 \leq t \leq 2\pi$ and $\phi_\nu(2\pi) > \phi_\nu(0)$. There is λ such that $\phi(t;c_0,\lambda)$ is periodic.

Proof We suppose that $c_0 > 0$; the same proof works when $c_0 < 0$ if we make the transformation $t \mapsto -t$. Again let $\phi_\lambda(t) = \phi_\lambda(t;c_0)$. If $\lambda > \lambda_0$, then by the proof (i) of Proposition 3.1, $\phi_\lambda(t) < \phi_\nu(t)$. Let $M = \max\limits_{0 \leq t \leq 2\pi} \phi_\nu(t)$; then $0 \leq \phi_\lambda(t) \leq M$ for $0 \leq t \leq 2\pi$. We then have

$$\dot\phi_\lambda = \alpha\phi_\lambda^3 + \beta\phi_\lambda^2 - \lambda_n\phi_\lambda \leq K - \lambda_n\phi_\lambda,$$

where $A = \max|\alpha|$, $B = \max|\beta|$ and $K = AM^3 + BM^2$. Writing μ for λ_n, it follows that

$$\phi_\lambda(t)e^{\mu t} - \phi_\lambda(0) \leq K\mu^{-1}(e^{\mu t}-1),$$

whence

$$\phi_\lambda(2\pi) \leq e^{-2\pi\mu}(\phi_\lambda(0) - K\mu^{-1}) + K\mu^{-1}.$$

As $\lambda \to \infty$, the right hand side tends to zero. So $\phi_\lambda(2\pi) < \phi_\lambda(0)$ if λ is sufficiently large. Now $\phi_\lambda(2\pi)$ depends continuously on λ; hence there is λ such that $\phi_\lambda(2\pi) = \phi_\lambda(0)$ - that is, $\phi(t;c_o,\lambda)$ is periodic.

Remarks (i) The hypothesis of Proposition 3.2 is certainly satisfied if c_0 is sufficiently small. (ii) If $\phi_\lambda(t)$ is defined for all values of $t \in [0,2\pi]$ and all $\lambda < \lambda_0$, then it can be shown that the result holds without the condition $\phi_\nu(2\pi) > \phi_\nu(0)$.

We now apply the above results to the two-dimensional system (1.2), noting that the transformation (2.1) is independent of λ.

Proposition 3.3 Suppose that x_o and ν are such that the orbit of (1.2) with $\lambda = \nu$ through $R_o : (x_o,0)$ is entirely contained in $\mathfrak{D}$, and that $\psi_\nu(2\pi) > \psi_\nu(0)$, where $\psi_\lambda(t)$ is the solution of the corresponding equation (2.4). There exists a unique λ such that (1.2) has a limit cycle through R_o. If, for $\lambda = 0$, the origin is a centre, then, for any $\lambda \neq 0$, there are no limit cycles in $\mathfrak{D}$.

Remark We note again that this result encompasses all small-amplitude limit cycles, and for $n = 2$, all limit cycles encircling the origin.

REFERENCES

[1] M.A.M. Alwash and N.G. Lloyd. Non-autonomous equations related to polynomial two-dimensional systems. Preprint, University College of Wales, Aberystwyth (1985).
[2] M.A.M. Alwash and N.G. Lloyd. Periodic solutions of a quartic non-autonomous equation. Preprint, University College of Wales, Aberystwyth (1985).
[3] M.A.M. Alwash. Bifurcation of periodic solutions of non-autonomous ordinary differential equations. Dissertation, University College of Wales, Aberystwyth (1985).
[4] R. Bamón. Solution of Dulac's problem for quadratic vector fields. Preprint, Instituto de Matemática Pura e Aplicada, Rio de Janeiro (1985).
[5] T.R. Blows and N.G. Lloyd. The number of small-amplitude limit cycles of Liénard equations. Math. Proc. Cambridge Philos. Soc. 95 (1984), 359-366.
[6] T.R. Blows and N.G. Lloyd. The number of limit cycles of certain polynomial differential equations. Proc. Roy. Soc. Edinburgh 98A (1984), 215-239.

[7] T.R. Blows, M.C. Kalenge and N.G. Lloyd. Bifurcating limit cycles of certain cubic systems. Preprint, University College of Wales, Aberystwyth (1985).

[8] C. Chicone. Limit cycles of a class of polynomial vector fields in the plane. J. Differential Equations, to appear.

[9] W.A. Coppel. A simple class of quadratic system. Research report No.7 (1985), Australian National University, Canberra.

[10] H. Dulac. Sur les cycles limites. Bull. Soc. Math. France 51 (1923), 45-188.

[11] M.C. Kalenge. On some polynomial systems in the plane. Dissertation, University College of Wales, Aberystwyth (1985).

[12] D.E. Koditschek and K.S. Narendra. Limit cycles of planar quadratic differential equations. J. Differential Equations 54 (1984), 181-195.

[13] T. Kohda, K. Imamura and Y. Oono. Small-amplitude periodic solutions of the quadratic Liénard equation. Trans. IECE Japan E 68 (1985), 154-158.

[14] A. Lins Neto. On the number of solutions of the equation $\frac{dx}{dt} = \sum_{j=0}^{n} a_j(t)x^j$, $0 \leq t \leq 1$, for which $x(0) = x(1)$. Inventiones Mathematicae 59 (1980), 67-76.

[15] A. Lins Neto, W. de Melo and C.C. Pugh. On Liénard's equation. In Geometry and Topology (Rio de Janeiro, 1976). Lecture Notes in Mathematics, no.597 (Springer-Verlag, 1977), 335-357.

[16] N.G. Lloyd. The number of periodic solutions of the equation $\dot{z} = z^N + p_1(t)z^{N-1} + \ldots + p_N(t)$. Proc. London Math. Soc. (3) 27 (1973), 667-700.

[17] N.G. Lloyd. A note on the number of limit cycles in certain two-dimensional systems. J. London Math. Soc. (2) 20 (1979), 277-286.

[18] N.G. Lloyd. Small-amplitude limit cycles of polynomial differential equations. In Ordinary Differential Equations and Operators, edited by W.N. Everitt and R.T. Lewis, Lecture Notes in Math., no.1032 (Springer-Verlag, 1982), 346-357.

[19] M. Mieussens. 'Sur les cycles limites des systems quadratiques.' C.R. Acad. Sci. Paris Ser. A, 291 (1980), 337-340.

[20] V.V. Nemystkii and V.V. Stepanov. Qualitative Theory of Differential Equations (Princeton University Press, 1960).

[21] Shi Songling. A concrete example of the existence of four limit cycles for plane quadratic systems. Scientia Sinica 23 (1980), 153-158.

CONVEXITY STRUCTURES AND KANNAN MAPS

G. Oldani - D. Roux
Dipartimento di Matematica
Universita degli Studi di Milano
via Cesare Saldini, 50
20133 MILANO (Italy)

ABSTRACT. A fixed point theorem about a "nonexpansive generalized Kannan map" is given in an abstract form, using the notion of convexity structure. This contains and extends various known theorems.

1. J. P. Penot ([4]) and W. A. Kirk ([2],[3]) using the notion of convexity structure, reformulate in a more abstract setting the well known theorem of Browder-Goede-Kirk about fixed points of nonexpansive mappings and some related results. With such a technique, proofs of other fixed point theorems can be carried out in an abstract framework in order to unify and generalize the results. In this paper we deal with so-called "generalized Kannan maps" in a metric space possessing convexity structure.

2. Let (X,d) be a metric space and a G convexity structure in (X,d) ([1]). For every subset $I \subseteq X$, let

$$\text{diam } I = \text{Sup } \{d(x,y) \ / \ x, y \in I\}$$

and

$$co_G \ I = \cap \{S \ / \ S \in G \text{ and } S \supseteq I\}.$$

We define G a <u>quasi-normal relative structure</u> (q.n.r. structure), if for every bounded non void non singleton $S \in G$, there exists z_S in X such that

([1]) i.e. a class of subsets of X such that

(a) $\phi \in G$, $X \in G$, $x \in G$ for every $x \in X$;

(b) if $\{S_\alpha\}_{\alpha \in A} \subseteq G$, then $\bigcap_{\alpha \in A} S_\alpha \in G$.

S. P. Singh (ed.), Nonlinear Functional Analysis and Its Applications, 327–333.

$$d(z_s,x) < \text{diam } S \quad \text{for every} \quad x \in S; \tag{2.1}$$

if $y \in X$ and $r > \text{diam } S$, then

$$d(x,y) < r \text{ for every } x \in S \Rightarrow d(z_s,y) \leq r. \tag{2.2}$$

In particular, if the convexity structure P(X) of the subsets of X is a q.n.r. structure, we say that X has q.n.r. structure. In this case, every convexity structure of X is a q.n.r. structure.

Examples of normed and metric spaces with q.n.r. structure can be found in [9] and [7].

In particular, L^p-spaces ($1 \leq p \leq \infty$), separable Banach spaces, spaces with quasi-normal structure, dual spaces of (complex) AL-spaces have q.n.r. structure.

Q.n.r. convexity structures can also exist in metric spaces which haven't q.n.r. structure.

For example, let X be the set of R^2-points of the form $(x,0)$ or $(0,x)$ with $x \geq 1$ and G the convexity structure whose elements are X and the convex subsets of X. X doesn't possess q.n.r. structure but G is a q.n.r. convexity structure of X.

For every $x \in X$ and $r \geq 0$, let

$$B(x,r) = \{y \in X \ / \ d(y,x) \leq r\}.$$

G is said to be an <u>admissible convexity structure</u> (²), if for every $x \in X$ and $r \geq 0$ we have $B(x,r) \in G$.

There exist q.n.r. convexity structures which are not admissible. It suffices to consider the above example.

3. Let us now consider a <u>generalized Kannan map</u> $T : X \to X$ which for every $x, y \in X$ satisfies

$$d(Tx,Ty) \leq a(x,y)d(x,Tx) + a(y,x)d(y,Ty) \tag{3.1}$$

where $a : X \times X \to R^+$ is such that

$$a(x,y) + a(y,x) \leq 1 \tag{3.2}$$

$$a(x,y) \to 1 \Rightarrow \text{Max } \{d(x,Tx),d(y,Ty)\} \to 0 \quad \text{or } \infty. \tag{3.3}$$

Maps satisfying (3.1) were first considered in [1] ($a(x,y) = a(y,x) = \alpha < 1/2$), in [5] ($\alpha = 1/2$), in [10] ($a(x,y) + a(y,x) \leq \beta < 1$), in [6] ($a(x,y) + a(y,x) < 1$) and in the general case in [7].

(²) see [4] and [2].

The following theorem holds.

THEOREM Let (X,d) be a metric space which possesses an admissible, countably compact (3) and q.n.r. convexity structure. Then every map $T : X \to X$ which satisfies (3.1) has a (unique) fixed point in X.

If we leave out the hypothesis that G is admissible, the theorem is no longer true.

Indeed, let X and G be as in the foregoing example; let

$$T(x) = \begin{cases} (\alpha,0) & \text{if } x = (0,\alpha) \\ (0,\alpha) & \text{if } x = (\alpha,0). \end{cases}$$

G is a countably compact q.n.r. convexity structure, $T : X \to X$ satisfies (3.1) with $a(x,y) = 1/2$ for every $x, y \in X$ and it has no fixed point in X.

Let us consider, in particular, a subset X of a normed space E and set

$$d(x,y) = \| x - y \| \quad \text{for every} \quad x, y \in X.$$

If X is a weakly compact set (a weak* closed set of a dual space), the convexity structure, whose elements are the weak closed (X and the weak* closed bounded) subsets of X, is an admissible (4) compact convexity structure.

Then we have the following

COROLLARY Let X be a weak compact (weak* closed) subset of a normed (dual) space. Every map $T : X \to X$ satisfying (3.1) has a (unique) fixed point in X.

Then the theorem of this paper unifies and generalizes to the metric spaces the results of [8] [11] and [7].

4. In the proof of the theorem we need three lemmas.

The first one points out a property of admissible convexity structures; the other lemmas concern the maps T satisfying (3.1).

Throughout the sequel, let G be a convexity structure of a metric space (X,d) and $T : X \to X$ satisfying (3.1).

Let us set for every $r \geq 0$

$$A_r = \{x \in X \mid d(x,Tx) \leq r\}$$

and, if $A_r \neq \phi$, $C_r = co_G \ TA_r$.

---

(3) i.e. such that each countable subfamily of G, which has the finite intersection property, has non void intersection.

(4) $B(x,r)$ is the intersection of X and the weak (weak*) closed ball of E with centre x and radius r.

LEMMA 1 <u>If G is an admissible convexity structure, then for every $I \subseteq X$.</u>

$$\operatorname{diam} \operatorname{co}_G I = \operatorname{diam} I. \tag{4.1}$$

<u>Proof</u>. It suffices to consider the case diam $I = \delta < \infty$. Set

$$A = \bigcap_{x \in I} B(x,\delta) \text{ and } C = \bigcap_{y \in A} B(x,\delta).$$

Since $A \supseteq I$, then $A \supseteq C$; therefore if $u, v \in C$, we have $d(u,v) \leq \delta$. Hence diam $C \leq \delta$.

Since for every $x \in I$ and $y \in A$ we have $d(x,y) \leq \delta$, then $C \supseteq I$. But $C \in G$, so

$$\operatorname{diam} \operatorname{co}_G I \leq \operatorname{diam} C \leq \delta$$

and (4.1) holds.

REMARK At least when X is a normed space and $d(x,y) = \| x - y\|$ for every $x, y \in X$, (4.1) implies G to be admissible.

Indeed, if there exist $x \in X$ and $r > 0$ such that $B(x,r) \notin G$, let

$$y \in \operatorname{co}_G B(x,r) - B(x,r) \neq \phi.$$

Obviously

$$d(y, x - r \frac{y - x}{\| y - x\|}) > 2r$$

and $B(x,r)$ does not satisfy (4.1).

LEMMA 2 <u>If G is an admissible convexity structure and $A_r \neq \phi$, then</u>

$$\operatorname{diam} C_r \leq r \quad \underline{\text{and}} \quad A_r \supseteq C_r.$$

<u>Proof</u>. Lemma 1, (3.1) and (3.2) give

$$\operatorname{diam} C_r = \operatorname{diam} TA_r \leq \sup_{x,y \in A} d(Tx,Ty) \leq r.$$

Now we have to prove that $d(x,Tx) \leq r$ for every $x \in C_r$.

Fix $x \in C_r$; if $d(x,Tx) > 0$, let $s = \operatorname{Sup} \{d(Tz,Tx) \, / \, z \in A_r\}$. Obviously $B(Tx,s) \supseteq TA_r$.

Since $B(Tx,s) \supseteq C_r$ and $x \in C_r$, we have $d(x,Tx) \leq s$.

For every $\varepsilon > 0$, there exists $z \in A_r$ such that $d(Tx,Tz) \leq s - \varepsilon$. Then we have

$$d(x,Tx) - \varepsilon \leq s - \varepsilon \leq d(Tx,Tz) \leq a(z,x)d(z,Tz) + a(x,z)d(x,Tx).$$

Since $d(x,Tx) > 0$, we have, remembering (3.2) and (3.3),

$$d(x,Tx) \leq \frac{a(z,x)\,d(z,Tz) + \varepsilon}{1 - a(x,z)}$$

with $1 - a(x,z) > \alpha > 0$.

Then $d(x,Tx) \leq r$. q.e.d.

LEMMA 3 Let G be an admissible q.n.r. convexity structure of X. If $\operatorname{diam} C_r > 0$, then $d(z_{C_r}, Tz_{C_r}) < r$.

Proof. Let us set $d(z_{C_r}, Tz_{C_r}) = p$.

If $p = r$, we have $z_{C_r} \in A_r$, $Tz_{C_r} \in C_r$.

Then $d(z_{C_r}, Tz_{C_r}) < r$, absurd.

If $p > r$, for every $x \in A_r$ we have

$$d(Tz_{C_r}, Tx) \leq a(z_{C_r}, x)p + (1 - a(z_{C_r}, x))r = r + a(z_{C_r}, x)(p - r)$$

$$\leq r + \sup_{x \in A_r} a(z_{C_r}, x)(p - r) < p' < p.$$

Then

$$d(C_r, Tz_{C_r}) \leq \operatorname{diam}(C_r \cup \{Tz_{C_r}\}) = \operatorname{diam} co_G (C_r \cup \{Tz_{C_r}\})$$

$$= \operatorname{diam} co_G (TA_r \cup \{Tz_{C_r}\}) < p'.$$

Since (Lemma 2) $\operatorname{diam} C_r \leq r < p'$, condition (2.2) gives

$$d(z_{C_r}, Tz_{C_r}) \leq p',$$

absurd and lemma 3 is proved.

5. Proof of the theorem. Let us set

$$r_0 = \operatorname{Inf}\{d(x,Tx),\ x \in X\}.$$

We have to prove that $A_{r_0} \neq \phi$ and $r_0 = 0$.

Let $\{r_n\}_{n=1}^{\infty} \subseteq R^+$ be a sequence such that $A_{r_n} \neq \phi$ for every n and $r_n \downarrow r_0$.

Then $C_{r_n} \neq \phi$ and $C_{r_{n+1}} \subseteq C_{r_n}$ for every n.

Since G is countably compact, $\bigcap_n C_{r_n} \neq \phi$.

Then lemma 2 gives $\bigcap_n A_{r_n} \supseteq \bigcap_n C_{r_n}$.

Since $A_{r_0} = \bigcap_n A_{r_n}$, we have $A_{r_0} \neq \phi$.

If $r_0 > 0$, lemma 2 gives diam $C_{r_0} = r_0$; hence, by lemma 3,

$$d(z_{r_0}, Tz_{r_0}) < r_0, \text{ absurd.}$$

This ends the proof.

References

1. R. Kannan, *Some Results on Fixed Points*, Bull. Calcutta Math. Soc. 60 (1968), 71-76.

2. W.A. Kirk, *Nonexpansive Mappings in Metric and Banach Spaces*, Rend. Sem. Mat. Milano 51 (1981), 133-144.

3. W.A. Kirk, *An Abstract Fixed Point Theorem for Nonexpansive Mappings*, Proc. Amer. Math. Soc. 82 (1981), 640-642.

4. J.P. Penot, *Fixed Point Theorems Without Convexity*, Analyse non convexe (1977, Pau), Bull. Soc. Mat. France, 60 (1979), 129-152.

5. S. Riech, *Kannan's Fixed Point Theorem*, Boll. Un. Mat. Ital. (4) 4 (1971), 1-11.

6. D. Roux - E. Maluta, *Contractive Kannan Maps in Compact Spaces*, Riv. Mat. Univ. Parma (4) 5 (1979), 141-145.

7. D. Roux - C. Zanco, *Kannan Maps in Normed Spaces*, Atti Accad. Naz. Lincei Cl. Sci. Fis. Natur., 65 (1978), 252-258.

8. P.M. Soardi, *Struttura Quasi-normale e Teoremi di Punto Unito*, Rend. Ist. Mat. Univ. Trieste, 4 (1972), 105-114.

9. P.M. Soardi, *Existence of Fixed Points of Nonexpansive Maps in Certain Banach Lattices*, Proc. Amer. Math. Soc. 73 (1979), 23-29.

10. R.M. Tiberio Bianchini, *Su Un Problema di S. Reich riguardante la Teoria dei Punti Fissi*, Boll. Un. Mat. Ital. (4) 5 (1972), 103-108.

11. C.S. Wong, *On Kannan Maps*, Proc. Amer. Math. Soc., 47 (1975), 105-111.

Work supported by Italian M.P.I. and C.N.R.

ON SOME CONVERSES OF GENERALIZED BANACH CONTRACTION PRINCIPLES

B. Palczewski and A. Miczko
Department of Mathematics
Technical University of Golausk
Majakowskiego 11/12
80952 Gdansk, POLAND

The purpose of this paper is to state some converses to generalized contraction principles for pairs of selfmappings on metric spaces.

We prove here (see Theorem 2.1) that if f_1, f_2 are known to be selfmappings (not necessarily continuous ones) on metric space (X,d) and there exist a point $\bar{x} \in X$ and some real numbers $\alpha_i \in (0,1)$, such that $d(\bar{x}, f_i x) \leq \alpha_i\, d(\bar{x}, x)$ for each $x \in X$, $i = 1,2$, then for each $\gamma \in (0, \frac{1}{3})$ there exists a metric d_γ, topologically equivalent to d, and complete if d is complete, such that $d_\gamma (f_1 x, f_2 y) \leq \gamma(d_\gamma (x,y) + d_\gamma (x, f_2 y) + d_\gamma (f_1 x, y))$, $x,y \in X$.

For continuous $f_1, f_2 : X \to X$ we get the following result (see Theorem 2.3) : if there exists $\bar{x} \in X$ and $\alpha > 0$ such that $d(\bar{x}, f_i x) \leq \alpha\, d(\bar{x}, x)$, $x \in X$, $i = 1,2$, then conditions (i) and (ii) are equivalent.

(i) for each $\lambda_i \in (0,1)$ there exists a metric d_{λ_i}, topologically equivalent to d, and complete if d is complete, that $d_{\lambda_i} (f_i x, f_i y) \leq \lambda_i\, d_{\lambda_i} (x,y)$, $x,y \in X$, $i = 1,2$.

(ii) for each $\gamma \in (0, \frac{1}{3})$ there exists a metric d_γ, topologically equivalent to d, and complete if d is complete, that $d_\gamma(f_1 x, f_2 y) \leq \gamma(d_\gamma(x,y) + d_\gamma(x, f_2 y) + d_\gamma(f_1 x, y))$, $x,y \in X$.

This paper also includes some converses of coincidence type for commuting selfmapping on metric spaces.

Our results generalize the well-known converses of Banach fixed point principle for continuous mappings stated by L. Janos [9], P. Meyers [16], and others (see example [1], [2], [4], [5], [7]).

S. P. Singh (ed.), Nonlinear Functional Analysis and Its Applications, 335–351.

Many authors formulate some interesting converses of Banach fixed-point theorem in uniform spaces (see [19]) or give converses to Banach theorem in generalized metric spaces the metrics of which admit values in partially ordered sets (see [13] and [25]) but will not be dealt with here.

§1. DEFINITIONS, NOTIONS AND LEMMAS

Let X be a nonempty set and let d and e be two metrics on X. We say that d is topologically equivalent to e, if the topologies τ_e and τ_d generated by e and d repectively are the same.

Metrics d and e on X are C-equivalent, if every (x_n) is a Cauchy sequence in (X,d) iff it is a Cauchy sequence in (X,e).

Remark 1.1. If d and e are metrics on X, then C-equivalence of d and e implies the topologically equivalence of this metrics.

Let X be a nonempty set, $f : X \to X$ and let x_0 be given. We say that (x_n) is (f,x_0) -orbit, if $x_n = f^n x_0$, $n = 0,1,\ldots$.

Let (X,d) be a metric space, f be a selfmapping on X and point $x_0 \in X$ be given. The (f,x_0) -orbit (x_n) is a Cauchy (f,x_0) -orbit if (x_n) is a Cauchy sequence. We say that (X,d) is (f,x_0) -orbitally complete, if a Cauchy (f,x_0) -orbit is convergent to $x \in X$ and (X,d) is f -orbitally complete, if it is (f,x_0) -orbitally complete for any $x_0 \in X$.

Remark 1.2. The above definition of f -orbitally completeness of (X,d) is a slight modification of the well-known definition of L. Ćirić's paper [3].

Lemma 1.1. (The Banach contraction principle) Let (X,d) be a metric space and let $f : X \to X$. Suppose that there exists $\alpha \in [0,1)$ that $d(fx,fy) \le \alpha\, d(x,y)$, $x,y \in X$. If there exists $x_0 \in X$ that (X,d) is (f,x_0) -orbitally complete then there exists the point $\bar{x} \in X$ that

(i) $\bar{x} = f\,\bar{x}$

(ii) $d(f^n x,\bar{x}) \to 0$ as $n \to \infty$ for each $x \in X$

(iii) there exists an open neighbourhood U of $\bar{x}$ such that $f^n(U) \to \{\bar{x}\}$, i.e. for each neighbourhood V of $\bar{x}$ there exists $n(V) \in N$, that $f^n(U) \subset V$ for $n > n(V)$

(iv) f is a continuous selfmapping

(v) for any neighbourhood W of $\bar{x}$ there exists some neighbourhood V of $\bar{x}$ such that $x \in V$ implies $f^n(x) \in W$, $n \in N$.

We say that the mapping $f : X \to X$ is a contraction on a metric space (X,d), and we write $f \in c(X,d)$, if for each $\lambda \in (0,1)$ there exists a metric d_λ, topologically equivalent to d, and complete if d is complete, that $d_\lambda(fx,fy) \le \lambda d_\lambda(x,y)$, $x,y \in X$.

Lemma 1.2. (see P. Meyers theorem [16]) Let X be a metrizable space whose topology is generated by d and let f be a continuous selfmapping on X. If there exists $\bar{x} \in X$ such that

(i) $\bar{x} = f\bar{x}$

(ii) $d(f^n x,\bar{x}) \to 0$ as $n \to \infty$ for each $x \in X$

(iii) there exists an open neighbourhood U of $\bar{x}$ such that

$f^n(U) \to \{x\}$ as $n \to \infty$,
then $f \in c(X,d)$.

Remark 1.3. From a larger collection of special papers we known a wider class of continuous selfmappings f on the complete metric space (X,d) fulfilling assumptions (i) - (iii) of P. Meyers theorem (see D. Xieping [24], Th. 7). On the other hand a large number of contractive mappings are not continuous ones.

Lemma 1.3. (The generalized Banach contraction principle for a pair of mappings) Let f_1,f_2 be selfmappings on a metric space (X,d) such that $d(f_1x,f_2y) \le \gamma(d(x,y) + d(x,f_2y) + d(f_1x,y))$, $x,y \in X$, where $\gamma \in [0,\frac{1}{3})$. If there exists $x_0 \in X$ that (X,d) is $(f_2 \circ f_1,x_0)$ or $(f_1 \circ f_2,f_1x_0)$ -orbitally complete, then there exists $\bar{x} \in X$ that $d(\bar{x},f_i x) \le K\, d(\bar{x},x)$, $x \in X$, where $K = 2\gamma(1 - \gamma)^{-1}$.

Proof. We define sequence $(x_n)_{n\in N_0}$ in the following way

$x_{2n} = (f_2 \circ f_1)^n x_0$ and $x_{2n+1} = (f_1 \circ f_2)^n f_1 x_0$, $n = 0,1,\ldots$. By standard arguments we find out that (x_n) is a Cauchy sequence. Thus $x_{2n} \to \bar{x}$ and $x_{2n+1} \to \bar{x}$ as $n \to \infty$ for some $\bar{x} \in X$. We have $d(\bar{x},f_2 x)$ $\le d(\bar{x},f_1 x_{2n}) + \gamma(d(x_{2n},x) + d(x_{2n},f_2 x) + d(x_{2n+1},x))$, $x \in X$, $n \in N_0$. Taking $n \to \infty$ we get $d(\bar{x},f_2 x) \le \gamma(2d(\bar{x},x) + d(\bar{x},f_2 x))$ and thus $d(\bar{x},f_2 x) \le 2\gamma(1 - \gamma)^{-1} d(\bar{x},x)$. In an analogical way, $d(\bar{x},f_1 x)$ $\le 2\gamma(1 - \gamma)^{-1} d(\bar{x},x)$.

From Lemma 1.3 we easily obtain the following

Lemma 1.4. Let (X,d) be a metric space and let $f : X \to X$ be such that $d(f x,f y) \le \gamma(d(x,y) + d(x,f y) + d(f x,y)$, $x, y \in X$, $\gamma \in [0,\frac{1}{3})$. If there exists $x_0 \in X$, that (X,d) is (f,x_0) -orbitally complete, then there is $\bar{x} \in X$, that $d(\bar{x}, f\,x) \le 2\gamma(1-\gamma)^{-1} d(\bar{x}, x)$ for any $x \in X$.

We say that a selfmapping f on a metric space (X,d) belongs to class $gc(X,d)$, i.e. f is a generalized contraction on (X,d), if for each $\lambda \in (0,1)$ there exists a metric d_λ, topologically equivalent to d, and complete if d is complete, such that $d_\lambda(f\,x, f\,y) \le \gamma(d_\lambda(x,y) + d_\lambda(x, f\,y) + d_\lambda(f\,x, y))$, $x, y \in X$, where $\gamma = \frac{1}{3}\lambda$.

Let f_1, f_2 be selfmappings on (X,d). We say that pair (f_1,f_2) belongs to the class of generalized contractions of pairs and we write $(f_1,f_2) \in gcp(X,d)$ if for each $\lambda \in (0,1)$ there exists a metric d_λ, topologically equivalent to d, and complete if d is complete, that $d_\lambda(f_1\,x, f_2\,y) \le \gamma(d_\lambda(x,y) + d_\lambda(x, f_2\,y) + d_\lambda(f_1\,x, y))$, $x,y \in X$, $\gamma = \frac{1}{3}\lambda$.

K. Goebel in [6] (see also R. Machucca [16] and M. Khan [12]) proved the coincidence theorem for two mappings from a nonempty set into a complete metric space. Now we give some versions of K. Goebel result.

Lemma 1.5. Let A and X be nonempty sets and let $f,g : A \to X$ be such that $f(A) \subset g(A)$. Suppose that there exist $\lambda \in (0,1)$ and a metric d on $g(A)$, that $d(f\,u, f\,v) \le \lambda d(g\,u, g\,v)$, $u,v \in A$. If $(g(A),d)$ is (h,x_0) -orbitally complete for some $x_0 \in g(A)$, where $h(x) = (f \circ g^{-1})(x)$, $x \in g(A)$, then there exist $\bar{x} \in g(A)$ and $\bar{u} \in A$ that

(i) $\bar{x} = f\,\bar{u} = g\,\bar{u}$

(ii) $d(h^n\,x, \bar{x}) \to 0$ for any $x \in g(A)$

(iii) there exists an open neighbourhood U of $\bar{x}$ in $g(A)$, that $h^n(U) \to \{\bar{x}\}$.

(iv) h is continuous

(v) for each neighbourhood W of $\bar{x}$ in $g(A)$ there exists a neighbourhood V of $\bar{x}$, that $h^n(x) \in W$, $n \in N$ for each $x \in W$.

Remark 1.4. Function h of Lemma 1.5 is well defined, because the set $\{f(g^{-1}(x))\}$ has exactly one element for any $x \in g(A)$ (see K. Goebel [6]).

Lemma 1.6. Let A_1, A_2 and X be nonempty sets and let $f_i, g_i : A_i \to X$ be such that $f_i(A_i) \subset Z$, $i = 1,2$, where $Z = g_1(A_1) \cap g_2(A_2)$. Suppose that there exist a metric d on Z and $\lambda \in [0,1)$ that $d(f_1\,u, f_2\,v) \le \gamma(d(g_1\,u, g_2\,v) + d(g_1\,u, f_2\,v) + d(f_1\,u, g_2\,v))$, $u \in A_1$, $v \in A_2$, $\gamma = \frac{1}{3}\lambda$. If there exist $x_0 \in Z$ and the choice functions $h_i : Z \to Z$,

$h_i(x) \in f_i(g_i^{-1}(x))$, $x \in Z$, $i = 1,2$, are such that (Z,d) is $(h_2 \circ h_1, x_0)$ or $(h_1 \circ h_2, h_1 x_0)$ -orbitally complete then there exists $\bar{x} \in Z$ that $d(\bar{x}, f_i u_i) \le 2\gamma(1 - \gamma)^{-1} d(\bar{x}, g_i u_i)$ for each $u_i \in A_i$, $i = 1,2$ and $\bar{x} = f_i \bar{a}_i = g_i \bar{a}_i$ for some $\bar{a}_i \in A_i$, $i = 1,2$.

Proof. For each of the pairs of choice functions h_0, $i = 1,2$, we obtain inequality $d(h_1 x, h_2 y) \le \gamma(d(x,y) + d(x, h_2 y) + d(h_1 x, y))$, $x,y \in Z$. From Lemma 1.3, there exists $\bar{x} \in Z$, that $d(\bar{x}, h_i x) \le 2\gamma(1 - \gamma)^{-1} d(\bar{x}, x)$, $x \in Z$, $i = 1,2$. Obviously $\bar{x} = h_1 \bar{x} = h_2 \bar{x}$ and $\bar{x}$ is a unique common fixed point fo h_1 and h_2. From relation $\bar{x} \in f_i(g_i^{-1}(\bar{x}))$ we get $f_i \bar{a}_i = g_i \bar{a}_i = \bar{x}$ for some $\bar{a}_i \in A_i$, $i = 1,2$.

Remark 1.5. From Lemma 1.6 we come to the following conclusion: Let A and X be nonempty sets and let $f,g : A \to X$, $f(A) \subset g(A)$. Suppose that there exist a metric d on $g(A)$ and $\lambda \in [0,1)$ that $d(f u, f v) \le \gamma((g u, g v) + d(g u, f v) + d(f u, g v))$, $u,v \in A$, $\gamma = \frac{1}{3}\lambda$. If there exist $x_0 \in g(A)$ and the function $h : g(A) \to g(A)$, $h(x) \in f(g^{-1}(x))$, $x \in g(A)$, that $(g(A),d)$ is (h,x_0) -orbitally complete, then there exists $\bar{x} \in g(A)$, that $d(\bar{x}, f u) \le 2\gamma(1 - \gamma)^{-1} d(\bar{x}, g u)$ for any $u \in A$ and $\bar{x} = f \bar{a} = g \bar{a}$ for some $\bar{a} \in A$ (compare M. Khan [12]).

Lemma 1.7. Let X be a nonempty set and let $f_i, g_i : X \to X$, $f_i \circ g_i = g_i \circ f_i$, $f_i(X) \subset Z$, $Z = g_1(X) \cap g_2(X)$. Suppose that there is a metric d on Z and $\lambda \in [0,1)$ that $d(f_1 x, g_2 y) \le \gamma(d(g_1 x, g_2 y) + d(g_1 x, f_2 y) + d(f_1 x, g_2 y))$, $x,y \in X$, $\gamma = \frac{1}{3}\lambda$. If there exists $x_0 \in Z$ and the functions $h_i : Z \to Z$, $h_i(x) \in f_i(g_i^{-1}(x))$, $x \in Z$, $i = 1,2$, that (Z,d) is $(h_2 \circ h_1, x_0)$ or $(h_1 \circ h_2, h_1 x_0)$ -orbitally complete, then there exists $\bar{x} \in Z$ that $\bar{x}$ is a unique common fixed point of f_1, f_2, g_1 and g_2 in X and $d(\bar{x}, f_1 x) \le 2\gamma(1 - \gamma)^{-1} d(\bar{x}, g_i x)$ for each $x \in X$, $i = 1,2$.

Proof. From Lemma 1.6, there exists $\bar{x} \in Z$, that $\bar{x} = f_i \bar{a}_i = g_i \bar{a}_i$ for some $\bar{a}_i \in X$, $i = 1,2$. We have $g_i \bar{x} = g_i \circ f_i \bar{a}_i = f_i \circ g_i \bar{a}_i = f_i \bar{x}$, $i = 1,2$. Putting $\bar{z}_i = g_i \bar{x}$, $i = 1,2$, we get $d(\bar{x}, \bar{z}_2) = d(f_1 \bar{x}, f_2 \bar{a}_2) \le \gamma(d(g_1 \bar{x}, g_2 \bar{a}_2) + d(g_1 \bar{x}, f_2 \bar{a}_2) + d(f_1 \bar{x}, g_2 \bar{a}_2))$

$= 3\gamma\, d(\bar{x},\bar{z}_2)$ and $\bar{x} = \bar{z}_2$. Obviously we also have $\bar{x} = \bar{z}_1$ and thus $\bar{x} = f_i\,\bar{x} = g_i\,\bar{x}$, $i = 1,2$. From the inequality $d(\bar{x},h_i\,x) \le 2\gamma(1-\gamma)^{-1}\, d(\bar{x},x)$, $x \in Z$, we receive $d(\bar{x},f_i\,x) \le 2\gamma(1-\gamma)^{-1} d(\bar{x},g_i\,x)$, $x \in X$.

Remark 1.6. From Lemma 1.7 on a metric space (X,d) for $f,g : X \to X$, $f(X) \subset g(X)$, $f \circ g = g \circ f$ on condition that $d(f\,x,f\,y) \le \gamma(d(g\,x,g\,y) + d(g\,x,f\,y) + d(f\,x,\,g\,y))$, $x,y \in X$, $\gamma \in [0,\frac{1}{3})$ we obtain an implication: if there exists $x_0 \in g(A)$ and the function $h : g(X) \to g(X)$, $h(x) \in f(g^{-1}(x))$, $x \in g(X)$, that $(g(X),d)$ is (h,x_0) -orbitally complete then there is $\bar{x} \in g(X)$, $\bar{x} = f\,\bar{x} = g\,\bar{x}$ and $d(\bar{x},f\,x) \le 2\gamma(1-\gamma)^{-1}\, d(\bar{x},g\,x)$ for any $x \in X$.

Remark 1.7. G. Jungck in [11] proved the coincidence version of fixed point theorem for commuting mappings with slightly different conditions, which are not subject to analysis in this paper.

§2. CONVERSES OF GENERALIZED BANACH CONTRACTION PRINCIPLES

At first we shall concentrate our attention on the converse of generalized Banach contraction principle for two not necessarity continuous mappings on a metric space.

Theorem 2.1. Let (X,d) be a metric space and let $f_1,f_2 : X \to X$. Suppose that there exist the point $\bar{x} \in X$ and real numbers $\alpha_i \in (0,1)$ that the inequality holds

$$d(\bar{x},\, f_i\,x) \le \alpha_i\, d(\bar{x},\, x) \tag{2.1}$$

for $x \in X$, $i = 1,2$. Then the pair (f_1,f_2) belongs to the class of generalized contractions of pairs in (X,d), i.e. for each $\lambda \in (0,1)$ there exists a metric d_λ, topologically equivalent to d, and complete if d is complete, that $d_\lambda(f_1\,x,f_2\,y) \le \gamma(d_\lambda(x,y) + d_\lambda(f_1\,x,y) + d_\lambda(x,f_2\,y))$, $x,y \in X$, $\gamma = \frac{1}{3}\lambda$.

Proof. a) We define two families of balls $(B_n(i))_n$, $i = 1,2$, as follows

$$B_n(i) = \{x \in X : d(\bar{x},x) \le \alpha_i^n\},$$

$n \in \mathcal{F} := \{0, \pm 1, \ldots\}$.

Let

$$n_i(x) = \max\{n : x \in B_n(i)\},\ x \neq \bar{x} \text{ and}$$

$$n_i(\bar{x}) = \infty,\ i = 1,2.$$

We define function μ in the following way

$$\mu(x,y) := \begin{cases} n_1(x) + n_1(y) + n_2(x) + n_2(y) & \text{for } x \neq \bar{x} \text{ and } y \neq \bar{x} \\ 2\min\{n_1(x) + n_2(x),\ n_1(y) + n_2(y)\} & \text{for } x = \bar{x} \text{ or } y = \bar{x}. \end{cases}$$

From the definition of μ, we get for $x \neq \bar{x}$ and $y \neq \bar{x}$ the inequality

$$\mu(f_1 x, f_2 y) \geq \max\{\mu(x,y),\ \mu(x,f_2 y),\ \mu(f_1 x,y)\} + 1.$$

For $\gamma = \frac{1}{3}\lambda$ we define

$$\Xi\gamma(x,y) := \begin{cases} 0, & x = y = \bar{x} \\ \gamma^{\mu(x,y)}\, d(x,y), & x,y \in X,\ x \neq \bar{x} \text{ or } y \neq \bar{x}. \end{cases}$$

If $x \neq \bar{x}$ and $y \neq \bar{x}$, then we can easily obtain the inequality $\Xi\gamma(f_1 x, f_2 y) \leq \gamma(\Xi\gamma(x,y) + \Xi\gamma(x,f_2 y) + \Xi\gamma(f_1 x,y))$. However if $x = \bar{x}$ and $y \neq \bar{x}$, then

$$\Xi\gamma(f_1\bar{x}, f_2 y) = \Xi\gamma(\bar{x}, f_2 y) = \gamma^{\mu(\bar{x},f_2 y)}\, d(\bar{x}, f_2 y)$$

$\leq \gamma^{\mu(x,y)}\gamma\, d(\bar{x},y) = \Xi\gamma(\bar{x},y)$ and again we receive

$$\Xi\gamma(f_1 \bar{x}, f_2 y) \leq \gamma(\Xi\gamma(\bar{x},y) + \Xi\gamma(\bar{x},f_2 y) + \Xi\gamma(f_1 \bar{x},y)).$$

Thus for each $x,y \in X$,

$$\Xi\gamma(f_1 x, f_2 y) \leq \gamma(\Xi\gamma(x,y) + \Xi\gamma(x,f_2 y) + \Xi\gamma(f_1 x,y)).$$

We have $\Xi\gamma(x,y) = \Xi\gamma(y,x)$ and $\Xi\gamma(x,y) = 0$ iff $x = y$, $x,y \in X$.

b) Now we introduce the functional for which the triangle inequality holds.

Let

$d_\lambda(x,y) = \inf\{L_\gamma(\sigma_{xy}) : \sigma_{xy} \in \Sigma_{xy}\}$, where Σ_{xy} denotes the

the set of chains $[x = x_0, \ldots, x_m = y]$ and

$$L_\gamma(\sigma_{xy}) = \sum_1^m \Xi\gamma(x_i, x_{i-1}).$$

We have $d_\lambda(x,y) = d_\lambda(y,x)$, $d_\lambda(x,x) = 0$ and $d_\lambda(x,y) \le d_\lambda(x,z) + d_\lambda(z,y)$ for $x,y \in X$.

(c) We will prove that from construction $d_\lambda(x,y) > 0$ for $x \ne y$, $x,y \in X$.

Let $y \ne \bar{x}$ and let, for example, $n_i(x) \le n_i(y)$, $i = 1,2$, for some $y \in X$. Then by simple calculations we arrive at

$$d_\lambda(x,y) \ge \gamma^{2\{n_1(y) + n_2(y)\}} \min\{d(x,y),\ d(y, B_{n_1(y)+1}(1)),$$

$$d(y, B_{n_2(y)+1}(2)),\ d(x, B_{n_1(y)+1}(1)),\ d(x, B_{n_2(y)+1}(2))\}$$

and hence $d_\lambda(x,y) > 0$, (by $d(x,A)$ we denote, as usual, the distance between the point x and the set A).

Analogically, if $y = \bar{x}$ then we have

$$d_\lambda(x,\bar{x}) \ge \gamma^{2\{n_1(x) + n_2(x)\}} \min\{d(x, B_{n_1(x)+1}(1)),$$

$$d(x, B_{n_2(x)+1}(2))\} > 0.$$ Thus in this case we also have

$$d_\lambda(x,y) > 0 \text{ for } x \ne y,\ x,y \in X.$$

(d) Metrics d_λ and d are topologically equivalent.

At first let $x \ne \bar{x}$ and let $x \in (B_{n_i(x)} - k_i(i))^0$ for some $k_i \in N$, $i = 1,2$, and moreover let $n_i(y) \ge n_i(x)$ for some $y \in X$, $i = 1,2$. We have the inequality

$$d_\lambda(x,y) \le \gamma^{2\{n_1(x) + x_2(x) - k_1 - k_2\}} \min\{d(x,y),$$

$$d(x, X\setminus(B_{n_1(x)} - k_1(1))^0),\ d(y, X\setminus(B_{n_2(x)} - k_2(2))^0),$$

$$d(y,X\setminus(B_{n_1(x)} - k_1(1))^0),\ d(y,X\setminus(B_{n_2(x)} - k_2(2))^0)\}.$$

Let $\varepsilon > 0$. If $d(x,y) < \delta$, where $\delta = \gamma^{-2\{n_1(x)+n_2(x)-k_1-k_2\}}$ $\min\{1,\ d(x,X\setminus(B_{n_1(x)-k_1}(1))^0),\ d(x,X\setminus(B_{n_2(x)-k_2}(2))^0),$ $d(y,X\setminus(B_{n_1(x)-k_1}(1))^0),\ d(y,X\setminus(B_{n_2(x)-k_2}(2))^0)\}$ then $d_\lambda(x,y) < \varepsilon$ and therefore, if $d(x_n,x) \to 0$, then $d_\lambda(x_n,x) \to 0$ as $n \to \infty$.

Let $n_i(y) \le n_i(x)$, $i = 1,2$, for some $x,y \in X$, $x \ne \bar{x}$. Then the inequality holds

$$d_\lambda(x,y) \ge \gamma^{2\{n_1(x) + n_2(y) + k_1 + k_2\}} \min\{d(x,y),$$

$$d(x,B_{n_1(x)+k_1}(1)),\ d(x,B_{n_2(x)+k_2}(2)),\ d(y,B_{n_1(x)+k_1}(1)),$$

$$d(y,B_{n_2(x)+k_2}(2))\}$$

for some $k_1,k_2 \in N$.

Let

$$0 < \varepsilon < \min\{d(x,B_{n_1(x)+k_1}(1)),\ d(x,B_{n_2(x)+k_2}(2)),$$

$$d(y,B_{n_1(x)+k_1}(1)),\ d(y,B_{n_2(x)+k_2}(2))\}.$$

Then, if $d_\lambda(x,y) < \delta = \varepsilon\gamma^{2\{n_1(x)+n_2(x)+k_1+k_2\}}$ then $d(x,y) < \varepsilon$. Thus $d_\gamma(x_n,x) \to 0$ implies $d(x_n,x) \to 0$ as $n \to \infty$.

Let now $x = \bar{x}$ and let, for example, $y \in B_0(i)$, $i = 1,2$. We have

$$d_\lambda(\bar{x},y) \le \Xi\gamma(\bar{x},y) \le d(\bar{x},y)$$

and so if $d(x_n,\bar{x}) \to 0$ then $d_\lambda(x_n,\bar{x}) \to 0$ as $n \to \infty$.

For each $\varepsilon > 0$ there exists n_i that $\alpha_i^{n_i} < \frac{\varepsilon}{2}$, $i = 1,2$. If $d(x,y) > \varepsilon$ then $d(y,B_{n_i}(i)) > \frac{\varepsilon}{2}$, $i = 1,2$ and $d_\lambda(x,y)$

$\geq \gamma^{n_0} \max \{d(y,B_{n_1}(1)), d(y,B_{n_2}(2))\} > \gamma^{n_0}\frac{\varepsilon}{2}$ for some $n_0 \in N$. If $d_\lambda(\bar{x},y) < \delta = \varepsilon\gamma^{-n_0}$, then $d(\bar{x},y) < \varepsilon$. Therefore d_λ is topologically equivalent to d.

(e) Let (X,d) be complete. We will prove that in that case (X,d_λ) is also complete.

Let (x_n) be a Cauchy sequence in (X,d_λ) and let us assume that (x_n) is not convergent in (X,d_λ).
Then we have $n_i(x_n) < a_i < \infty$ for each $n \geq 0$, i.e. for each $n \geq 0$, $x_n \notin B_{a_i}(i)$, $i = 1,2$.

Let $b_i = \alpha_i^{a_i} - \alpha_i^{a_i+1}$, $i = 1,2$ and $b = \min \{b_1,b_2\}$. For sufficiently large n,

$$d_\lambda(x_n,x_{n+j}) < b\gamma^{2\{a_1+a_2+2\}}.$$

It is easy to verify that

$$d_\lambda(x_n,x_{n+j}) \geq \gamma^{2\{a_1+a_2+2\}} \min \{d(x_n,x_{n+j}), b\}.$$

In that way

$$\gamma^{-2(a_1+a_2+2)} d_\lambda(x_n,x_{n+j}) \geq d(x_n,x_{n+j})$$

and (x_n) is a Cauchy sequence in (X,d). Then $d(x_n,x) \to 0$ for some $x \in X$ and from the topological equivalence of d and d_λ, $d_\lambda(x_n,x) \to 0$. This contradication proves that (X,d_λ) is complete if (X,d) is complete.

From Theorem 2.1 we get the converse of the generalized Banach fixed point principle for one selfmapping on a metric space.

Theorem 2.2. Let $\underline{f}$ be a selfmapping on a metric space (X,d). Suppose that there exists $\bar{x} \in X$ and $\alpha \in (0,1)$ that the inequality holds

$$d(\bar{x},f\,x) \leq \alpha\, d(\bar{x},x) \tag{2.2}$$

for every $x \in X$. Then f is a generalized contraction on (X,d), i.e. for each $\lambda \in (0,1)$ there exists a metric d_λ, topologically equivalent to d, and complete if d is complete, that

$d_\lambda(f\,x, f\,y) \le \gamma(d_\lambda(x,y) + d_\lambda(x, f\,y) + d_\lambda(f\,x, y))$, $x, y \in X$,

$\gamma = \frac{1}{3}\lambda$.

Now we shall give the converse of the generalized Banach contraction principle for two continuous mappings in a metric space.

Theorem 2.3. Let (X,d) be a metric space and let $f_1, f_2 : X \to X$ be continuous mappings. Suppose that there exist the point $\bar{x}$ in X and a real number $\alpha > 0$ the inequality holds

$$d(\bar{x}, f_i\,x) \le \alpha\, d(\bar{x}, x),\ i = 1,2, \tag{2.3}$$

for each $x \in X$. Then the following conditions are equivalent

(i) mappings f_1 and f_2 are contractions on (X,d), i.e. for each $\lambda \in (0,1)$ there exist the metrics d_1, d_2, topologically equivalent to d, and complete if d is complete, such that $d_i(f_i\,x, f_i\,y) \le \lambda d_i(x,y)$, $x, y \in X$, $i = 1,2$.

(ii) the pair (f_1, f_2) fulfils a generalized contraction on (X,d), i.e. for each $\lambda \in (0,1)$ there exists a metric d_λ, topologically equivalent to d, and complete if d is complete, that $d_\lambda(f_1\,x, f_2\,y)$ $\le \gamma(d_\lambda(x,y) + d_\lambda(x, f_2\,y) + d_\lambda(f_1\,x, y))$, $x, y \in X$, $\gamma = \frac{1}{3}\lambda$.

Proof. 1° If (f_1, f_2) fulfils the condition (ii) then we have the inequality $d_\lambda(\bar{x}, f_1\,x) \le 2\gamma(1-\gamma)^{-1}\, d_\lambda(\bar{x}, x)$, $x \in X$, $i = 1,2$. Thus all assumptions of P. Meyers theorem [17] hold and $f_i \in c(X,d)$, $i = 1,2$.

2° Let $f_i \in c(X,d)$, $i = 1,2$.

(a) There exists an open neighbourhood U_i of $\bar{x}$ that $f_i^n(U_i) \to \{\bar{x}\}$, $i = 1,2$. Following P. Meyers [16] we can prove that there exist W_1 and W_2 (open in the topology generated by d) that $f_i(W_i) \subset W_i$, $i = 1,2$. We define the family $(K_n)_{n\in z}$ of closed subsets of X as follows

$$K_n := \begin{cases} \overline{f_1^n(K_0)} \cup \overline{f_2^n(K_0)}, & n \ge 1 \\ f_1^{-n}(K_0) \cap f_2^{-n}(K_0), & n \le 0, \end{cases}$$

where $K_0 = \overline{W_1 \cap W_2}$.

Let us define

$$n(x) := \begin{cases} \max\{n : x \in K_n\} & \text{for } x \neq \bar{x} \\ \infty & \text{for } x = \bar{x} \end{cases}$$

and

$$\mu(x,y) = \begin{cases} n(x) + n(y) \text{ for } x,y \in X,\ x \neq \bar{x} \text{ and } y \neq \bar{x} \\ 2 \min\{n(x), n(y)\} \text{ for } x = \bar{x} \text{ or } y = \bar{x} \end{cases}$$

From the continuity of f_1 and f_2 and from the definition of μ, we easily find that

$$\mu(f_1\ x, f_2\ y) \geq \max\{\mu(x,y),\ \mu(x,f_2\ y),\ \mu(f_1\ x,y)\} + 1$$

for $x,y \neq \bar{x}$.

(b) Let $M = \max\{1,\alpha\}$ and $1 = \frac{1}{3}\lambda\frac{1}{M}$.

Define

$$\Xi_1(x,y) = \begin{cases} 1^{\mu(x,y)} d(x,y) \text{ if } x,y \in X,\ x \neq x \text{ or } y \neq \bar{x} \\ 0 \qquad \text{if } x = y = \bar{x} \end{cases} .$$

We obtain

$$\Xi_1(f_1\ x,f_2\ y) = 1^{\mu(f_1\ x; f_2\ y)} d(f_1\ x,f_2\ y)$$

$$\leq 1^{\mu(f_1\ x,f_2\ y)} (d(x,y) + d(x,f_2\ y) + d(f_1\ x,y)),\ x,y \in X,$$

$x \neq \bar{x}$ and $y \neq \bar{x}$.

Thus for $x \neq \bar{x}$ and $y \neq \bar{x}$ the inequality holds

$$\Xi_1(f_1\ x,f_2\ y) \leq 1(\Xi_1(x,y) + \Xi_1(x,f_2\ y) + \Xi_1(f_1\ x,y))$$

$$\leq 1\,M(\Xi_1(x,y) + \Xi_1(x,f_2\ y) + \Xi_1(f_1\ x,y)).$$

If, for example, $x = \bar{x}$ and $y \neq \bar{x}$, then

$$1^{(f_1 \bar{x}, f_2 y)} = 1^{\mu(f_1 \bar{x}, f_2 y)} d(f_1 \bar{x}, f_2 y)$$

$$\leq 1^{\mu(x, f_2 y)} \propto d(\bar{x}, y) \leq 1 \, M \, \Xi_1(\overline{xy})$$

$$\leq 1 \cdot M(\Xi_1(\bar{x}, y) + \Xi_1(\bar{x}, f_2 y) + \Xi_1(\bar{x}, y)).$$

Thus in general

$$\Xi_1(f_1 x, f_2 y) \leq \gamma(\Xi_1(x,y) + \Xi_1(x, f_2 y) + \Xi_1(f_1 x, y)),$$

$$x, y \in X, \ \gamma = \frac{1}{3} \lambda.$$

We have $\Xi_1(x,y) = 0$ iff $x = y$ and $\Xi_1(x,y) = \Xi_1(x,y)$, $x,y \in X$.

(c) Similarly to Theorem 2.1, we define the desidered metric d_λ in the following way

$$d_\lambda(x,y) = \inf \{L_1(\sigma_{xy}) : \sigma_{xy} \in \Sigma_{xy}\}$$

(see proof of Theorem 2.1).

The proof that d_λ is a metric, that d_λ is topologically equivalent to d, and that d_λ is complete if d is complete is nearly identical to the proof of Theorem 2.1 (see also the proof of P. Meyers theorem [16]. So we can ignore it.

§3. COINCIDENCE TYPE CONVERSES FOR COMMUTING SELFMAPPINGS ON METRIC SPACES

Theorem 3.1. Let X be a nonempty set and $f_i, g_i : X \to X$, $f_i(X) \subset Z$, $f_i \circ g_i = g_i \circ f_i$, $i = 1,2$, $Z = g_1(X) \cap g_2(X)$. Suppose that (Z,d) is a metric space and that there exist $x_0 \in Z$ and the choice functions $h_i : Z \to Z$, $h_i(x) \in f_i(g_i^{-1}(x))$, $x \in X$, $i = 1,2$, that (X,d) is $(h_2 \circ h_1, x_0)$ or $(h_1 \circ h_2, h_1 x_0)$ -orbitally complete. Then for each $\lambda \in (0,1)$ there exists a metric d_λ, topologically equivalent to d, and complete if d is complete such that $d_\lambda(f_1 x, f_2 y) \leq \gamma(d_\lambda(g_1 x, g_2 y) + d_\lambda(g_1 x, f_2 y) + d_\lambda(f_1 x, g_2 y))$, $x,y \in X$, $\gamma = \frac{1}{3}\lambda$, iff there exists $\bar{x} \in Z$, that $d(\bar{x}, f_i x) \leq \alpha_i d(\bar{x}, g_i x)$ for $x \in X$, $\alpha_i \in (0,1)$ and $\bar{x} = f_i \bar{x} = g_i \bar{x}$, $i = 1,2$.

Proof. 1° If such a metric d_λ exists, then from Lemma 1.7 there exists $\bar{x} \in Z$ that $\bar{x} = f_i\,\bar{x} = g_i\,\bar{x}$, and $d_\lambda(\bar{x}, f_i\,x) \le 2\gamma(1-\gamma)^{-1}\,d_\lambda(\bar{x}, g_i\,x)$ for each $x \in X$, $i = 1,2$.

2° If the general assumptions of this theorem hold and $d(\bar{x}, f_i\,x) \le \alpha_i\,d(\bar{x}, x)$ for each $x \in X$, then the assumptions of Theorem 2.1 are right for h_1 and h_2. Therefore for each $\lambda \in (0,1)$ there exists a metric d_λ, topologically equivalent to d, and complete if d is complete, that $d_\lambda(h_1\,x, h_2\,y) \le \gamma(d_\lambda(x,y) + d_\lambda(x, h_2\,y) + d_\lambda(h_1\,x, y))$ for $x,y \in X$. In this way $d_\lambda(f_1\,x, f_2\,y) \le \gamma(d_\lambda(g_1\,x, g_2\,y) + d_\lambda(g_1\,x, f_2\,y) + d_\lambda(f_1\,x, g_2\,y))$ for $x,y \in X$, $\gamma = \frac{1}{3}\lambda$.

From the above converse we get the following theorem

Theorem 3.2. Let X be a nonempty set and let f,g be selfmappings on X, such that $f(X) \subset g(X)$ and $f \circ g = g \circ f$.

Let (g(X,d) be a metric space and let us assume that there exist $x_0 \in g(X)$ and the choice function $h : g(X) \to g(X)$, $h(x) \in f(g^{-1}(x))$, $x \in g(X)$, that $(g(X),d)$ is (h,x_0) -orbitally complete. Then for each $\lambda \in (0,1)$ there exists a metric d_λ, topologically equivalent to d, and complete if d is complete such that $d_\lambda(f\,x, f\,y) \le \gamma(d_\lambda(g\,x, g\,y) + d_\lambda(g\,x, f\,y) + d_\lambda(f\,x, g\,y)$, $x,y \in X$, $3\gamma = \lambda$, iff there exists $\bar{x} \in g(X)$ that $d(\bar{x}, f\,x) \le \lambda d(\bar{x}, x)$ for each $x \in X$ and $\bar{x} = f\,\bar{x} = g\,\bar{x}$, $\lambda \in (0,1)$.

On the base of Theorem 2.3 we obtain the following result

Theorem 3.3. Let X be a nonempty set and let $f_i, g_i : X \to X$ be such that $f_i(X) \subset Z$ and $f_i \circ g_i = g_i \circ f_i$, $i = 1,2$, $Z = g_1(X) \cap g_2(X)$. Suppose that there exist a metric d on Z and the point $\bar{x} \in Z$ and moreover a real number $\alpha > 0$ that $d(\bar{x}, f_i\,x) \le \alpha\,d(\bar{x}, g_i\,x)$ for any $x \in X$, $i = 1,2$. If there exists the continuous choice functions $h_i : Z \to Z$, $h_i(x) \in f_i(g_i^{-1}(x))$, $x \in Z$, then the following conditions are equivalent

(i) for each $\lambda \in (0,1)$ there exists a metric d_i topologically equivalent to d, and complete if d is complete such that $d_i(f_i\,x, f_i\,y) \le \lambda\,d_i(g_i\,x, g_i\,y)$, $x,y \in X$, $i = 1,2$.

(ii) for each $\lambda \in (0,1)$ there exists a metric d_λ, topologically equivalent to d, and complete if d is complete, that $d_\lambda(f_1\,x, f_2\,y) \le \gamma(d_\lambda(g_1\,x, g_2\,y) + d_\lambda(g_1\,x, f_2\,y) + d_\lambda(f_1\,x, g_2 y))$, $x,y \in X$, $\gamma = \frac{1}{3}\lambda$.

From P. Meyers theorem [16] and Lemma 1.6 we get

Theorem 3.4. Let X be a nonempty set and let f,g be selfmappings on X, $f(X) \subset g(X)$ and $f \circ g = g \circ f$. Let (g(X,d) be a metric space and let there exist $x_0 \in g(X)$ and $h : g(X) \to g(X)$, $h(x) \in f(g^{-1}(x))$, $x \in X$, that $(g(X),d)$ is (h,x_0) -orbitally complete. Then for each $\lambda \in (0,1)$ there exists a metric d_λ, topologically equivalent to d, and complete if d is complete such that $d_\lambda(f\,x,f\,y) \le \lambda d_\lambda(g\,x,g\,y)$, $x,y \in X$ iff h is continuous with respect to d and there exists $\bar{x} \in g(X)$, that

(i) $\bar{x} = f\,\bar{x} = g\,\bar{x}$

(ii) $d(h^n\,x,\bar{x}) \to 0$ for each $x \in g(X)$

(iii) there exists an open neighbourhood U of $\bar{x}$ in (X,d), that $f^n(U) \to \{\bar{x}\}$ as $n \to \infty$.

Remark 3.1. A. C. Babu in [1] gives the following theorem: Let X be a metrizable topological space whose topology is generated by a metric d. Let g be homeomorphism of X into itself and f a continuous self -map of X which commutes with g. Then for each $\lambda \in (0,1)$ there exists a metric d_λ, topologically equivalent to d, and complete if d is complete, such that $d_\lambda(f\,x,f\,y) \le \lambda d_\lambda(g\,x,g\,y)$, $x,y \in X$ iff

(a) there exists a point $\bar{x} \in X$ such that $f\,\bar{x} = g\,\bar{x} = \bar{x}$

(b) $g^{-n}\,f^n\,x \to \bar{x}$ for all $x \in X$

It is clear that for the truth of necessary condition of this assertion we must assume, for example, that (X,d) is $g^{-1} \circ f$ -orbitally complete. On the other hand the assumptions that f is continuous and g is homeomorphism is very strong.

Remark 3.2. B. Rhoades in [21] and [22] gives some very important and interesting remarks about the comparisons of various definitions of contractive mappings.

In paper [19] we are formulating some remarks about the comparisons of contractive mappings taking advantage for this purpose of the converses to various Banach fixed point principles (compare also D. Xieping [24], Theorem 7).

REFERENCES

1. A.C. Babu, *A converse to a generalized Banach contraction principle*, Publ, de L'Inst. Math. (N.S) 32 (46) (1982), 5-6.

2. C. Bessaga, *On the converse of the Banach fixed -point principle*, Coll. Math. VII, 1 (1959), 41-43.

3. L. Ćirić, *Generalized contractions and fixed -point theorems*, Publ. Inst. Math. Beograd (N.S) 12 (26) (1971), 19-26.

4. K. P. Chew, K. K. Tan, *Remetrization and family of commuting contractive type mappings*, Fixed Point Theory and its Applications, Acad. Press, New York - London (1976), 41-50.

5. M. Edelstein, *A short proof of a theorem of L. Janos*, Proc. Amer. Math. Soc. 20 (1969), 509-510.

6. K. Goebel, *A coincidence theorem*, Bull, de L'Acad, Polon. Des Sciences, Serie des sciences math. astr. et phys. XVI, 9, (1968), 733-736.

7. A. J. Goldman, P. R. Meyers, *Simultaneous contractification*, J. res. nat. Bur. Stand. (1968), 301-305.

8. V. I. Istratescu, *Fixed point theory*, Mathematics and Its Applications, 7, D. Reidel Publ. Comp. (1981).

9. L. Janos, *A converse of Banach's contraction theorem*, Proc. Amer. Math. Soc. 18 (1967), 287-289.

10. L. Janos, *Topological homothieties on compact Hausdorff spaces*, Proc. Amer. Math. Soc. 21 (1969), 562-568.

11. G. Jungck, *Commuting mappings and fixed points*, Amer. Math. Monthly 83 (1976), 261-268.

12. M. S. Khan, *Remarks on some fixed point theorems*, Demostratio Mathematica, XV, 2 (1982), 375-379.

13. M. Kwapisz, *Some remarks on abstract form of iterative methods in functional equation theory*, Ann. Soc. Math. Polon. Coment. Math. XIV (1984) 281-294.

14. S. Leader, *A topological characterization of Banach contractions*, Pac. J. Math. 69, 2 (1977), 461-466.

15. R. Machuca, *A coincidence theorem*, Ann. Math. Month. 74 (1967), 469.

16. P. R. Meyers, *A converse to Banach's contraction theorem*, J. Res. Nat. Bur, Standards (1967) 73-76.

17. P. R. Meyers, *Contractifiable Semigroups*, J. Res. Nat. Stand. 4 (1970), 315-322.

18. A. Miczko, B. Palczewski, *Some remarks on the Sehgal generalized contraction mappings*, Zeszyty Naukowe Politechniki Gdańskiej, Math. XII, (1982), 21-32.

19. A. Miczki, B. Palczewski, *A comparison of some definitions of contractive mappings* (in preparation).

20. P. Morales, *Topological contraction principle*, Fund. Math. CX (1980), 135-144.

21. B. E. Rhoades, *A comparison of various definitions of contractive mappings*, Trans. Amer. Soc. 226 (1977), 256-290.

22. B. E. Rhoades, *Contractive definitions revisited*, Contemporary Mathematics, 21 (1983), 189-205.

23. M. Turinici, *Nonlinear contractions and applications to Volterra Functional equations*, An. St. Univ. Iasi, S.I. 23 (1977), 43-50.

24. D. Xieping, *Fixed point theorems of generalized contractive type mappings II*, Chin. Ann. of Math. 4B (2) (1983), 153-163.

MULTIPLICITY RESULTS FOR SUPERLINEAR ELLIPTIC EQUATIONS

Bernhard Ruf
Forschungsinstitut Für Mathematik, ETH Zürich
8092 Zürich
Switzerland

INTRODUCTION.

In this note we will present some recent progress in the question of the existence of multiple solutions of superlinear elliptic boundary value problems. More precisely, we will study equations of the form

$$(1) \qquad \begin{cases} -\Delta u - f(u) = h \ , \text{ in } \Omega \subset \mathbb{R}^n \\ \qquad\qquad u = 0 \ , \text{ on } \partial\Omega \end{cases}$$

where Ω is a bounded region in $\mathbb{R}^n$, and $f \in C(\mathbb{R})$ satisfies

$$(2) \qquad \begin{aligned} f^+ &:= \lim_{s\to+\infty} \inf \frac{f(s)}{s} = +\infty \\ f^- &:= \lim_{s\to-\infty} \sup \frac{f(s)}{s} < +\infty \end{aligned}$$

We are interested to find (estimates for) the number of solutions that exist for equation (1) in dependence of a given function h.

The particular feature of equation (1) with assumption (2) is the fact that the nonlinearity f interferes strongly with the spectrum of the Laplacian. In fact, if λ_i, $i \in \mathbb{N}$, denote the eigenvalues of the Laplacian with eigenfunctions e_i, $i \in \mathbb{N}$,

$$\begin{cases} -\Delta e_i = \lambda_i e_i \ , \text{ in } \Omega \\ \quad e_i = 0 \qquad , \text{ on } \partial\Omega \end{cases}$$

then we have

$$(3) \qquad f^- < \lambda_k < f^+$$

S. P. Singh (ed.), Nonlinear Functional Analysis and Its Applications, 353–367.

for infinitely many λ_k's . We say that f crosses the eigenvalue λ_k if (3) holds. It is this interference of nonlinearity and spectrum which gives rise to an interesting and complicated solution behaviour.

AMBROSETTI-PRODI TYPE RESULTS

The situation which has been studied most extensively is when f crosses the first eigenvalue λ_1 of the Laplacian. In 1973 Ambrosetti-Prodi [3] obtained a very precise result in the case where f crosses exactly the first eigenvalue λ_1 , i.e.

(4) $$0 < f^- < \lambda_1 < f^+ < \lambda_2 ,$$

and if in addition

(5) $$f \in C^2 (\mathbb{R}) , f''(t) > 0, \forall t \in \mathbb{R} .$$

By a further characterization of Berger-Podolak [6] their result can be expressed as follows.

Theorem 1. Assume (4) and (5). Then, for every h_1 with $\int_\Omega h_1 \cdot e_1 dx = 0$ given, there exists a number $T(h_1)$ such that (1) has for $h = h_1 + te_1$ with

- $t < T(h_1)$: exactly two solutions
- $t = T(h_1)$: exactly one solution
- $t > T(h_1)$: no solution .

In subsequent papers by Hess-Ruf [12], Amann-Hess [2], Dancer, [9], Berestycky-Lions [5], deFigueiredo [10], deFigueiredo-Solimini [11] similar results were obtained under more and more general assumptions on the nonlinearity leading to the following general theorem, see [11].

Theorem 2. Assume that $f^- < \lambda_1 < f^+$ and

$$\frac{f(s)}{s^p} \xrightarrow[s \to +\infty]{} 0, \text{ for some } 1 < p < \frac{N+2}{N-2} .$$

Then, for h_1 with $\int h_1 e_1 = 0$ given there exist constants

$T_2(h_1) \le T_0(h_1)$ s.th. (1) has for $h = h_1 + te_1$ with

- $t < T_2(h_1)$: at least two solutions
- $t > T_0(h_1)$: no solution

This result seems to indicate that the general structure of either zero or two solutions is present under (essentially) the sole condition that the first eigenvalue is crossed by the non linearity. However, we note that the above theorem gives only a lower estimate on the number of solutions for $t < T_2(h_1)$. We will expose below a result (for a corresponding Sturm-Liouville problem) which shows that for the superlinear problem one should in fact expect many more solutions.

Idea of Proof: We indicate here the idea of a proof for theorem 2 in the following simpler "model problem"

$$(6) \qquad \begin{cases} -\Delta u = (u^+)^p + \lambda u + te_1 & , \Omega \\ u = 0 & , \partial\Omega \end{cases}$$

where $1 < p < \frac{N+2}{N-2}$, $\lambda < \lambda_1$.

(a) Nonexistence: Assume $t > 0$, and assume that u is a solution of (6). Since $\lambda < \lambda_1$, we conclude from the maximum principle that $u > 0$, since

$$-\Delta u - \lambda u = (u^+)^p + te_1 > 0 .$$

Multiplying (6) with e_1 (with $e_1 > 0$, $\|e_1\| = 1$) we get

$$(\lambda_1 - \lambda)(u, e_1) = ((u^+)^p, e_1) + t$$

$$\ge (\lambda_1 u - c, e_1) + t .$$

Hence we conclude that $t \le d$, i.e. there exists no solution for $t > d$.

(b) The existence of a first solution is immediate in our case: take $u_1 = \frac{t}{\lambda_1 - \lambda} e_1$, with $t < 0$. Then $u < 0$, since $\lambda_1 > \lambda$. Hence $(-\Delta - \lambda)u_1 - (u_1^+)^p = (\lambda_1 - \lambda)u_1 = te_1$. This solution is a strict local minimum of the functional $I : H_0^1(\Omega) \to \mathbb{R}$ given by

$$I(u) = \frac{1}{2}\int|\nabla u|^2 - \frac{\lambda}{2}\int|u|^2 - \frac{1}{p+1}\int|u^+|^{p+1} - \int te_1u$$

since $I'(u)v = \int\nabla u\nabla v - \lambda\int uv - \int(u^+)^pv - \int te_1v = 0 \ , \ \forall\, v \in H_0^1$

and $I''(u)[v,v] = \int|\nabla v|^2 - \lambda\int|v|^2 > 0 \ , \qquad \forall\, v \neq 0$.

Also under general assumptions on f one can obtain a strict local minimum of the corresponding functional by using the method of upper and lower solution, see [11].

The second solution is now obtained by the Mountain-Pass theorem of Ambrosetti-Rabinowitz [4]. In fact, one notes that

$$I(\alpha e_1) = \alpha^2\,\frac{\lambda_1 - \lambda}{2} - \frac{\alpha^{p+1}}{p+1}\int e_1^{p+1} - t\alpha \xrightarrow[\alpha\to+\infty]{} -\infty$$

Let now α_0 s.th. $I(\alpha_0e_1) < I(u_1)$, and define the class of paths

$$\Gamma = \{\gamma:[0,1] \to H_0^1 \ , \ \gamma(0) = u_1 \ , \ \gamma(1) = \alpha_0e_1\} \ .$$

Then set

(7) $$c = \inf_{\Gamma}\max_{\gamma([0,1])} I(u) \ .$$

By the Mountain-Pass theorem [4] , c is a critical value giving rise to a second solution for equation (1) , see fig. 1.

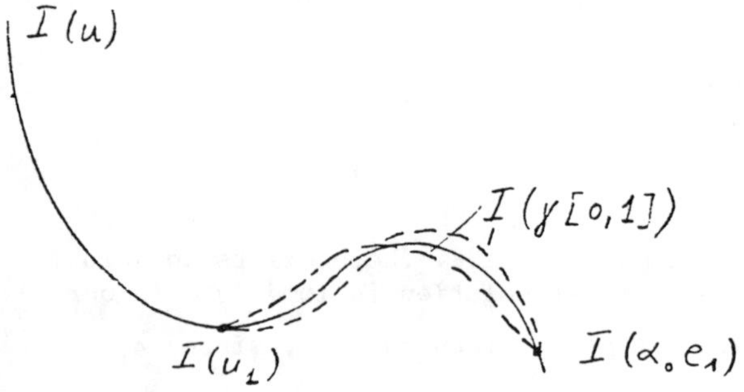

Figure 1

EXISTENCE OF MANY SOLUTIONS FOR A SUPERLINEAR STURM-LIOUVILLE PROBLEM

We now consider the following superlinear Sturm-Liouville problem

$$\begin{cases} -u'' - f(u) = t \quad \text{in } (0,1) \\ u(0) = u(1) = 0. \end{cases} \tag{8}$$

We will show that this equation with the sole assumption that $f \in C^1(\mathbb{R})$ satisfies

$$f'(+\infty) = +\infty \ , \ f'(-\infty) < \infty \tag{9}$$

has, with t decreasing, more and more solutions. This seems to be the first general result which establishes many solutions for equations of the form (8). C. Scovel [22] obtained similar results for the very specific equation,

$$-u'' = 6u^2 + t, \ u(0) = u(1) = 0.$$

Theorem 3. (Ruf-Solimini [19]). Let $f \in C^1(\mathbb{R})$ satisfy (8). Then there exists a sequence $T_k \underset{k\to+\infty}{\to} -\infty$ such that for $t < T_k$ equation (8) has at least k solutions.

Remark. This result leads to the following questions:
(a) Is the statement of theorem 3 true for the nonautonomous equation

$$\begin{cases} -u'' - f(x,u) = t\cdot p(x) \\ u(0) = u(1) = 0 \end{cases}$$

with $\quad f'(x,+\infty) = +\infty, \ \forall x \in [0,1]$

$$f'(x,-\infty) < +\infty,$$

$$p(x) > 0 \quad \text{in } [0,1].$$

(b) Do similar results hold for the corresponding partial differential equation?

Idea of proof. The proof of this result relies again on variational methods. In fact by similar arguments as above one obtains again a local minimum corresponding to a negative solution and a saddle point by mountain-pass arguments for the functional

$$I(u) = \frac{1}{2}\int_0^1 |u'|^2 - \int_0^1 F(u) - t\int_0^1 u,$$

where $F(t) = \int_0^t f(s)ds$ is the primitive of f.

The key idea is now to show that for t sufficiently negative the mountain-pass solution has exactly one sign change. Then one can do the following: One considers equation (8) on the intervals $(0, \frac{1}{j})$, $1 \le j \le k-1$, and obtains for t sufficiently negative (say $t < T_k$) again a mountain-pass solution with exactly one sign change. Since furthermore $|u'(0)| = |u'(\frac{1}{j})|$ this follows by integrating $-\frac{1}{2}(|u'|^2)' = f(u)u' + tu'$, which yields

$$-\frac{1}{2}|u'(\tfrac{1}{j})|^2 + \frac{1}{2}|u'(0)|^2 = F(u(\tfrac{1}{j})) - F(u(0)) + tu(\tfrac{1}{j}) - tu(0) = 0),$$

we can therefore join these solutions to obtain k solutions with 0,1,3,...,2k-1 sign changes, respectively.

To prove that the mountain-pass solution has exactly one sign change one proceeds in two steps.

(a) One first shows that for t sufficiently negative the mountain-pass solution, lets call it u_2, cannot be positive. This is done by direct estimates. Hence one concludes that u_2 must have a sign change.

(b) To prove that u_2 has at most one sign change one uses strongly its variational characterization. In fact, assume u_2 has more than one node. Then we can rearrange u_2 in the following way. Let $I^{\pm} = \{u \gtrless 0\}$, and let $\beta \in (0,1)$ such that $\mu([0,\beta]) = \mu(I^+)$ and $\mu([\beta,1]) = \mu(I^-)$. For a positive function v, we denote by $\sigma(v)$ the Steiner symmetrization of v. We now assign to u the rearranged function $\tilde{u}$ given by $\tilde{u}^+ = \sigma(u^+)$ on $[0,\beta]$ and $-\tilde{u}^- = \sigma(-u^-)$ on $[\beta,1]$.

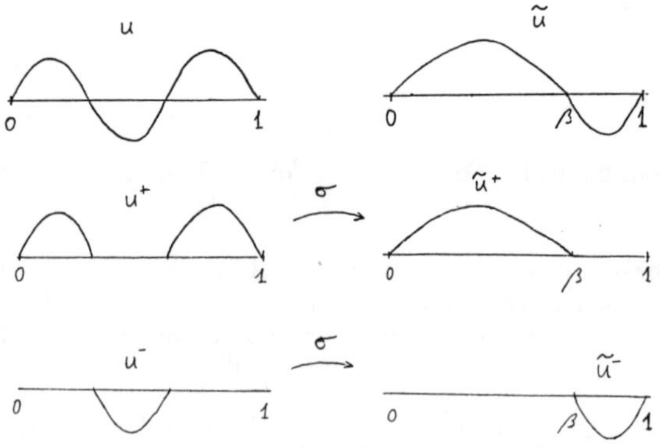

Figure 2

Note that $u \in H^1(0,1)$ implies $\tilde{u} \in H^1(0,1)$ and

$$I(\tilde{u}) = \frac{1}{2}\int|\tilde{u}'|^2 - \int F(\tilde{u}) - t\int\tilde{u}$$

$$\qquad\qquad\qquad\qquad \| \qquad\quad \|$$

$$\leq \frac{1}{2}\int|u'|^2 - \int F(u) - t\int u = I(u)$$

by the properties of the Steiner symmetrization. Clearly, it is not sufficient to do this for the solution, but we have to do it for the paths on which we minimax. More precisely, by the minimax characterization we know that there exists a sequence $(\gamma_n) \subset \Gamma$ such that

$$\max_{\gamma_n} I(u) \to c \qquad \text{(see (7))}.$$

Each of these paths γ_n we now replace by paths $\tilde{\gamma}_n = \{\tilde{u} | u \in \gamma_n, \tilde{u}$ constructed as above$\}$. There are some technical problems to show that these $\tilde{\gamma}_n$ are again continuous paths. Crucial to prove this is the fact that the Steiner symmetrization $\sigma\colon\ H^1 \to H^1$ is continuous. This has been shown recently by Coron [8] in one dimension. It is not known whether this is also true in higher dimensions.

From the above we then have

$$\max_{\tilde{\gamma}_n} I(u) \leq \max_{\gamma_n} I \underset{n\to\infty}{\to} c.$$

By the mountain-pass theorem we can find a sequence (u_n) with $I(u_n) = \max_{\gamma_n} I(u)$ and $u_n \to u$ such that $I'(u) = 0$ and $I(u) = c$. Clearly, u has exactly one sign change.

Remark. Note that the mountainpass solution we find for (8) is in fact a periodic solution, i.e. a solution of the equation

$$-u'' = I(u) + t,\ u(0) = u(1),\ u'(0) = u'(1).$$

The statement that for t large negative the mountainpass solution has precisely, one change then means in fact that in the mountainpass has minimal period. This result should be viewed in relation with the question of the minimality of period of periodic solution for certain Hamiltonian systems; see e.g. the recent result of Hofer-Kekland [13] who show that mountain-pass solutions of convex autonomous Hamiltonian systems have minimal period.

SUPERLINEAR NONLINEARITIES WITH PARTIAL INTERFERENCE WITH THE SPECTRUM

We note that a superlinear function with $f^+ = +\infty$, $f^- < \lambda_1$ crosses all eigenvalues of the linear operator. The crossing of the first eigenvalue was crucial in the existence as well as in the nonexistence proof of theorem 2. We now turn to problems where $\lambda_1 < f^- < c$, i.e. where f interferes only partially with the spectrum. We will find that now there do exist solutions for data function $h = h_1 + te_1$ with t large positive. For simplicity, we state the theorem in model form.

Theorem 4. Ruf-Srikanth [20]. Let $1 < p < \frac{N+2}{N-2}$, $\lambda > \lambda_1$, $\lambda \neq \lambda_k$, $k \in \mathbb{N}$

$$\begin{cases} -\Delta u = (u^+)^p + \lambda u + te_1, & \Omega \\ u = 0 & , \partial\Omega \end{cases} \tag{10}$$

Then, for any $t > 0$, equation (10) has at least 2 solutions.

Note that $f(s) = \lambda s + (s^+)^p$ satisfies $f^+ = +\infty$, $f^- = \lambda > \lambda_1$, i.e. we are in the situation described above.

Idea of proof. One works again with variational methods. It is again clear that $u_1 = \frac{t}{\lambda_1 - \lambda} e_1$ is a (negative) solution for (10). However, because $\lambda > \lambda_1$, this solution is no longer a minimum of the corresponding functional, and this means that we cannot apply the mountain-pass theorem to obtain a second solution. To overcome this difficulty we apply "linking arguments" devised by Rabinowitz [16] for variational problems.

First, we introduce the following functional, with $\alpha = \frac{t}{\lambda_1 - \lambda} < 0$,

$$I(u) = \frac{1}{2}\int |\nabla u|^2 - \frac{1}{p+1}\int ((u+\alpha e_1)^+)^{p+1} - \frac{\lambda}{2}\int |u|^2$$

Note that critical points of I are solutions of

$$-\Delta y - ((y+\alpha e_1)^+)^p - \lambda y, \text{ in } \Omega;\ y = 0 \text{ on } \partial\Omega. \tag{11}$$

Equation (11) is equivalent to equation (10) by the relation $u = y+\alpha e_1$. Furthermore, note that the solution $u_1 = \frac{t}{\lambda_1-\lambda} e_1$ of (10) corresponds to the trivial solution $y \equiv 0$ of (11), with critical level $I(0) = 0$.

Now, let $k = \max\{n \in \mathbb{N} | \lambda_n < \lambda\}$, and set $E_k = \mathrm{span}\{e_1,\ldots, e_k | e_j$: j-th eigenfunction of Laplacian$\}$, and let X be the orthogonal complement of E_k in $H_0^1(\Omega)$. We consider the following two sets:

$B_\rho(0) \subset X$: ball of radius ρ with center 0 in X

$Q_v := \{m+\alpha v | m \in E_k, \|m\| \leq \rho, 0 \leq \alpha \leq 2\rho\}$,

where $v \in X$ is fixed with norm $\|v\| = 1$.

The sets B_ρ and Q_v "link" in the sense of Rabinowitz [16].

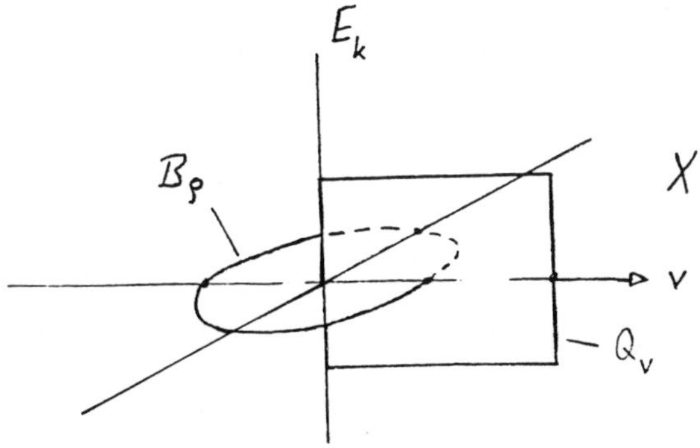

Figure 3

From the figure one sees that the sets ∂B_ρ and ∂Q_v link like two members of a chain. Crucial for a minimax procedure are now the following two statements:

(i) There exist numbers $\rho > 0$ and $\delta > 0$ such that

$$\inf_{\partial B_\rho(0)} I(u) \geq \delta > 0$$

(ii) There exists a $v \in X$, $\|v\| = 1$, such that $I/\partial Q_v \leq 0$.

Statement (i) is easily seen to hold for our functional I, since for $x \in X$ with $\int |\nabla x|^2 = \|x\|^2 = 1$

$$I(\rho x) = \frac{1}{2}\int|\nabla\rho x|^2 - \frac{\lambda}{2}\int|\rho x|^2 - \frac{1}{p+1}\int((\rho x+\alpha e_1)^+)^{p+1}$$

$$\geq \frac{1}{2}(1 - \frac{\lambda}{\lambda_{k+1}})\rho^2\|x\|^2 - \frac{\rho^{p+1}}{p+1}\int|x^+|^{p+1}$$

$$\geq \frac{1}{2}(1 - \frac{\lambda}{\lambda_{k+1}})\rho^2\|x\|^2 - \frac{\rho^{p+1}}{p+1}\|x\|^{\alpha(p+1)}$$

with $\alpha(p+1) < 2$, by the Sobolev imbedding theorem. Clearly, for $\rho > 0$ small enough, this last expression is positive for all $x \in X$, $\|x\| = 1$. For statement (ii) one needs two ingredients. First, one shows by a direct estimate as for (i) that $I/_{\partial Q_v \cap E_k} \leq 0$, again using that for ρ small the quadratic term of I is dominant. Second, one constructs a function $v \in X$ such that $I(\alpha v) \leq 0$ for all α larger than some $\alpha_0 > 0$. The construction of this function v is somewhat tedious and technical. Finally, one shows that in fact the estimate can be extended to hold on all of ∂Q_v, i.e. $I/_{\partial Q_v} \leq 0$.

Having properties (i) and (ii) one now proceeds as follows. One defines

$$\Sigma = \{S=h(Q)\,|\,h:Q_v \to H_0^1 \text{ continuous, } h/_{\partial Q_v} = id\}$$

and sets

$$c = \inf_{\Sigma} \max_{S} I(u).$$

One now verifies, using (i) and the linking property of the sets $B_\rho(0)$ and Q_v that $c \geq \delta > 0$: In fact, it is clear that for any $S \in \Sigma$ one has $S \cap \partial B_\rho(0) \neq \emptyset$, and therefore $\max_S I(u) \geq \inf_{\partial B_\rho} I(u) \geq \delta > 0$, for all $S \in \Sigma$ i.e. $\inf_\Sigma \max_S I(u) \geq \delta > 0$.

This allows now to prove by standard variational methods (using the fact that the functional I is of class $C^1(H_0^1,\mathbb{R})$ and satisfies a compactness condition, the so-called Palais-Smale condition) that c is a critical value of I. Essential for this is that the class of sets Σ is invariant under the gradient flow associated with I, (see [16]). Since $c = I(u_2) > 0$, u_2 is a second solution for (10).

RELATING THE NUMBER OF SOLUTIONS AND THE NUMBER OF UN-CROSSED EIGENVALUES

In several recent works on nonlinear problems relations between the number of solutions and the number of eigenvalues with which the nonlinearity interferes have been established (see e.g. Castro-Lazer [7], Amann [1], Lazer-McKenna [14], Ruf [17,18]). We will now present a result for equation (10) in one dimension, which establishes - quite surprisingly - a relation between the number of solutions and the number of eigenvalues which are not crossed.

Theorem 5. (Ruf-Srikanth [21]). Let $p > 1$, $\lambda \in (\lambda_k, \lambda_{k+1})$. Then

$$\begin{cases} -u'' - (u^+)^p - \lambda u = te_1, \text{ in } (0,1) \\ u(0) = u(1) = 0 \end{cases} \tag{12}$$

has for any $t > 0$ at least $2k + 2$ solutions.

Remarks. (a) We again restrict ourselves to the model equation (12). The result also holds under more general assumptions.

(b) Note that $f(s) := (s^+)^p + \lambda s$ satisfies $f^+ = +\infty$, $f^- \in (\lambda_k, \lambda_{k+1})$, i.e. f crosses all but the first k eigenvalues.

Idea of proof. This proof is not variational, but relies on bifurcation theory. This approach was first applied to such type of problems in [17,18].

As before we transform (12) into the equivalent equation

$$\begin{cases} -y'' - ((y+\alpha e_1)^+)^p - \lambda y = 0 \\ y(0) = y(1) = 0 \end{cases} \tag{13}$$

with $\alpha = \frac{t}{\lambda_1 - \lambda} < 0$. Now we note that the term $((y+\alpha e_1)^+)^p$ is of small order near zero in the C^1-norm, that is

$$\frac{((y+\alpha e_1)^+)^p}{\|y\|_{C^1}} \xrightarrow{C^1} 0, \text{ as } \|y\|_{C^1} \to 0.$$

In fact, this term vanishes identically for $\|y\|_{C^1}$ small enough.

Clearly, (13) has the set of trivial solutions $(\lambda, 0) \in \mathbb{R} \times C^{2,\tau}$, $0 < \tau < 1$. We can therefore consider (13) as a bifurcation problem in $(\lambda,u) \in \mathbb{R} \times C^{2,\tau}$. By the global bifurcation theorem of Rabinowitz [16] we know that all eigenvalues λ_k, $k \in \mathbb{N}$, are bifurcation points of global bifurcation branches $S_k^{\pm}$, where S_k^{+} (S_k^{-}) contains the solutions with k-1 nodes and with positive (negative) derivative in zero. Because of the nodal properties of solutions of Sturm-Liouville problems, these branches are mutually disjoint. Hence, they connect $(\lambda_k,0)$ to infinity in $\mathbb{R} \times C^{2,\alpha}$. By careful estimates one can in fact obtain the asymptotic behaviour of all the branches. The bifurcation diagram looks as follows:

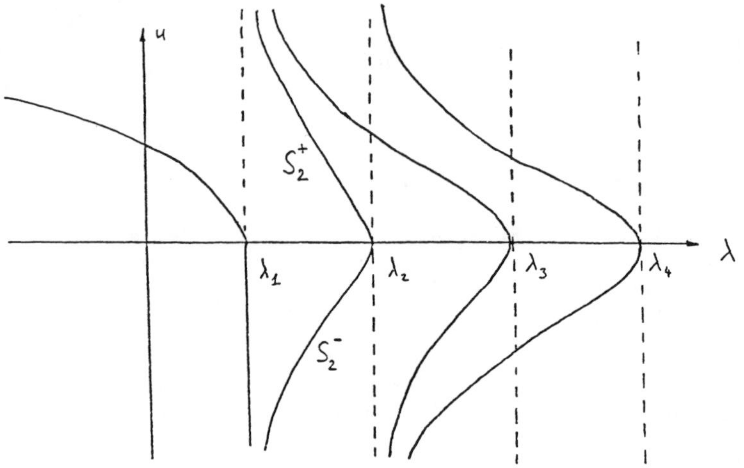

Figure 4

The main estimate to obtain the asymptotic behaviour is the following: For $k \geq 2$ and any sequence $(\eta_n,u_n) \in S_k^{\pm}$ with $\|(\eta_n,u_n)\|_{\mathbb{R}\times C^{2,\tau}} \to \infty$ for $n \to \infty$ holds:

$$\eta_n \leq \lambda_k \text{ and } |\{x|u_n(x) \geq 0\}| \underset{n\to\infty}{\to} 0$$

The following example shows how this estimate is used to derive the asymptotic behaviour. Suppose $(\eta_n,u_n) \in S_2^+$, $\|(\eta_n,u_n)\| \to \infty$. Then, since u_n has exactly one sign change for all $n \in \mathbb{N}$, we conclude from the above estimate that $|\{x|u_n(x) < 0\}| \to 1$. But u_n^- solves the equation

$$\begin{cases} -(u_n^-)'' = \eta_n u_n^- & \text{on } (a_n,1) \\ u_n(a_n) = u_n(1) = 0 \end{cases}$$

with $a_n \to 0$ for $n \to \infty$, from which we conclude that $\eta_n \underset{n\to\infty}{\to} \lambda_1$.

Finally, it is easy to obtain from the above diagram the multiplicity result of the theorem: one simply has to count the number of bifurcation branches which intersect the hyperplane $\{\lambda\} \times C^{2,\alpha}$ in $\mathbb{R} \times C^{2,\alpha}$.

REMARKS AND OPEN PROBLEMS

We remark that the nonlinearity $f(s) = \lambda s + (s^+)^p$, $p > 1$, $\lambda \in (\lambda_k, \lambda_{k+1})$ of theorem 5 also satisfies the assumptions of theorem 3. These two theorems give therefore quite a complicated solution structure for the Sturm-Liouville problem in dependence of the data function h:

- if $h = te_1$, $t > 0$, there exist at least $2k + 2$ solutions
- if $h = t$ with $t < T_n$, there exist at least n solutions (with n as large as one chooses).

This of course opens many questions, e.g. the following:
- exactness of the number of solutions for $t > 0$;
- number of solutions for other "directions" of h;
- are similar results true for the corresponding partial differential equations.

REFERENCES

1. Amann, H., *Saddle points and multiple solutions of differential equations*, Math. Z. 169 (1979), 127-166.

2. Amann, H., Hess, P., *A multiplicity result for a class of elliptic boundary value problems*, Proc. Roy, Soc. ed. 84-A (1979), 145-151.

3. Ambrosetti, A., Prodi, G., *On the inversion of some differentiable mappings with singularities between Banach spaces*, Ann. Math. Pura Appl. 93 (1973), 231-247.

4. Ambrosetti, A., Rabinowitz, P. H., *Dual variational methods in critical point theory and applications*, J. Functional Anal. 14 (1973), 349-381.

5. Berestycki, H., Lions, P. L., *Sharp existence results for a class of semilinear problems*, Bol. Soc. Bras, Mat. 12 (1981), 9-20.

6. Berger, M., Podolak, E., *On the solutions of a nonlinear Dirichlet problem*, Indiana Univ. Math. J. 24 (1975), 837-845.

7. Castro, A., Lazer, A. C., *Critical point theory and the number of solutions of a nonlinear Dirichlet problem*

8. Coron, J. M., *The continuity of the rearrangement in $W^{1,p}(\mathbb{R})$*, Ann. Scuola Norm. Pisa, 9, Ser. IV (1984), 57-85.

9. Dancer, E. N., *On the range of certain weakly nonlinear elliptic partial differential equations*, J. Math. Pures et Appl. 57 (1978), 351-366.

10. de Figueiredo, D. G., *On the superlinear Ambrosetti-Prodi problem*, MRC Tech. Rep #2252 May 1983.

11. de Figueiredo, D. G., Solimini, S., *A variational approach to superlinear elliptic problems*, MRC Tech. Rep # 2568, Sept. 1983.

12. Hess, P., Ruf, B., *On a superlinear elliptic boundary value problem*, Math, Z. 164, 1978.

13. Hofer, H., Ekeland, I., *Periodic Solutions with prescribed minimal period for convex autonomous Hamiltontian systems*, to appear.

14. Lazer, A. C., McKenna, P. J., *On a conjecture related to the number of solutions of a nonlinear Dirichlet problem*, Proc, Roy. Soc. Ed.

15. Rabinowitz, P. H., *Nonlinear Sturm-Liouville problems for second order ordinary differential equations*, Comm. Pure Appl. Math. 23 (1970), 936-961.

16. Rabinowitz, P. H., *Some aspects of critical point theory*, MRC Tech. Rep # 2465, Nov. 1982.

17. Ruf, B., *Remarks and generalizations related to a recent multiplicity result of A. Lazer and P. McKenna*, Nonlin. An. T.M.A., to appear.

18. Ruf, B., *Multiplicity and eigenvalue intersecting nonlinearities*, SISSA-preprint, Trieste, 1984.

19. Ruf. B., Solimini, S., *On a superlinear Sturm-Liouville problem with arbitratily many solutions*, SIAM J. on Math. An., to appear.

20. Ruf, B., Srikanth, P. N., *Multiplicty results for ODE's with nonlinearities crossing all but a finite number of eigenvalues*, Nonlinear An. T.M.A., to appear.

21. Ruf, B., Srikanth, P. N., *Multiplicity results for superlinear elliptic problems with partial interference with the spectrum*, J. Math. Anal. and Appl., to appear.

22. Scovel, C., *PhD Thesis*, Courant Institute, New York, 1983.

A NOTE ON PERIODIC SOLUTIONS OF HEAT EQUATION WITH A SUPERLINEAR TERM

Luis Sanchez
CMAF
Av. Prof. Gama Pinto, 2
1699 LISBOA CODEX
PORTUGAL

ABSTRACT. We present a simple proof of the existence of a periodic solution for a heat equation containing a nonlinear term that behaves like a certain power of the unknown function. The proof uses the Leray-Schauder principle.

1. INTRODUCTION AND STATEMENT OF THE RESULT

Consider the problem

$$
\begin{aligned}
& u_t - u_{xx} - u + g(t,x,u) = f(t,x) \\
& u(t+T,x) = u(t,x) \\
& u(t,0) = 0 = u(t,\pi), \quad 0 \leq x \leq \pi, \quad t \in \mathbb{R}
\end{aligned} \tag{1}
$$

where $T > 0$, g and f are T-periodic as functions of the variable t and g behaves, in a sense to be specified, like a power of u large $|u|$.

Problems of this type have been widely treated in the literature: we mention the results contained in [1], [2], [4], [5] and the references in [6]. Here we show that, for some power-like behaviour of g, a very simple proof of the existence of a solution of (1) can be given by using the Leray-Schauder principle.

Let us introduce the following notation. Let Q be the strip $\{(t,x) : t \in \mathbb{R} \text{ and } 0 < x < \pi\}$, $\Omega = (0,T) \times (0,\pi)$ and let H be the space of real functions $u(t,x)$, defined in Q, which are T-periodic in t and such that $u|_\Omega \in L^2(\Omega)$. For α,β nonnegative integers, let $H^{\alpha,\beta}$ denote the space of functions $u : Q \to \mathbb{R}$, T-periodic in t, such that $D_t^i u$, $D_x^j u$ belong to H for integers i,j such that $0 \leq i \leq \alpha, 0 \leq j \leq \beta$ (where D_t, D_x denote generalized derivatives). The closure in $H^1 \equiv H^{1,1}$ of the space of C^∞ functions, T-periodic in t and vanishing in a neighbourhood of the boundary of Q, is denoted by H_0^1. All these are Hilbert spaces with natural norms (see [6] for details). We mention that a function $\underline{u}$ belongs to $H^{1,2} \cap H_0^1$ if, and

S. P. Singh (ed.), Nonlinear Functional Analysis and Its Applications, 369–374.

only if, its Fourier series

$$\sum_{k,j} u_{kj}\, e^{ik\omega t} \sin jx\,,\ \overline{u}_{kj} = u_{-k,j}\,;\ \omega = 2\pi/T$$

$(k \in Z\,;\ j = 1,2,3,\ldots)$ has $\sum_{k,j} |u_{kj}|^2 (k^2 + j^4) < +\infty$; the square-root of this sum is an admissible norm for $H^{1,2} \cap H^1_o$.

Le L be the linear operator in H defined by $Lu = u_t - u_{xx} - u$ if $u \in H^{1,2} \cap H^1_o$ ($u_t = D_t u$, $u_{xx} = D^2_x u$). Solutions of the problem (1), are understood in the sense that

$$Lu + Gu = f \tag{2}$$

We assume that $f \in H$. G is the Niemytski operator associated to g, i.e. $(Gu)(t,x) = g(t,x,u(t,x))$.

We assume that g is a Charatheodory function in $Q \times \mathbb{R}$ (respect. $Q \times \mathbb{R}^2$), i.e. that it is measurable in (t,x) for fixed $\underline{u}$ and continuous in $\underline{u}$ for almost every (t,x) in Q. Then the following theorem holds:

THEOREM. Let g be a Caratheodory function, T-periodic in t, and suppose that for some $A \in]0,5[$ and positive constants c_i $(i = 1,2,3,4)$ we have

$$c_1 |u|^{A+1} - c_2 \le u\, g(t,x,u) \le c_3 |u|^{A+1} + c_4,\quad (t,x,u) \in Q \times \mathbb{R}\,.$$

Then, for each $f \in H$, equation (2) has a solution $u \in H^{1,2} \cap H^1_o$.

REMARK. The nonlinear term in (2) has a meaning, because $H^{1,2} \cap H^1_o$ is continuously imbedded in $C(\overline{Q})$ (cf. [6]).

We give a list of auxiliary results in §2 and the proof of the theorem in §3.

2. SOME AUXILIARY RESULTS

It is well known that the nullspace of L, N(L), is the one-dimensional space of multiples of the function sin x, and the range of L, R(L), is the orthogonal $N(L)^{\perp}$ in H. The restriction $L|_{H^{1,2} \cap H^1_o \cap R(L)}$ has a continuous inverse

$$R : R(L) \to H^{1,2} \cap H^1_o$$

so that K is compact as an operator in R(L).

We shall suppose from now on that $T = 2\pi$, so that the notation is slightly simplified. This involves no loss of generality.

If $f \in R(L)$, let

$$f = \sum_{(k,j) \in I} f_{kj}\, e^{ikt} \sin jx$$

where $I = \{(k,j) \in \mathbb{Z} \times \mathbb{N} : (k,j) \neq (0,1)\}$. Then Kf is given by

$$Kf = \sum_{(k,j) \in I} \frac{f_{kj}}{ik + j^2 - 1}\, e^{ikt} \sin jx \tag{3}$$

Let us introduce, for each function $f \in H$, its odd, 2π- periodic extension $\tilde{f}$ with respect to x,

$$\tilde{f}(t,x) = \begin{cases} f(t,x) & \text{if } (t,x) \in \Omega \\ -f(t,-x) & \text{if } (t,-x) \in \Omega . \end{cases}$$

Then we can rewrite (3) as a convolution,

$$(Kf)(t,x) = \frac{1}{4\pi^2} \int_{[o,2\pi]^2} h(t-t', x-x')\tilde{f}(t',x')\,dt'dx' \tag{4}$$

where h is the function defined in the torus $[o,2\pi]^2$ by

$$h(t,x) = \sum_{\substack{(k,j) \in \mathbb{Z}^2 \\ (k,|j|) \in I}} \frac{1}{ik + j^2 - 1}\, e^{i(kt+jx)} .$$

Now, since the sum $\sum_{(k,|j|) \in I} \left| \frac{1}{ik + j^2 - 1} \right|^{\beta}$ is finite for $\beta > 3/2$, the Hausdorff-Young theorem on Fourier series (see for instance [7]) implies that $h \in L^{\alpha}((o,2\pi)^2)$ for all $\alpha < 3$. Then, by well-known properties of convolutions, we obtain the following.

Fact 1. The linear mapping $f \to Kf$ given by (4) is bounded from $L^p(\Omega)$ into $L^q(\Omega)$, where

(i) q is any number such that

$$1 \leq q < 3p/(3-2p)$$

if $1 \leq p < 3/2$,

(ii) q is any number ≥ 1 if $p = 3/2$,

(iii) $q = \infty$ if $p > 3/2$.

Let us write

$$(f,g) = \int_{\Omega} fg \; dt \; dx$$

if the integral exists. In particular, (,) is the usual inner product of H. The following simple property of K is easily established using, for example, Fourier series.

Fact 2. For any $f \in R(L)$ we have $(Kf,f) \geq 0$.

Now let P_R and P_N be the orthogonal projections onto R(L) and N(L) respectively and let $| \; |_p$ denote the usual norm of $L^p(\Omega)$, $p \geq 1$. Clearly, P_R and P_N can be considered as linear operators in $L^p(\Omega)$ by setting

$$P_N f = \frac{2}{\pi T} (f, \sin x) \sin x; \quad P_R f = f - P_N f, \; f \in L^p(\Omega).$$

Moreover, since there obviously exists a constant $c = c(p)$ such that

$$|P_N f|_p \leq c|f|_p \;, \; f \in L^p(\Omega)$$

we see that the following is true:

Fact 3. For every $p \geq 1$, P_R and P_N are continuous with respect to the norm of $L^p(\Omega)$.

3. PROOF OF THE THEOREM

We prove that equation (2) has a solution by considering the equivalent equation

$$\tilde{u} + P_N G(u) + K P_R G(u) = K\tilde{f} + \bar{f} \tag{5}$$

where $\tilde{u} = P_R u$. $\tilde{f} = P_R f$, $\bar{f} = P_N f$. We consider equation (5) in the Banach space $X = L^r(\Omega)$ were $r = \max(1.2A)$. By the hypotheses on g, and since $H^{1,2} \cap H_0^1$ is compactly imbedded in X, the left-hand side of (5) is a compact perturbation of a Fredholm operator of index zero. Hence, the existence of a solution of (5) will follow, according to Mawhin's version ([3]) of the Leray-Schauder continuation principle, if we show that the set of solutions of the family of equations

$$\tilde{u} + (1-t)\bar{u} + t \; P_N G(u) + t \; KP_R G(u) = tK\tilde{f} + t\bar{f}, \; 0 \leq t \leq 1 \tag{6}$$

is bounded in X. (Here $\bar{u} = P_N \; u$).

We prove the theorem by showing that solutions of (6) form a bounded set, not only in X, but also in $H^{1,2} \cap H^1_o$. Let $u = \overline{u} + \tilde{u}$ be such that (6) holds for some $t \in [0,1]$. Then we have

$$(\tilde{u}, P_R G(u)) + t(K P_R G(u), P_R G(u)) = t(K\tilde{f}, P_R G(u)),$$

$$(1-t)(\overline{u},\overline{u}) + (\overline{u}, P_N G(u)) = t(\overline{f}, P_N G(u)).$$

Since $(u, G(u)) = (\tilde{u}, P_R G(u)) + (\overline{u}, P_N G(u))$ it follows from Fact 2 that

$$(u, G(u)) \leq t(K\tilde{f}, P_R G(u)) + (\overline{f}, P_N G(u)).$$

Noting that $K\tilde{f}$ and $\overline{f}$ are continuous on $\overline{Q}$, we obtain

$$(u, G(u)) \leq c\ |G(u)|_1$$

where $\underline{c}$ denotes a constant (independent of $\underline{t}$ and $\underline{u}$). Using the growth hypothesis on g we immediately get $|u|_{A+1} \leq c$ (in the sequel $\underline{c}$ denotes some constant independent of $\underline{t}$ and u). Again by the hypothesis,

$$|G(u)|_{A+1)/A} \leq c$$

and, since $\overline{u}$ is bounded in any norm (because of Fact 3) we derive from (6) and Facts 1 and 3;

$$|\tilde{u}|_r \leq c, \quad (c = c(r))$$

where $r < 3q_1/(3-2q_1)$ if $q_1 \equiv (A+1)/A < 3/2$, r is an arbitrary number if $q_1 = 3/2$ and $r = \infty$ if $q_1 > 3/2$. If $q_1 \geq 3/2$ the proof is finished because we immediately get the boundedness of G(u) in H and hence that of u in $H^{1,2} \cap H^1_o$, because of the properties of K. If not, we repeat the above procedure to improve the estimate on $\tilde{u}$. We claim that, after a finite number of steps, an inequality of the form

$$|G(u)|_q \leq c \quad \text{where} \quad q \geq 3/2$$

must be reached, so that the proof ends as before. To prove this assertion we argue by contradiction. Suppose that there exists an infinite sequence of numbers (q_n) such that

$$q_1 = \frac{A+1}{A}, \quad 0 < \frac{1}{A} \cdot \frac{3q_n}{3-2q_n} - q_{n+1} < \frac{1}{n},$$

and $q_n < 3/2$ for all $n \in \mathbb{N}$. Then such a sequence can be chosen to be strictly increasing. This follows by induction from the following facts:

(i) $q_1 < \frac{1}{A} \cdot \frac{3q_1}{3-2q_1}$ since by hypothesis, we have $0 < A < 5$; (ii) if, for some $n \in \mathbb{N}$, we have $q_n < \frac{1}{A} \cdot \frac{3q_n}{3-2q_n}$, then $q_n < q_{n+1}$ gives

$q_{n+1} < \frac{1}{A} \cdot \frac{3q_n}{3-2q_n} < \frac{1}{A} \cdot \frac{3q_{n+1}}{3-2q_{n+1}}$ because the function $3x/(3-2x)$ is strictly increasing. If we then let $q = \lim_{n\to\infty} q_n$, we obtain $q=3q/(A(3-2q))$, so that

$$q = 3(A-1)/(2A) < q_1$$

a contradiction. This ends the proof.

REMARK. If g is a smooth function, say, of class C^2, and $g(t,0,0) = g(t,\pi,0) = 0$, then the Niemytski operator induced by g acts continuously on $H^{1,2} \cap H^1_o$. In this case, equation (5) can be viewed as an equation in $H^{1,2} \cap H^1_0$ and the proof of theorem 1 works in this space, since K is also compact in $R(L) \cap H^{1,2} \cap H^1_0$. If further, $f \in H^{1,2} \cap H^1_0$, then, by the regularity properties of K (see [6]) the solution obtained in the theorem is in $H^{2,4}$ and therefore is a classical solution.

REFERENCES

[1] *Brézis-Nirenberg*, Characterization of the ranges of some nonlinear operators and application to boundary value problems, Ann. Scuola Norm. Sup. Pisa 5 (1978), 225-326.

[2] *Fucik*, Solvability of nonlinear equations and boundary value problems. Reidel, Dordrecht, 1980.

[3] *Mawhin*, Topological degree methods in nonlinear boundary value problems. NSF-CBMS Regional Conference Series in Maths. No 140 American Mathematical Society, Providence, R.I. (1979).

[4] *Stastnova-Fucik*, Note to periodic solvability of the boundary value problem for nonlinear heat equation, Comment. Math. Univ. Caroline 18 (1977), 735-740.

[5] *Stastnova-Fucik*, Weak periodic solutions of the boundary value problem for nonlinear heat equation, Applikace Matematiky 24 (1979), 284-303.

[6] *Vejvoda*, Partial Differential equations: time periodic solutions, M. Nijhoff, the Hague 1982.

[7] *Zygmund*, Fourier Series, Dover 1935.

FIXED POINT SETS ON PAIRS OF SPACES

Helga Schirmer
Department of Mathematics and Statistics
Carleton University
Ottawa K1S 5B6
Canada

ABSTRACT. This paper deals with some questions concerning the existence of a fixed point, minimal fixed point sets, and prescribed fixed point sets on A, X - A and $C\ell$ (X - A) for a given selfmap f: (X,A) → (X,A) of a pair of spaces (X,A) and maps homotopic to f. Some new answers are presented, and existing results as well as open questions are discussed. The main tool is the relative Nielsen number, which was recently introduced for selfmaps of a pair of compact metrizable ANR's, and which is here generalized to "admissible" selfmaps of pairs of arbitrary metrizable ANR's.

1. INTRODUCTION

We will discuss in this paper questions concerning the existence of fixed points, minimum number of fixed points and realization of possible fixed point sets on A, X - A and the closure $C\ell$ (X - A) for a selfmap f: (X,A) → (X,A) of a pair of metric ANR's. Sufficient conditions for the existence of a fixed point on $C\ell$ (X - A) for compact selfmaps of such pairs were first given, in terms of Lefschetz numbers, by C. Bowszyc [1], and extended to the class of maps of compact attraction by L. Górniewicz and A. Granas [2]. Our results here are obtained with the help of the relative Nielsen number introduced in [6] and some generalizations of it. Under suitable assumptions on (X,A) it is possible to obtain very precise information about the fixed points which can occur on A, but the results concerning fixed points on X - A and $C\ell$ (X - A) are less complete. We have, therefore, included examples and several open questions.

The relative Nielsen numbers which we use here were introduced in [6] for selfmaps of pairs of compact metric ANR's only, and we first extend some results from [6] to "admissible" selfmaps of pairs of not necessarily compact ANR's. Admissible selfmaps include those considered in [1] and [2]. The definition of relative Nielsen numbers for selfmaps of pairs of non-compact ANR's is modelled on the definition of Nielsen fixed point classes for selfmaps of a non-compact ANR by K. Scholz in [8] and sketched in §2, where we also introduce a variant

S. P. Singh (ed.), Nonlinear Functional Analysis and Its Applications, 375–386.

of the relative Nielsen number which will be helpful later on.

The various Nielsen numbers of § 2 are used in § 3 to obtain lower bounds for the number of fixed points on A and Cℓ (X- A) for admissible maps f: (X,A) → (X,A) of a pair of metric ANR's. In § 4 we use results from [6] and [7] to describe minimal fixed point sets on A, X - A and Cℓ (X - A) for all maps in the homotopy class of a given map f: (X,A) → (X,A) for a suitable pair of compact polyhedra (X,A). The information is very precise if f is a deformation (i.e. homotopic to the identity map of (X,A)), but in general the interplay between fixed points on X - A and the boundary Bd A causes some problems.

In § 5 we review some results from [7] concerning the realization of a prescribed set $K \subset X$ as the fixed point set of a selfmap of a pair (X,A). Finally, in § 6, we discuss the existence of selfmaps of (X,A) with prescribed fixed points and fixed point indices on X - A, and show that the extension of the results from [4] for the case $A = \emptyset$ is not straightforward.

I wish to thank L. Górniewicz who made me aware of [1] and [2].

2. RELATIVE NIELSEN NUMBERS FOR PAIRS OF NON-COMPACT ANR'S.

Let f: (X,A) → (X,A) be a map of a pair of spaces (X,A). We write $\bar{f}$: A → A for the restriction of f to A, and f: X → X for the map f: (X,A) → (X,A) if the condition that $f(A) \subset A$ is immaterial. Hence homotopies of f: (X,A) → (X,A) are maps of the form H: $(X \times I, A \times I) \to (X,A)$, and homotopies of f: X → X are maps of the form H: $X \times I \to X$. Relative Nielsen numbers for maps f:(X,A) → (X,A), which are invariant under homotopy and a lower bound for the number of fixed points of f, were introduced in [6] for the case where X and A are compact metric ANR's. They are defined in terms of the fixed point classes of X → X and $\bar{f}$: A → A. As the theory of fixed point classes was extended to selfmaps of non-compact ANR's by K. Scholz [8], we can do the same for the various fixed point classes associated with a selfmap of a pair of non-compact ANR's. This extension is quite similar to that in [8], and we give therefore only an outline. We write Fix f = $\{x \in X \mid f(x) = x\}$ for the fixed point set of f: X → X, and Bd Y, Cℓ Y and Int Y for the boundary, closure and interior in X of a subspace $Y \subset X$. By $Bd_A Z$ we mean the boundary in A of a subspace $Z \subset A$. The index of the fixed point class F of f: X → X is written as ind (X,f,F). Other notation is as in [8]. A class $\mathfrak{F}$ of selfmaps f: (X,A) → (X,A) of pairs of spaces (X,A) will be called <u>admissible</u> if, for each $f \in \mathfrak{F}$, the maps f: X → X and $\bar{f}$: A → A induced by f: (X,A) → (X,A) are admissible in the sense of [8], p. 82, i.e. if

(i) f: X → X and $\bar{f}$: A → A have a generalized Lefschetz number,

(ii) Fix f is compact in X and Fix $\bar{f}$ is compact in A,

(iii) X and A are metrizable ANR's.

An $\mathfrak{F}$ - homotopy is a map H: $(X \times I, A \times I) \to (X,A)$ so that both H: $X \times I \to X$ and H: $A \times I \to A$ are $\mathfrak{F}$ - homotopies in the sense of [8],

p. 81. The class $\mathfrak{F}$ admits an index if, for each $f \in \mathfrak{F}$, an index which satisfies the usual five axioms [8], p. 82 exists for all $f: X \to X$ and $\bar{f}: A \to A$ induced by $f \in \mathfrak{F}$. We shall, in the rest of this section and in § 3, assume that $\mathfrak{F}$ is an admissible class of selfmaps of pairs which admits an index. Examples of such classes include the class of compact selfmaps of pairs of metrizable ANR's considered by Bowszyc [1] and the more general class of maps of compact attraction between pairs of such spaces considered in [2].

If $f \in \mathfrak{F}$, then the Nielsen numbers $N(f)$ and $N(\bar{f})$ can be defined as in [8] as the number of essential fixed point classes of $f: X \to X$ and $\bar{f}: A \to A$. Hence we can, as in [6], Definition 2.1, define $N(f,\bar{f})$ as the number of essential common fixed point classes of f and $\bar{f}$, i.e. as the number of those essential fixed point classes of $f: X \to X$ which contain an essential fixed point class of $\bar{f}: A \to A$. We write, as in [6],

$$N(f;X,A) = N(f) + N(\bar{f}) - N(f, \bar{f})$$

for the relative Nielsen number of a map $f: (X, A) \to (X, A)$ with $f \in \mathfrak{F}$. As $N(f)$ and $N(\bar{f})$ are invariant under $\mathfrak{F}$- homotopies, the invariance of $N(f,\bar{f})$ under $\mathfrak{F}$- homotopies can be proved as in [6], proof of Theorem 3.3. Hence $N(f;X,A)$ is invariant under $\mathfrak{F}$- homotopies. It follows from the definition that $N(f;X,A)$ is a positive integer with $N(f;X,A) \geq N(f)$ and $N(f;X,A) \geq N(\bar{f})$ (see [6], Theorem 3.2). The following result can be proved by the same counting argument as the one used in [6], proof of Theorem 3.1.

Theorem 2.1. If $f: (X,A) \to (X,A)$ is map in $\mathfrak{F}$, then any map $\mathfrak{F}$- homotopic to f has at least $N(f;X,A)$ fixed points.

In order to obtain lower bounds for the number of fixed points on $C\ell\,(X - A)$ we shall in § 3 make use of a variant of $N(f,\bar{f})$, which we now introduce. If $f \in \mathfrak{F}$ and F is a fixed point class of $f: X \to X$, then $F \cap A$ is the union of fixed point classes of $\bar{f}: A \to A$ [6], Lemma 2.2, and hence the index of $F \cap A$ with respect to $\bar{f}: A \to A$ is well defined as the sum of the indices of the fixed point classes of $\bar{f}: A \to A$ contained in $F \cap A$. We say that the fixed point class F of $f: X \to X$ assumes its index in A if

$$\text{ind}\,(X,f,F) = \text{ind}\,(A,\bar{f},F \cap A),$$

and write $N(f,A)$ for the number of essential fixed point classes of $f: X \to X$ which assume their index in A. Hence $N(f,A)$ is an integer with $0 \leq N(f,A) \leq N(f,\bar{f})$ and it is $\mathfrak{F}$- homotopy invariant according to the next theorem.

Theorem 2.2 If $f,g: (X,A) \to (X,A)$ are $\mathfrak{F}$- homotopic maps in $\mathfrak{F}$, then $N(f,A) = N(g,A)$.

Proof. Let H be an $\mathcal{F}$ - homotopy from f to g and let F be an essential fixed point class of f: X → X with

$$\text{ind }(X,f,X) = \text{ind }(A,\bar{f},F \cap A).$$

Then

$$F \cap A = \cup \,(\bar{F}_j \mid j = 1,2,\ldots, k) \cup \bar{F}' ,$$

where each $\bar{F}_j$ is an essential fixed point class of $\bar{f}$: A → A and ind $(A,\bar{f},\bar{F}') = 0$. As in [6], proof of Theorem 3.3, the $\mathcal{F}$ - homotopy H relates the essential fixed point class F of f: X → X to an essential fixed point class G of g: X → X and the essential fixed point classes $\bar{F}_j$ of $\bar{f}$: A → A to essential fixed point classes $\bar{G}_j$ of $\bar{g}$: A → A in such a way that

$$G \cap A = \cup(\bar{G}_j \mid j = 1,2,\ldots,k) \cup \bar{G}',$$

with ind $(A,\bar{g},\bar{G}_j)$ = ind $(A,\bar{f},\bar{F}_j)$ and ind $(A,\bar{g},\bar{G}') = 0$. Hence

$$\begin{aligned}\text{ind }(X,g,G) &= \text{ind }(X,f,F) = \text{ind }(A,\bar{f},F \cap A)\\ &= \Sigma\ (\text{ind }(A,\bar{f},\bar{F}_j) \mid j = 1,2,\ldots,k)\\ &= \Sigma\ (\text{ind }(A,\bar{g},\bar{G}_j) \mid j = 1,2,\ldots,k)\\ &= \text{ind }(A,\bar{g},G \cap A).\end{aligned}$$

So H relates essential fixed point classes of f which assume their index in A to essential fixed point classes of g which assume their index in A, and Theorem 2.2 follows.

We finish with an example which will be used in § 3.

Example 2.3. Let $X = B^4$ be a closed 4-ball in a Euclidean space and let B_1^4 and B_2^4 be two disjoint smaller 4-balls contained in Int X. Let $A_1 \subset B_1^4$ be a subspace which is homeomorphic to $S^2 \times I$, where S^2 is a 2-sphere, and let $A_2 \subset B_2^4$ be a subspace which is homeomorphic to $P \times I$, where P is a pair of pants (i.e. a disk with two holes). Let χ denote the Euler characteristic. If $A = A_1 \cup A_2$ and if f is the identity map of (X,A), then the one fixed point class of f: X → X has index ind (X,f,F) = $\chi(X) = 1$, and

$$\begin{aligned}\text{ind }(A,\bar{f},F \cap A) &= \text{ind }(A,\bar{f},A_1) + \text{ind }(A,\bar{f},A_2)\\ &= \chi(A_1) + \chi(A_2) = 2 - 1 = 1.\end{aligned}$$

Therefore $N(f,A) = N(f,\bar{f}) = N(f) = 1$ and $N(\bar{f}) = N(f;X,A) = 2$.

On the other hand, the identity maps g_j of (X,A_j), for j = 1,2, have Nielsen numbers $N(g_j,A_j) = 0$ and $N(g_j,\bar{g}_j) = N(g_j) = N(\bar{g}_j) = N(g_j;X,A_j) = 1$.

3. LOWER BOUNDS FOR THE NUMBER OF FIXED POINTS FOR MAPS OF PAIRS OF ANR'S.

Sufficient conditions for the existence of a fixed point on the complement of A for a given map f: (X,A) → (X,A) in terms of Lefschetz numbers were found by C. Bowszyc. One of his main results, which was extended somewhat in [2], is

Theorem 3.1. ([1], Theorems 4.4 and 4.5) Let (X,A) be a pair of metrizable ANR's, let A be either closed or open and let f: (X,A) → (X,A) be a compact map. If $L(f) \neq L(\bar{f})$, then f has a fixed point on $C\ell$ (X-A)

We will show in this section that the Nielsen numbers of § 2 can be used to obtain lower bounds for the number of fixed points on A and $C\ell$ (X - A). Again $\mathfrak{F}$ is an admissable class of maps of pairs which admits an index.

Theorem 3.2. Let f: (X,A) → (X,A) be a map in $\mathfrak{F}$. Then f has

(i) at least $N(\bar{f})$ fixed points on A,

(ii) at least N(f) - N(f,A) fixed points on $C\ell$ (X - A).

Proof. (i) is clear by definition of $N(\bar{f})$. To prove (ii), we let F be an essential fixed point class of f: X → X which does not assume its index in A. Then $F \neq 0$ and

$$\text{ind}\ (A,\bar{f},F \cap A) \neq \text{ind}\ (X,f,F).$$

But this implies $F \cap C\ell\ (X - A) \neq \emptyset$, as otherwise $F \subset \text{Int}\ A$ and thus

$$\text{ind}\ (X,f,F) = \text{ind}\ (X,f,F \cap A) = \text{ind}\ A,\bar{f},F \cap A).$$

Therefore we can select N(f) - N(f,A) distinct fixed points in $C\ell$(X - A), one for each such fixed point class.

Remark 3.3. According to 2.1 the map of Theorem 3.2 has at least N(f;X,A) fixed points in all. But Theorem 3.2 (ii) cannot be sharpened to guarantee at least $N(f;X,A) - N(\bar{f}) = N(f) - N(f,\bar{f}) \geq N(f) - N(f,A)$ fixed points on X - A. To see this, let $X = B^2 = \{x|\ ||x|| = 1\}$ be the unit disk in Euclidean 2-space and let A be the annulus $\{x|\ \frac{1}{2} \leq ||x|| \leq 1\}$. If f: (X,A) → (X,A) is a deformation, then N(f) = 1 and $N(f, \bar{f}) = N(f,A) = 0$. Theorem 4.4 (ii) below will show that there exists a deformation of (X,A) which has no fixed point on X - A, but one on Bd A.

As $N(f) - N(f,A) \leq N(f;X,A) - N(\bar{f})$, we obtain as a corollary to Theorem 3.2 a lower bound for the number of fixed points on $C\ell$ (X - A) which can be easier to compute.

Corollary 3.4. Any map f: (X,A) → (X,A) in $\mathfrak{F}$ has at least $N(f;X,A)-N(\bar{f})$ fixed points on $C\ell$ (X - A).

Remark 3.5. The lower bound of Theorem 3.2 (ii) is in general a better lower bound than the one of Corollary 3.4. This can be seen from Example 2.3, with $(X,A) = (X,A_1)$ or $(X,A) = (X,A_2)$. Theorem 4.4 (iii) will show that there exists a deformation of (X,A_j), for $j = 1,2$, which has precisely $N(f) - N(f,A_j) = 1$ fixed points on $C\ell\ (X - A_j)$. So in this case $N(f) - N(f,A_j)$ is the best possible lower bound, but not $N(f;X,A_j) - N(f) = 0$.

Theorem 3.1 is false if $N(f)$ and $N(\bar{f})$ are used instead of $L(f)$ and $L(\bar{f})$. This can again be seen from Example 2.3, as Theorem 4.4 (iii) will show that there exists a deformation of $(X, A_1 \cup A_2)$ which has no fixed point on $X - \mathrm{Int}\ (A_1 \cup A_2)$ although $N(f) \neq N(\bar{f})$. But we can obtain a partial analogue of Theorem 3.1 in terms of Nielsen numbers.

Theorem 3.6. If $f: (X, A) \to (X,A)$ is a map in $\mathfrak{F}$ and if $N(f) \geq N(\bar{f})$, then f has a fixed point on $C\ell\ (X - A)$.

Proof. This follows immediately from Corollary 3.4, as $N(f,\bar{f}) \leq N(\bar{f})$ and thus in this case $N(f;X,A) - N(\bar{f}) = N(f) - N(f,\bar{f}) > 0$.

Remark 3.7. The condition $L(f) \neq L(\bar{f})$ need not imply $N(f) - N(f,A) > 0$, as can be seen by taking $X = S^1$ as a circle, A as an arc in X and $f: (X,A) \to (X,A)$ as a deformation. In this case $L(f) = 0 \neq L(\bar{f}) = 1$, but $N(f) = 0$ and hence $N(f) - N(f,A) = 0$.

Problem 1. Find an example of a map $f: (X,A) \to (X,A)$ in $\mathfrak{F}$ with $N(f) - N(f,A) > 0$ but $L(f) = L(\bar{f})$, and hence show that the conditions of Theorem 3.1 and Theorem 3.2 (ii) are independent of each other.

4. MINIMAL FIXED POINT SETS FOR PAIRS OF COMPACT POLYHEDRA.

The lower bound $N(\bar{f})$ of Theorem 3.2 (i) for the number of fixed points on A is often sharp for all maps in the $\mathfrak{F}$ - homotopy class of $f: (X,A) \to (X,A)$ (i.e. there exists a map $g: (X,A) \to (X,A)$ which is $\mathfrak{F}$ - homotopic to f and has precisely $N(\bar{f})$ fixed points on A), but in general this is not the case for the lower bounds of Theorem 3.2 (ii) or its Corollary 3.4 for the number of fixed points on $C\ell\ (X - A)$. Maps with a minimal fixed point set on $X - A$ as well as on A were constructed in [6] for a large class of pairs of compact polyhedra. To describe them, we need some definitions from [6]. A compact polyhedron X is called a Nielsen space if every map $f: X \to X$ is homotopic to a map $g: X \to X$ which has $N(f)$ fixed points, and if these fixed points can lie anywhere in X. A subspace A of X can be by-passed if every path in X with endpoints in $X - A$ is homotopic (relative to its endpoints) to a path in $X - A$. An example of a Nielsen space X with a subspace A which can be by-passed is a compact triangulable manifold of dimension dim $M \geq 3$ with its boundary $A = \mathrm{Bd}\ M$, or with a subpolyhedron of dimension $\leq$ dim $M - 2$. The following theorem contains the best information to date concerning maps of pairs of spaces with minimal fixed point sets.

Theorem 4.1. Let (X,A) be a pair of compact polyhedra so that

(i) X is connected,

(ii) X - A has no local cutpoint and is not a 2-manifold,

(iii) every component of A is a Nielsen space,

(iv) A can be by-passed.

Then any map $f: (X,A) \to (X,A)$ is homotopic to a map $g: (X,A) \to (X,A)$ which has precisely $N(f;X,A)$ fixed points. Of these $N(\bar{f})$ fixed points lie on A, and $N(f;X,A) - N(\bar{f})$ fixed points lie on X - A.

Proof. Theorem 4.1 follows from an inspection of the (lengthy) proof of [6], Theorem 6.2.

Theorems 3.2 and 4.1 show that $N(\bar{f})$ is for such polyhedral pairs a sharp lower bound for the number of fixed points on A. In the proof of [6], Theorem 6.2 all $N(\bar{f})$ fixed points on A are actually chosen on Bd A, and hence g has $N(f;X,A)$ fixed points on $C\ell$ (X - A). The proof can, however, be carried out as long as at least one point is chosen on Bd A for every common fixed point class of f and $\bar{f}$. But it may not be possible to choose $N(\bar{f}) - N(f,\bar{f})$ (let alone $N(\bar{f}) - N(f,\bar{f}) + N(f,A)$) fixed points in Int A. (See e.g. the example in Remark 3.7). By combining the information of Theorems 3.2 (ii) and 4.1 we can only conclude that under the assumptions of Theorem 4.1 a minimal fixed point set on $C\ell$ (X - A) of a selfmap g of (X,A) in the homotopy class of f consists of at least $N(f) - N(f,A)$ and at most $N(f;X,A)$ fixed points. That both of these possibilities can occur is shown by the next two examples.

Example 4.2. Let $X = B^n$ be a closed n-ball in a Euclidean space ($n \geq 3$) and let A be a smaller closed n-ball contained in Int X. Then any map $f: (X,A) \to (X,A)$ is homotopic to the constant map $g: (X,A) \to (X,A)$ given by $g(x) = a$ for all x, where $a \in$ Int A. Hence $N(f) = N(f,A) = N(f;X,A) = 1$, and a minimal fixed point set on $C\ell$ (X - A) consists of $N(f) - N(f,A) = 0$ points.

Example 4.3. If again X is an n-ball but A a subspace homeomorphic to an (n - 1) - ball and contained in Bd X (with $n \geq 3$), then $N(f) = N(f,A) = 1$, but any map $g: (X,A) \to (X,A)$ with a minimal fixed point set on $C\ell$ (X - A) has $N(f;X,A) = 1$ fixed point on $C\ell$ (X - A). (In this example A = Bd A, so no fixed points can lie in Int A. But it is also possible to construct examples where Int A $\neq \emptyset$ but where a minimal fixed point set on $C\ell$ (X - A) consists of $N(f;X,A) > N(f) - N(f,A)$ points.)

In the last two examples $f: (X,A) \to (X,A)$ was a deformation, i.e. homotopic to the identity map of (X,A). For such maps precise information about the location of minimal fixed point sets was obtained in [7], Theorem 4.1. The conditions on (X,A) can in this case be somewhat relaxed, as we only need that (X,A) is a 2-dimensionally connected pair of polyhedra. This means that X is connected, X - A is 2-dimensionally connected (see [7], § 3, or [5], p.421) and that each component of A admits a small deformation (precisely: a proximity map, see [7], § 2, or [5], p. 421) which is fixed point free if $\chi(X) = 0$ and has precisely one fixed point in an arbitrary location if $\chi(X) \neq 0$. A pair consisting of a compact triangulable manifold X and its boundary A is 2-dimensionally connected if dim $X \geq 2$.

Theorem 4.4. ([7], Theorem 4.1). If (X,A) is a 2-dimensionally connected pair of compact polyhedra, then there exists a deformation of (X,A) with N(id;X,A) fixed points. They can be located as follows:

(i) If $\chi(X) = 0$ and $\chi(A_j) = 0$ for each component A_j of A, then f is fixed point free,

(ii) If $\chi(X) \neq 0$ but $\chi(A_j) = 0$ for each component A_j of A, then f has exactly one fixed point which can be located anywhere in $C\ell$ (X - A),

(iii) if $\chi(A_j) \neq 0$ for $k \geq 1$ components $A_1, A_2, \ldots, A_k$ of A, then f has exactly k fixed points $a_1, a_2, \ldots, a_k$. Each A_j, for $j = 1,2,\ldots,k$, must contain one fixed point a_j. If $\chi(X) = \chi(A)$, then each a_j can be located anywhere in A_j, but if $\chi(X) \neq \chi(A)$, then at least one a_j must lie on Bd A.

It is easy to find the various Nielsen numbers in these three cases, as it is known [6], Theorem 2.6 that

$$N(id;X,A) = \begin{cases} N(\overline{id}) & \text{if } N(\overline{id}) \neq 0, \\ N(id) & \text{if } N(\overline{id}) = 0. \end{cases}$$

A detailed calculation shows that a minimal fixed point set on $C\ell$ (X-A) consists in cases (i) and (ii) of N(id;X,A) = N(id) - N(id,A) points. But in case (iii) the number of points of a minimal fixed point set on $C\ell$ (X - A) can lie anywhere between N(id) - N(id,A) and N(id;X,A). This shows that the next problem may in general not have a "nice" answer.

Problem 2. Find a sharp lower bound for the minimum number of fixed points on $C\ell$ (X - A) for a map f: (X,A) $\to$ (X,A) of a pair of polyhedra which satisfies the assumptions of Theorem 4.1.

But there are some cases where Problem 2 has an easy solution, and we describe two in the next two results. Corollary 4.5 follows immediately from Theorem 4.4, and sharpens Theorem 3.1 for deformations of some polyhedral pairs.

Corollary 4.5. Let (X,A) be a 2-dimensionally connected pair of compact polyhedra and let $\text{Int}\, A_j \neq \emptyset$ for each component of A with $\chi(A_j) \neq 0$. If $L(id) \neq L(\overline{id})$, then there exists a deformation of (X,A) with precisely one fixed point on $C\ell\,(X - A)$.

Theorem 4.6. Let (X,A) satisfy the assumptions of Theorem 4.1. Then any map $f: (X,A) \to (X,A)$ with $N(\bar{f}) = 0$ is homotopic to a map $g: (X,A) \to (X,A)$ which has precisely $N(f) - N(f,A) = N(f)$ fixed points on $C\ell\,(X - A)$.

Proof. If $N(\bar{f}) = 0$, then $N(f,\bar{f}) = N(f,A) = 0$, and hence $N(f;X,A) = N(f) = N(f) - N(f,A)$, so the existence of g follows from Theorem 4.1.

We finish this section with some further problems. The first two should be fairly easy, although the answers may be complicated. The third is hard, as it is open even if $A = \emptyset$.

Problem 3. Extend Theorem 4.4 to deformations of non-compact but locally finite 2-dimensionally connected pairs of polyhedra, using (e.g.) the methods of G. -H. Shi [10].

Problem 4. Extend Theorem 4.4 to deformations of compact but not 2-dimensionally connected pairs of polyhedra. This was done in the case $A = \emptyset$ by G. -H. Shi [9], but his results have recently been obtained in a much more elegant form by K. Scholz [oral communication].

Problem 5. Extend Theorems 4.1 and 4.4 to pairs of compact metric ANR's.

5. PRESCRIBED FIXED POINT SETS ON PAIRS OF POLYHEDRA.

The deformations with minimal fixed point sets of Theorem 4.4 can be used to obtain maps with prescribed fixed point sets. It has been known for some time that every 2-dimensionally connected compact polyhedron X has the complete invariance property (CIP), which means that every closed non-empty subset K of X can be realized as the fixed point set Fix f of a map $f: X \to X$. The construction of f is carried out with the help of a small deformation with a minimal fixed point set. (See also [5] for a survey of the CIP). But it is not true that K can always be realized as the fixed point set of a map of pairs $f: (X,A) \to (X,A)$. The following result was proved in [7], Theorem 5.1.

Theorem 5.1. Let (X,A) be a 2-dimensionally connected pair of compact polyhedra, let A_j be the components of A and let K be a closed subset of X. Then there exists a deformation $f: (X,A) \to (X,A)$ with K as its fixed point set if and only if

(i) $K \cap A_j \neq \emptyset$ if $\chi(A_j) \neq 0$,

(ii) $K \cap C\ell\,(X - A) \neq \emptyset$ if $\chi(X) \neq \chi(A)$.

It was also pointed out in [7], Remark 5.2, that the conditions (i) and (ii) can in general not be omitted even if maps of pairs in an arbitrary homotopy class are used.

An easy consequence of Theorem 5.1 is a sharpened version of Corollary 4.5.

Theorem 5.2. Let (X,A) be a 2-dimensionally connected pair of compact polyhedra and let Int $A_j \neq \emptyset$ for each component of A with $\chi(A_j) \neq 0$. If K_o is a closed subset of $C\ell$ (X - A) which is non-empty if $L(id) \neq L(\overline{id})$, then there exists a deformation f: (X,A) → (X,A) with Fix ∩ $C\ell$ (X - A) = K_o.

Proof. Let A_j, with j = 1,2,...,k, be the components of A with $\chi(A_j) \neq 0$. We pick k points $a_j \in$ Int A_j and let $K = K_o \cup \{a_1, a_2, \ldots, a_k\}$. Then a deformation f of (X,A) with fixed point set K exixts by Theorem 5.1, and it satisfies Theorem 5.2.

The CIP was extended from compact 2-dimensionally connected polyhedra to all locally finite polyhedra in [3]. Therefore a solution to the next problem should not be very hard, but as the case A = ∅ uses results from [9] and [10], it may be necessary to attack Problems 3 and 4 first.

Problem 6. Extend Theorem 5.1 to pairs of locally finite polyhedra.

6. PRESCRIBED FIXED POINTS AND INDICES ON THE COMPLEMENT OF A SUBSPACE.

Let points c_k of a polyhedron X and integers i_k be given. Necessary and sufficient conditions are given in [4] for the existence of a selfmap of X within a given homotopy class which has the c_k as its fixed points and the i_k as the corresponding fixed point indices. The methods of [4] can easily be used to obtain sufficient conditions for the existence of a selfmap of a polyhedral pair (X,A) within a given homotopy class which has prescribed fixed points c_k and fixed point indices i_k on X - A. The "splitting" and "moving" of fixed points [4], Lemmas 1 and 2 can be carried out in X - A without changing the map on A if X - A is 2-dimensionally connected. If (X,A) is a polyhedral pair which satisfies the assumptions of Theorem 4.4, then it is a routine exercise to adapt [4], Theorem 2 in order to obtain sufficient conditions on the i_k which ensure the existence of a map g: (X,A) → (X,A) within a given homotopy class and has on X - A the prescribed fixed points c_k and fixed point indices i_k. But these conditions are not likely to be necessary, for we can show that in the special case where f is a deformation any set of isolated fixed points and fixed point indices on X - A can be prescribed.

Theorem 6.1. Let (X,A) be a 2-dimensionally connected pair of compact polyhedra, where A ≠ ∅ and A ≠ X, Given points $c_k \in X - A$ and integers i_k, there exists a deformation g: (X,A) → (X,A) which has on X - A the fixed points c_k with fixed point indices i_k.

We sketch a proof: Let K ⊂ A be a set which consists of one point

$a_j \in A_j$ for each component A_j of A and of one point $a \in$ Bd A. Then, according to Theorem 5.1 and [7], Theorem 5.1 there exists a small deformation (= proximity map) f: (X,A) → (X,A) with Fix f = K. Hence the technique of [7], § 5 can be used to homotope f to a small deformation f' of (X,A) with Fix f' = Fix f ∪ $\{x_0\}$, where x_0 is a point in a maximal simplex of X - A, with ind (X,f',K) = ind (X,f,K) and therefore ind $(X,f',x_0) = 0$. As in [4], Proof of Lemma 1 the fixed point x_0 of f' can be split in order to obtain a deformation f" of (X,A) with Fix f" = K ∪ $\{y_0, b_1, b_2, \ldots, b_m\}$, where y_0, b_1, $b_2, \ldots, b_m$ lie in the carrier simplex of x_0, in such a way that

$$\text{ind } (X,f'',b_k) = i_k \text{ for all given } i_k,$$

$$\text{ind } (X,f'',y_0) = \Sigma\, (i_k \mid \text{all given } i_k).$$

The fixed points b_k can be moved in the 2-dimensionally connected space X - A to the given fixed points c_k without changing the indices i_k as in [4], Proof of Lemma 2, and y_0 can be deleted as a fixed point by uniting it with the point $a \in$ Bd A as in [7], Lemma 3.2.

We end with some problems.

Problem 7. The crux of the proof of Theorem 6.1 is the fact that any fixed point in X - A of a deformation f: (X,A) → (X,A) of a 2-dimensionally connected pair of compact polyhedra can be moved to Bd A. Is this still true if f: (X,A) → (X,A) is a map in an arbitrary homotopy class?

Problem 8. Depending on the answer to Problem 7, find the best possible extension of [4], Theorem 2 to maps of pairs of polyhedra.

Problem 9. Find extensions of Theorem 6.1 and of the answer to Problem 8 if fixed points c_k and indices i_k on $C\ell$ (X - A) (rather than X - A), or on A as well as on X - A, are prescribed.

References

[1] C. Bowszyc, 'Fixed point theorems for the pairs of spaces', Bull. Acad. Polon. Sci. 16 (1968), 845-850.

[2] L. Górniewicz and A. Granas, 'On a theorem of C. Bowszyc concerning the relative version of the Lefschetz fixed point theorem', preprint.

[3] Boju Jiang and H. Schirmer, 'Fixed point sets of continuous selfmaps on polyhedra', Fixed Point Theory (Proceedings, Sherbrooke, Quebec, 1980), Springer Verlag Berlin 1981, Lecture Notes in Mathematics v. 886, 171-177.

[4] H. Schirmer, 'Mappings of polyhedra with prescribed fixed points and fixed point indices', Pacific J. Math. 63 (1976), 521-530.

[5] H. Schirmer, 'Fixed point sets of continuous selfmaps', Fixed Point Theory (Proceedings, Sherbrooke, Quebec, 1980), Springer Verlag Berlin, 1981, Lecture Notes in Mathematics v. 886, 417-428.

[6] H. Schirmer, 'A relative Nielsen number', to appear in Pacific J. Math.

[7] H. Schirmer, 'Fixed point sets of deformations of pairs of spaces', to appear.

[8] K. Scholz, 'The Nielsen fixed point theory for noncompact spaces', Rocky Mountain J. Math. 4 (1974), 81-87.

[9] G. -H. Shi, 'Least number of fixed points of the identity class', Acta Math. Sinica 18 (1975), 192-202.

[10] G. -H. Shi, 'On the fewest number of fixed points for infinite complexes', Pacific J. Math. 103 (1982), 377-387.

FIXED POINT FREE DEFORMATIONS ON COMPACT POLYHEDRA

U. Kurt Scholz
Department of Mathematics
College of St. Thomas
St. Paul, Minnesota 55105
U.S.A.

ABSTRACT. A necessary and sufficient condition is obtained for deforming the identity map on a connected compact polyhedron X to a fixed point free map. The Euler characteristic of the space is required to be zero and non-negative on certain subpolyhedra. The result uses many of the ideas of Shi [4] where the problem of determining the minimum number of fixed points of deformations is placed in a combinatorial setting and certain "welding vertices" and transformations come into play.

1. INTRODUCTION

A 2-dimensionally connected compact polyhedron X admits a fixed point free deformation iff $\chi(X) = 0$ ([1], [3], [4], [5]). Simple examples show that this is false for connected compact polyhedra in general (e.g., X is the 2-sphere with a "figure eight" attached at one point). The problem arises when X may be disconnected by removing a finite number of points, producing a component whose closure has negative Euler characteristic.

We adopt the notation of [5] throughout this paper. Following Shi [4], a *part* P of a simplicial complex K is either a maximal 2-dimensionally connected subcomplex or a maximal 1-dimensional simplex. A subcomplex L which is a union of such parts is called a *finitely attached* subcomplex. Such subcomplexes have the property that the set $W(L) = |L| \cap |\overline{K-L}|$ is finite and we refer to the members of this set as the *welding vertices* of L. The union of all such welding vertices of the parts is the *welding set* of K, denoted by $M(K)$.

2. THE MAIN RESULT

If X is a compact polyhedron, then a particular triangulation results in a similar parts decomposition, although a different triangulation could produce different 1-dimensional parts. In this paper we will consider only the case where $X = |K|$, a finite simplicial complex.

S. P. Singh (ed.), Nonlinear Functional Analysis and Its Applications, 387–392.

Let $\chi(X)$ denote the Euler characteristic of a space X. A deformation on X is a map homotopic to the identity map.

The main result of this paper is

THEOREM 2.1: *A finite simplicial complex K admits a fixed point free deformation iff*

(*) $\chi(K) = 0$ *and* $\chi(L) \geq 0$ *for each finitely attached subcomplex L.*

The proof will occupy the remainder of this paper. At this point, however, note that if K consists of only one part then the theorem easily reduces to the known result mentioned in the introduction. We assume from here on that $M(K) \neq \emptyset$.

Example: Let Y be a 2-dimensionally connected compact polyhedron with $\chi(Y) = m > 0$. Let $|K|$ be homeomorphic to the space obtained by attaching m 1-cells to Y. Decomposing $|K|$ into the corresponding part P and complexes $P_1,\ldots,P_m$ we can easily see that if no component of $|K|-|P|$ has negative Euler characteristic then the conditions of the theorem are satisfied.

3. STAR TRANSFORMATIONS

A *star transformation* on K is a function $g\colon M(K) \to |K|$ which assigns to each welding vertex v a point in $St_K v$, the open star of v.

For a map $g\colon X \to X$ on a connected compact polyhedron with isolated fixed points $x_1,\ldots,x_m$, we denote the fixed point index of g at x_j by $i(g,x_j)$ (see, e.g., [1], Chapter 4). We require only a few of the well known properties namely that the sum

$$\sum_{j=1}^{m} i(g,x_j) = L(g),$$

the Lefschetz number of g (normalization) and if g is constant in a neighborhood of the fixed point x_j then $i(g,x_j) = 1$ (a consequence of the localization and normalization properties).

In combination with the fixed point index, star transformations play a fundamental role in minimizing the number of fixed points of deformations in the following manner:

A star transformation is easily seen to be extendable to a deformation $g\colon |K| \to |K|$ and, using the "Hopf construction" ([1] page 117), we may assume g has only finitely many fixed points. Let P be some part of K and suppose g maps $n(P)$ of the welding vertices of P into $|K|-|P|$ and the rest into $|P|$ leaving no fixed points on $W(P)$. In this setting, we easily have $|P| \cup g(|P|)$ retractible onto $|P|$ obtained by collapsing portions of the stars of the welding vertices which were mapped outside of $|P|$. Then the restriction of g to $|P|$ followed by such a retraction produces a map $g'\colon |P| \to |P|$ which is homotopic to the identity map on $|P|$.

Now g' has exactly $n(P)$ fixed points on $W(P)$ as a result of the retraction. Moreover, it is evident that at these vertices, g' is a locally constant map. Thus the fixed point index at each of these fixed vertices is +1. In the interior of $|P|$, g and g' have the same fixed points and it is not difficult to see that both maps also have the same fixed point index at these points. Letting $x_1,\ldots,x_m$ be the fixed points of g' in the interior of $|P|$ and summing up the indices of all of the fixed points of g' yields the Lefschetz number of g', which is the Euler characteristic of P since g is a deformation. We have

$$\sum_{j=1}^{m} i(g',x_j) + n(P) = \chi(P)$$

so that the sum of the indices of the *interior* fixed points of g' (and hence g) is $\chi(P)-n(P)$.

THEOREM 3.1: (Compare [4], Lemma 2) *If K is a connected finite simplicial complex and g is a fixed point free star transformation such that for each part P, g maps $\chi(P) \geq 0$ welding points of P into the complement of $|P|$ then $|K|$ admits a fixed point free deformation.*

Proof: From the above discussion we see that we can extend g to a deformation of $|K|$ with finitely many fixed points such that for each part P, the sum of the indices of the fixed points in P is zero. Eliminating the fixed points inside the parts is a standard procedure: if P is a maximal 1-simplex then this is obvious; otherwise, if P is 2-dimensionally connected, fixed points may be "combined" and finally eliminated ([1], [3], [6]).

The existence of the star transformation in this theorem requires that we have $0 \leq \chi(P) \leq \mathrm{card}(W(P))$ (cardinality of $W(P)$). Since a fixed point free deformation is obtained, it is also necessary that $\chi(K) = 0$. However, these conditions are not sufficient since, for example, there could be a welding vertex v contained in exactly two parts each with Euler characteristic 0. This would be a fixed vertex.

The arguments leading to the above lemma are carried out somewhat differently, but in greater generality in [4, Lemma 2].

By a *good* star transformation [4], we mean a star transformation g such that for each part P, g is fixed point free on $W(P)$ *only if* $0 \leq \chi(P) \leq \mathrm{card}(W(P))$ *and* g maps *exactly* $\chi(P)$ welding points of P into the complement of $|P|$. Thus, for example, if $\chi(P) < 0$ or $\chi(P) > \mathrm{card}(W(P))$ then g is required to have fixed points on $W(P)$.

THEOREM 3.2: (Shi, [4]) *Let K be a finite connected simplicial complex with $M(K) \neq \emptyset$. Then the least number of fixed points of the deformations on $|K|$ is equal to the least number of fixed points of the good star transformations.*

Note that the star transformations described in Theorem 3.1 is a good star transformation and hence Theorem 3.1 follows from Theorem 3.2.

4. THE PROOF

Assume throughout the rest of this paper that K is a connected finite simplicial complex. In view of Theorem 3.2, we obtain

LEMMA 4.1: *$|K|$ admits a fixed point free deformation iff there exists a good star transformation on K with no fixed points.*

Let L be a finitely attached subcomplex of K. Define $C(L) = \text{card}(W(L)) - \chi(L)$, called the *capacity* of L.

The next lemma summarizes some of the technical details we need.

LEMMA 4.2: *Suppose K satisfies (*) of Theorem 2.1 and let $P_1,\dots,P_m$ be its parts. Then*

1) $C(L) \geq 0$ for each finitely attached subcomplex

2) if $L = \bigcup_{j=1}^{k} P_{i_j}$ then $\chi(L) = \text{card}\left(\bigcup_{j=1}^{k} W(P_{i_j})\right) - \sum_{j=1}^{k} C(P_{i_j})$

3) $\sum_{i=1}^{m} (C(P_i) = card(M(K))$

The tedious calculations are straightforward and will be omitted. Note that 3) is simply 2) with $L = K$ and using $\chi(K) = 0$.

LEMMA 4.3: *If there is a good star transformation on K* without fixed points then $\chi(L) \geq 0$ for each finitely attached subcomplex L.

Proof: If g is fixed point free then writing $L = \bigcup_{j=1}^{k} P_{i_j}$, we note that $\sum_{j=1}^{k} C(P_{i_j})$ welding points of $P_{i_1},\dots,P_{i_k}$ are mapped into the interior of $|L|$. Therefore, $0 \leq \text{card}\left(\bigcup_{j=1}^{k} W(P_{i_j})\right) - \sum_{j=1}^{k} C(P_{i_j})$ and hence by Lemma 4.2, $\chi(L) \geq 0$.

We are now prepared to prove one part of the main theorem.

COROLLARY 4.4: *If $|K|$ admits a fixed point free deformation then $\chi(K) = 0$ and $\chi(L) \geq 0$ for each finitely attached subcomplex L.*

Proof: $\chi(K) = 0$ is obvious by the Lefschetz fixed point theorem since the Lefschetz number of a deformation is the Euler characteristic. By Theorem 3.2, there exists a fixed point free good star transformation and now the result follows from the previous lemma.

The existence of good star transformations without fixed points reduces to a well known result in combinatorics. It is one of many theorems based on P. Hall's "marriage theorem" which is prevalent in Transversal theory of combinatorics.

LEMMA 4.5: ([2], page 44, Theorem 3.3.1). *Let* $A_1, A_2, \ldots, A_n$ *be a collection of subsets of a set E and let* $m_1, m_2, \ldots, m_n$ *be non-negative integers. Then there exist pairwise disjoint sets* $X_1, X_2, \ldots, X_n$ *such that* $X_i \subseteq A_i$, $card(X_i) = m_i$ $(i=1,\ldots,n)$ *iff for each subset* $J \subseteq \{1,2,\ldots,n\}$ $card(\bigcup_{i\in J} A_i) \geq \sum_{i\in J} m_i$.

The proof of our theorem now simply consists of tying together the last two lemmas. Suppose the connected complex K satisfies (*) of Theorem 2.1. *Let* $P_1,\ldots,P_n$ *be the parts of K. For any finitely attached subcomplex* $L = \bigcup_{j=1}^{k} (P_{i_j})$, we have

$$card\left(\bigcup_{j=1}^{k} W(P_{i_j})\right) = \sum_{j=1}^{k} C(P_{i_j}) + \chi(L) \geq \sum_{j=1}^{k} C(P_{i_j})$$

since $\chi(L) \geq 0$.

Using the sets $W(P_1),\ldots,W(P_n)$ and the non-negative integers $C(P_1),\ldots,C(P_n)$, the previous lemma shows that there exists pairwise disjoint sets $X_1,\ldots,X_n$ with $X_i \subseteq W(P_i)$ and card $(X_i) = C(P_i)$ for i = $1,\ldots,n$. Moreover, since $\sum_{i=1}^{n} C(P_i) = card(M(K))$, we conclude that $\bigcup_{i=1}^{n} X_i = M(K)$.

Finally, to obtain our good star transformation g having no fixed points, let $v \in M(K)$. Then $v \in X_i$ for a unique i and we let $g(v)$ be an arbitrary point in the interior of $|P_i|$ which is in the star of v.

References

[1] R.F. Brown, *The Lefschetz Fixed Point Theorem,* Scott, Foresman and Co., (1971).

[2] L. Mirsky, *Transversal Theory,* Academic Press, (1971).

[3] G.-H. Shi, *'On the least number of fixed points and Nielsen numbers', Chinese Math.* **8** (1966) 234-243.

[4] ________, *'The least number of fixed points of the identity mapping class,'* Acta Math Sinica **18** (1975), 192-202.

[5] E.H. Spanier, *Algebraic Topology,* McGraw Hill (1966).

[6] F. Wecken, *'Fixpunktklassen III,'* Math. Ann. **118** (1942), 544-577.

MINIMAX PRINCIPLES FOR A CLASS OF LOWER SEMICONTINUOUS FUNCTIONS AND APPLICATIONS TO NONLINEAR BOUNDARY VALUE PROBLEMS

Andrzej Szulkin
Department of Mathematics
University of Stockholm
113 85 Stockholm, Sweden

1. INTRODUCTION

Let X be a real Banach space, and $\psi : X \to (-\infty,+\infty]$ a convex, proper (i.e., $\psi \not\equiv +\infty$) and lower semicontinuous function. Recall that $D(\psi) = \{u \in X : \psi(u) < +\infty\}$ is called the effective domain of ψ, and the (possibly empty) set

$$\partial\psi(u) = \{u^* \in X^* : \psi(v) - \psi(u) \geqq \langle u^*, v - u\rangle \ \forall v \in X\}$$

is the subdifferential of ψ at u [2,5]. Here X* denotes the dual of X, and $\langle , \rangle$ is the duality pairing between X* and X.

Suppose that I is a function satisfying the following hypothesis:

$I = \Phi + \psi$, where $\Phi \in C^1(X, \mathbb{R})$, and $\psi : X \to (-\infty,+\infty]$ is convex, proper and lower semicontinuous. (H)

A point $u \in X$ is said to be a critical point of I if $u \in D(\psi)$ and

$$\langle\Phi'(u), v - u\rangle + \psi(v) - \psi(u) \geqq 0 \quad \forall v \in X.$$

Equivalently, u is critical if and only if $\Phi'(u) + \partial\psi(u) \ni 0$. Note that if $\psi \equiv 0$, this definition coincides with the usual one. A number $c \in \mathbb{R}$ will be called a critical value if $I^{-1}(c)$ contains a critical point.

It is easy to see that each local minimum of I is a critical point. In order to obtain further critical points, we introduce the following compactness condition:

If (u_n) is a sequence such that $I(u_n) \to c \in \mathbb{R}$, and $\Phi'(u_n) + \partial\psi(u_n) \ni z_n$, where $z_n \to 0$, then (u_n) possesses a convergent subsequence. (PS)

S. P. Singh (ed.), Nonlinear Functional Analysis and Its Applications, 393–399.

Note that this is a natural extension of (a version of) the usual Palais-Smale condition to the present context. It is shown in [8] that several minimax principles of Ambrosetti-Rabinowitz type (see, e.g., [1,6,7]) remain valid for functions satisfying (H) and (PS), and may therefore be used to prove the existence of critical points other than local minima.

In §2 we are concerned with the problem of existence of critical points of functions satisfying (H) and (PS), and in §3 we present some applications to boundary value problems. The results discussed here may be found in [8] in full detail.

2. EXISTENCE OF CRITICAL POINTS

Denote $B_\rho = \{u \in X : \|u\| < \rho\}$.

Theorem 1. Suppose that $I : X \to (-\infty,+\infty]$ is a function satisfying (H), (PS) and

(i) $I(0) = 0$ and there exist α, $\rho > 0$ such that $I|_{\partial B_\rho} \geq \alpha$,

(ii) $I(e) \leq 0$ for some $e \notin \overline{B}_\rho$.

Then I has a critical value $c \geq \alpha$ which may be characterized by

$$c = \inf_{f\in\Gamma} \sup_{t\in[0,1]} I(f(t)),$$

where $\Gamma = \{f \in C([0,1],X) : f(0) = 0,\ f(1) = e\}$.

Idea of proof. Assume that c is not a critical value. For f, $g \in \Gamma$, let $d(f,g) = \sup_t \| f(t) - g(t) \|$, and $\Pi(f) = \sup_t I(f(t))$. Let $\varepsilon > 0$ be given. Since (Γ,d) is a complete metric space, one may use Ekeland's variational principle [4, p. 444] in order to obtain an $f \in \Gamma$ such that $\Pi(f) \leq c + \varepsilon$ and

$$\Pi(g) - \Pi(f) \geq - \varepsilon\, d(f,g) \quad \forall g \in \Gamma.$$

It is shown in [8] that if ε is small enough, there exists a deformation α_s, $0 < s \leq s_0$, such that $\| u - \alpha_s(u)\| \leq s$ and $\Pi(\alpha_s \circ f) - \Pi(f) \leq - 2\varepsilon s$. Hence

$$-2\varepsilon s \geq \Pi(\alpha_s \circ f) - \Pi(f) \geq - \varepsilon\, d(f,\alpha_s \circ f) \geq -\varepsilon s,$$

a contradiction. Unfortunately, the deformation α_s is such that $f \in \Gamma$ does not imply that $\alpha_s \circ f \in \Gamma$. This difficulty is overcome in [8] by constructing an auxiliary family of mappings Γ_1 such that (Γ_1,d) is still complete, but $\alpha_s \circ f \in \Gamma_1$ whenever $f \in \Gamma_1$. □

Denote by Σ the collection of all symmetric subsets of $X - \{0\}$ which are closed in X. A nonempty set $A \in \Sigma$ is said to have genus k (denoted $\gamma(A) = k$) if k is the smallest integer for which there exists an odd continuous mapping from A to $\mathbb{R}^k - \{0\}$. If such an integer does not exist, $\gamma(A) = +\infty$, and if $A = \phi$, $\gamma(A) = 0$. Properties of genus may be found, e.g., in [1,7]. Let $\Gamma_j = \{A \in \Sigma : A \text{ is compact and } \gamma(A) \geq j\}$.

Theorem 2. Suppose that $I : X \to (-\infty,+\infty]$ satisfies (H) and (PS), $I(0) = 0$, and Φ, ψ are even. Let

$$c_j = \inf_{A \in \Gamma_j} \sup_{u \in A} I(u).$$

If $-\infty < c_j < 0$ for $j = 1, \ldots, k$, then I has at least k distinct pairs of nontrivial critical points.

Idea of proof. We prove only that c_j is a critical value (this will give the correct number of critical points if all c_j are distinct). Let $\Pi(A) = \sup_A I(u)$. The collection $\Gamma = \{A \in \Gamma_j : \Pi(A) \leq c_j + \varepsilon\}$ may be metrized by the Hausdorff metric dist, given by

$$\text{dist}(A,B) = \max\{\sup_{a \in A} d(a,B), \sup_{b \in B} d(b,A)\},$$

where $d(a,B)$ denotes the distance from the point a to the set B, and one can show that if $c_j + \varepsilon < 0$, then (Γ,dist) is a complete metric space. Finally, assuming that c_j is not a critical value, one obtains a contradiction in the same way as in the proof of Theorem 1. □

Theorem 3. Suppose that $I : X \to (-\infty,+\infty]$ satisfies (H) and (PS), $I(0) = 0$, and Φ, ψ are even. Assume also that

(i) there exists a subspace X_1 of X of finite codimension, and numbers $\alpha, \rho > 0$ such that $I|_{\partial B_\rho \cap X_1} \geq \alpha$,

(ii) there is a finite dimensional subspace X_2 of X, $\dim X_2 > \operatorname{codim} X_1$, such that $I(u) \to -\infty$ as $\|u\| \to \infty$, $u \in X_2$.

Then I has at least $\dim X_2 - \operatorname{codim} X_1$ distinct pairs of nontrivial critical points.

Again, the proof is effected by constructing suitable collections of subsets, which are complete in the Hausdorff metric, and using Ekeland's variational principle and the deformation α_s.

Corollary. Suppose that the hypotheses of Theorem 3 are satisfied, with (ii) replaced by

(ii') for any positive integer k, there is a k-dimensional subspace X_2 of X such that $I(u) \to -\infty$ as $\|u\| \to \infty$, $u \in X_2$.

Then I has infinitely many distinct pairs of nontrivial critical points.

3. APPLICATIONS

Let $\Omega \subset \mathbb{R}^N$ be a bounded domain with smooth boundary Γ, and let $H^m(\Omega) \equiv H^m$ and $H_0^m(\Omega) \equiv H_0^m$ be the usual Sobolev spaces. Furthermore, let λ_k denote the k-th eigenvalue and e_k the corresponding normalized eigenfunction of $(-\Delta)^m$ in H^m or in H_0^m.

Theorem 4. Let $\lambda \in \mathbb{R}$, $g \in L^2$, $g < 0$ a.e. in Ω, and let

$$\mathbb{K} = \{u \in H_0^1 : u \geqq 0 \text{ a.e. in } \Omega\}.$$

Suppose that $2 < p < p^*$, where $p^* = 2N/(N-2)$ if $N > 2$ and $p^* = +\infty$ if $N = 2$. Then the variational inequality

$$u \in \mathbb{K} : \int \nabla u \cdot \nabla(v-u)dx - \lambda\int_\Omega u(v-u)dx - \int_\Omega^{p-1}(v-u)dx$$

$$\geq \int_\Omega g(v-u)dx \;\forall v \in \mathbb{K}$$

has a nontrivial solution (in addition to the trivial one $u = 0$).

The proof is achieved by verifying the hypotheses of Theorem 1 with $X = H_0^1$, $I = \Phi + \psi$, and

$$\Phi(u) = \tfrac{1}{2}\int_\Omega |\nabla u|^2 dx - \tfrac{1}{2}\lambda\int_\Omega u^2 dx - p^{-1}\int_\Omega u^p dx - \int_\Omega gu\, dx,$$

$$\psi(u) = \begin{cases} 0 \text{ if } u \in \mathbb{K} \\ +\infty \text{ if not.} \end{cases}$$

Theorem 5. If $\lambda_k < \lambda < \lambda_{k+1}$, the inequality

$$u \in H^2 \cap H_0^1 : \int_\Omega (-\Delta u)(v-u)dx + \int_\Omega |\nabla v|dx - \int_\Omega |\nabla u|dx \geq \lambda \int_\Omega u(v-u)dx \quad \forall v \in H_0^1$$

possesses at least k distinct pairs of nontrivial solutions.

The proof uses Theorem 3 with $X = H_0^1$, $X_1 = X$, $X_2 = \text{span}\{e_1,\dots,e_k\}$, $\Phi(u) = -\frac{1}{2}\lambda\int_\Omega u^2 dx$ and

$$\psi(u) = \tfrac{1}{2}\int_\Omega |\nabla u|^2 dx + \int_\Omega |\nabla u| dx.$$

Theorem 6. Let $f(t)$ be an odd C^1 function such that $f(0) = f'(0) = 0$, f is nondecreasing and $f'(t) \to \infty$ as $|t| \to \infty$. If $\lambda > \lambda_k$, then the boundary value problem

$$\begin{cases} (-\Delta)^m u + f(u) = \lambda u & \text{in } \Omega \\ u \in H_0^m \end{cases}$$

has at least k distinct pairs of nontrivial solutions u such that $uf(u) \in L^1$.

Note that there is no growth restriction on f.

Idea of proof. Let $F(t) = \int_0^t f(s)ds$, $X = H_0^m$, $\Phi(u) = -\frac{1}{2}\lambda\int_\Omega u^2 dx$ and

$$\psi(u) = \begin{cases} \frac{1}{2}\int_\Omega |\nabla^m u|^2 dx + \int_\Omega F(u)dx & \text{if } F(u) \in L^1 \\ +\infty & \text{otherwise.} \end{cases}$$

Using the fact that $\lambda > \lambda_k$, one shows that $I = \Phi + \psi$ is negative on small spheres about the origin in $\text{span}\{e_1,\dots,e_j\}$, $j = 1,\dots,k$. Hence the numbers c_j in Theorem 2 are negative. Since I also satisfies (H) and (PS), it has at least k pairs of nontrivial critical points. Let $\tilde{\psi} : L^2 \to (-\infty,+\infty]$ be given by $\tilde{\psi}(u) = \psi(u)$ if $u \in D(\psi)$, $\tilde{\psi}(u) = +\infty$ otherwise. Then u is a critical point of I if and only if $\lambda u \in \partial\tilde{\psi}(u)$. Define an operator $A : D(A) \subset L^2 \to L^2$ by $Au = (-\Delta)^m u + f(u)$ with

$$D(A) = \{u \in H_0^m : (-\Delta)^m u + f(u) \in L^2,\ uf(u) \in L^1\}.$$

It is easy to see that A is monotone and $A \subset \partial\tilde{\psi}$ (in the sense of graph inclusion). By [3, Corollary IV.3], A is maximal monotone. Hence $A = \partial\tilde{\psi}$ and $\lambda u = Au$. So each critical point of I is a solution of the boundary value problem. □

Theorem 7. Let $2 < p \leqq (2N - 2)/(N - 2)$ if $N > 2$ and $2 < p < +\infty$ if $N = 2$. Suppose that either

(i) $B(t) = r^{-1}|t|^r$, $1 \leqq r < p$, or

(ii) $B : \mathbb{R} \to [0,+\infty]$ is an even, lower semicontinuous and convex function such that $B(0) = 0$ and the effective domain of B is a proper subset of $\mathbb{R}$.

Then the boundary value problem

$$\begin{cases} -\Delta u - \lambda u - |u|^{p-2}u = 0 & \text{in } \Omega \\ -\partial u/\partial n \in \beta(u) & \text{on } \Gamma, \end{cases}$$

where $\beta = \partial B$ and $\partial u/\partial n$ is the outward normal derivative, has infinitely many distinct pairs of nontrivial solutions $u \in H^2$.

Idea of proof. Let $X = H^1$, $\Phi(u) = -\frac{1}{2}(\lambda + 1)\int_\Omega u^2 dx - p^{-1}\int_\Omega |u|^p dx$, and

$$\psi(u) = \begin{cases} \frac{1}{2}\int_\Omega (|\nabla u|^2 + u^2)dx + \int_\Gamma B(u)d\sigma & \text{if } B(u) \in L^1(\Gamma) \\ +\infty & \text{otherwise.} \end{cases}$$

Then one uses Corollary to Theorem 3 with $X_2 = \text{span}\{\phi_1,\ldots,\phi_k\}$, where k is arbitrary and $\phi_1,\ldots,\phi_k$ are linearly independent functions in $C_0^\infty(\Omega)$. This gives infinitely many critical points of I. Using [2, Proposition II.2.9], one shows that each critical point is a solution of the boundary value problem. □

In [8] Theorems 6 and 7 are stated in more general form.

References

1. A. Ambrosetti and P. H. Rabinowitz, *Dual Variational Methods in Critical Point Theory and Applications*, J. Func. Anal. 14 (1973), 349-381.

2. V. Barbu, *Nonlinear Semigroups and Differential Equations in Banach Spaces*, Editura Academiei, Bucarest, Nordhoff, Leyden, 1976.

3. H. Brézis and L. Nirenberg, *Characterizations of the Ranges of Some Nonlinear Operators and Applications to Boundary Value Problems*, Ann. Scuola Norm. Sup. Pisa, Ser. IV, 5 (1978), 225-326.

4. I. Ekeland, *Nonconvex Minimization Problems*, Bull. Amer. Math. Soc. 1 (1979), 443-474.

5. I. Ekeland and R. Temam, *Convex Analysis and Variational Problems*, North-Holland, Amsterdam, American Elsevier, New York, 1976.

6. L. Nirenberg, *Variational and Topological Methods in Nonlinear Problems*, Bull. Amer. Math. Soc. 4 (1981), 267-302.

7. P. H. Rabinowitz, *Variational Methods for Nonlinear Eigenvalue Problems*, Proc. Sym. on Eigenvalues of Nonlinear Problems, Edizioni Cremonese, Rome, 1974, pp. 143-195.

8. A. Szulkin, *Minimax Principles for Lower Semicontinuous Functions and Applications to Nonlinear Boundary Value Problems*, Ann. I.H.P., Analyse non linéaire, to appear.

FIXED POINT THEOREMS AND COINCIDENCE THEOREMS FOR UPPER HEMI-CONTINUOUS MAPPINGS.

Kok-Keong Tan
Department of Mathematics, Statistics and Computing Science
Dalhousie University
Halifax, Nova Scotia
Canada B3H 4H8

ABSTRACT. By applying a generalization of Fan-Glicksberg's fixed point theorem to upper hemi-continuous mappings in non-compact settings, we obtain a new coincidence theorem from which several fixed point theorems and a minimax inequality are given as applications.

1. INTRODUCTION

The classical Fan-Glicksberg fixed point theorem [7, 13], which is the infinite-dimensional generalization of a well-known fixed point theorem of Kakutani [15], asserts that if f is an upper semi-continuous set-valued mapping defined on a non-empty compact convex set X in a Hausdorff locally convex space E such that $f(x)$ is a nonempty closed convex subset of X for each $x \in X$, then f has a fixed point in X; that is, there exists a point $x \in X$ such that $x \in f(x)$. Since 1961, Ky Fan has developed sharper methods to give a systematic and unified treatment of the interconnection between extensions of Fan-Glicksberg's fixed point theorem, minimax theorems, minimax inequalities and equilibrium point theorem, etc., see [8,9,10,11,12]. In this paper, we first present a generalization of Fan-Glicksberg's fixed point theorem to upper hemi-continuous mappings in non-compact settings in [17] from which a new coincidence theorem is obtained. As applications, several fixed point theorems and a minimax inequality are given, generalizing results in [16].

2. UPPER HEMI-CONTINUOUS MAPPINGS.

For a non-empty set X, we shall denote by 2^X the collection of all non-empty subsets of X. If E is a topological vector space, we shall denote by E' the vector space of all continuous linear functionals on E and by $\langle w,x \rangle$ for $w \in E'$ and $x \in E$ the pairing between E' and E. If X is a topological space and $A \subset X$, we shall denote by $cl(A)$ the closure of A and ∂A the boundary of A.

Let Y be a topological space, E be a topological vector space and $f : Y \to 2^E$. Then f is said to be <u>upper semi-continuous</u> on Y

S. P. Singh (ed.), Nonlinear Functional Analysis and Its Applications, 401–408.

if for each $y_0 \in Y$ and for each open set U in E containing $f(y_0)$, there exists an open neighborhood N of y_0 in Y such that $f(y) \subset U$ for all $y \in N$. According to Ky Fan [9, 10], we say that f is <u>upper demi-continuous</u> on Y if for every $y_0 \in Y$ and for every open half-space H in E containing $f(y_0)$, there exists an open neighborhood N of y_0 in Y such that $f(y) \subset H$ for all $y \in N$. Recall that an open half-space H in E is a set of the form $H := \{x \in E : \phi(x) < t\}$ for some non-zero $\phi \in E'$ and for some $t \in \mathbb{R}$.

<u>Definition 2.1.</u> Let Y be a topological space, E be a real topological vector space and $f : Y \to 2^E$. Then f is said to be <u>upper hemi-continuous</u> [2, Definition 1, p. 356] if for each $p \in E'$ and for each $\lambda \in \mathbb{R}$, the set $\{y \in Y : \sup_{u \in f(y)} \langle p,u \rangle < \lambda\}$ is open in Y.

It is obvious that every (set-valued) upper semi-continuous mapping f is upper demi-continuous and that if a set-valued mapping f is compact-valued, then f is upper hemi-continuous if and only if f is upper demi-continuous. Moreover, we have the following:

<u>Proposition 2.2</u>, [17].

(a) <u>Every upper demi-continuous (set-valued) mapping is upper hemi-continuous.</u>

(b) <u>Let</u> Y <u>be a topological space,</u> Z <u>be a non-empty compact subset of a real Hausdorff locally convex topological vector space and</u> $f : Y \to 2^Z$. <u>Then</u> f <u>is upper semi-continuous if and only if</u> f <u>is upper demi-continuous if and only if</u> f <u>is upper hemi-continuous.</u>

3. A GENERALIZATION OF FAN-GLICKSBERG's FIXED POINT THEOREM.

For a vector space E and a non-empty subset X of E, we define

$$I_X(y) := \{x \in E: \text{there exists } u \in X \text{ and } r > 0 \text{ such that } x = y + r(u - y)\},$$

$$O_X(y) := \{x \in E: \text{there exist } u \in X \text{ and } r > 0 \text{ such that } x = y - r(u - y)\}.$$

$I_X(y)$ and $O_X(y)$ are called the <u>inward</u> and <u>outward</u> sets of X at y, respectively. The definitions are due to Halpern (cf. [14]).

In [12, Theorem 6], Ky Fan obtained the following minimax inequality which is a slight improvement of a result in [1] which in turn generalizes the well-known Ky Fan minimax principle [10]:

Theorem 3.1. Let X be a non-empty convex set in a Hausdorff topological vector space. Let ψ be a real-valued function defined on $X \times X$ such that

(a) For each fixed $x \in X$, $\psi(x,y)$ is a lower semi-continuous function of y on X.

(b) For each fixed $y \in X$, $\psi(x,y)$ is a quasi-concave function of x on X.

(c) $\psi(x,x) \le 0$ for all $x \in X$.

(d) X has a non-empty compact convex subset X_0 such that the set $\{y \in X;\ \psi(x,y) \le 0 \text{ for all } x \in X_0\}$ is compact.

Then there exists a point $\hat{y} \in X$ such that $\psi(x,\hat{y}) \le 0$ for all $x \in X$.

By applying the above minimax inequality, by using a method given in a recent paper of Ky Fan [12, proof of Theorem 8], and by adopting a similar construction used by Browder [6, proof of Theorem 8], we have the following result:

Theorem 3.2, [17]. Let X be a paracompact convex subset of a real Hausdorff locally convex topological vector space E, X_0 a non-empty compact convex subset of X and K a non-empty compact subset of X. Let f be an upper hemi-continuous set-valued mapping defined on X with each $f(x)$ a non-empty closed convex subset of E such that

(i) For each $x \in K \cap \partial X$, $f(x) \cap \mathrm{cl}(I_X(x)) \neq \emptyset$ (respectively, $f(x) \cap \mathrm{cl}(O_X(x)) \neq \emptyset$).

(ii) For each $x \in X \backslash K$, $f(x) \cap \mathrm{cl}(I_{X_0}(x)) \neq \emptyset$ (respectively, $f(x) \cap \mathrm{cl}(O_{X_0}(x)) \neq \emptyset$).

Then f has a fixed point.

The above result generalizes the Fan-Glicksberg's fixed point theorem to upper hemi-continuous mappings in a non-compact setting. In case E is a normed space, the above result generalizes Bohnenblust-Karlin's fixed point theorem [3] and furthermore, when f is single-valued, it feneralized two earlier results of Browder [5, Corollaries 2 and 2'].

4. A COINCIDENCE THEOREM AND APPLICATIONS.

The following definition is due to Browder [6]:

Definition 4.1. Let X, Y be non-empty sets, $T : X \to 2^Y$ and $S : Y \to 2^X$. Then T and S are said to have a coincidence if there exists

$(x,y) \in X \times Y$ such that $y \in T(x)$ and $x \in S(y)$.

By applying Theorem 3.2, we have the following coincidence theorem which generalizes Theorem 2.2 in [16]:-

Theorem 4.2. *Let E be a real Hausdorff locally convex topological vector space, $C \subset E$ be non-empty paracompact convex, $X_0 \subset C$ be non-empty compact convex, $K \subset C$ be non-empty compact, F be a Hausdorff locally convex topological vector space and $D \subset F$ be non-empty. Let $T : C \to 2^D$ be such that (i) $T(x)$ is convex for all $x \in C$ and (ii) $T^{-1}(y)$ is open in C for all $y \in D$ and $S : D \to 2^C$ be upper hemi-continuous such that (iii) $S(y)$ is closed convex for all $y \in D$ and (iv) for each $x \in C \setminus K$ and for each $y \in T(x)$, $S(y) \cap cl(I_{X_0}(x)) \neq \emptyset$ (respectively, $S(y) \cap cl(O_{X_0}(x)) \neq \emptyset$). Then T and S have a coincidence.*

Proof. Since C is paracompact and $\{T^{-1}(y) : y \in D\}$ is an open cover of C , there is a continuous partition of unity $\{g_y : y \in D\}$ of C subordinated to the cover $\{T^{-1}(y) : y \in D\}$; i.e. $\{g_y : y \in D\}$ is a family of continuous non-negative real valued functions on C such that $\sum_{y\in D} g_y(x) = 1$ for all $x \in C$, supp $g_y \subset T^{-1}(y)$ for all $y \in D$ and $\{\text{supp } g_y : y \in D\}$ is locally finite. Define $\xi : C \to D$ by

$$\xi(x) = \sum_{y\in D} g_y(x)y \ , \ x \in C \ ,$$

then ξ is continuous on C . Let $x \in C$. If $g_y(x) \neq 0$, then $x \in \text{supp } g_y \subset T^{-1}(y)$ so that $y \in T(x)$; hence $\xi(x) = \sum_{y\in D} g_y(x)y \in T(x)$ since $T(x)$ is convex . Define $R : C \to 2^C$ by

$$R(x) = S(\xi(x)) \ , \ x \in C \ .$$

Then R is upper hemi-continuous on C such that for each $x \in C$, $R(x)$ is a closed convex subset of C so that $R(x) \cap cl(I_C(x)) \supset R(x) \cap C = R(x) \neq \emptyset$ (respectively, $R(x) \cap cl(O_C(x)) \supset R(x) \cap C = R(x) \neq \emptyset$) . Moreover, for each $x \in C\setminus K$, $S(y) \cap cl(I_{X_0}(x)) \neq \emptyset$ (respectively, $S(y) \cap cl(O_{X_0}(x)) \neq \emptyset$) for each $y \in T(x)$ so that $R(x) \cap cl(I_{X_0}(x)) = S(\xi(x)) \cap cl(I_{X_0}(x)) \neq \emptyset$ (respectively, $R(x) \cap$

$\mathrm{cl}(O_{X_0}(x)) = S(\xi(x)) \cap \mathrm{cl}(O_{X_0}(x)) \neq \emptyset)$. Hence by Theorem 3.2, there exists an $\hat{x} \in C$ such that $\hat{x} \in R(\hat{x})$. Let $\hat{y} = \xi(\hat{x})$. Then $\hat{x} \in R(\hat{x}) = S(\xi(\hat{x})) = S(\hat{y})$ and $\hat{y} = \xi(\hat{x}) \in T(\hat{x})$ so that T and S have a coincidence. □

We note that except S is required to be "closed-valued", Theorem 4.2 is a generalization of Theorem 3 in [6] to upper hemi-continuous mappings on non-compact sets.

As an immediate consequence of Theorem 4.2, we have the following result which slightly generalizes Theorem 3.1 in [16] which in turn generalizes Proposition 1 in [5]:

Corollary 4.3. Let E be a Hausdorff locally convex topological vector space, $C \subset E$ be non-empty paracompact convex, $X_0 \subset C$ be non-empty compact convex, $K \subset C$ be non-empty compact and $T : C \to 2^C$ be such that (i) $T(x)$ is convex for all $x \in C$; (ii) $T^{-1}(y)$ is open in C for all $y \in C$ and (iii) for each $x \in C \backslash K$, $T(x) \subset \mathrm{cl}(I_{X_0}(x))$ (respectively, $T(x) \subset \mathrm{cl}(O_{X_0}(x)))$. Then T has a fixed point in C .

Proof: Let $S : C \to 2^C$ be defined by $S(x) = \{x\}$ for all $x \in C$, then S is upper hemi-continuous such that for each $x \in C \backslash K$ and for each $y \in T(x)$, $S(y) = \{y\} \subset T(x) \subset \mathrm{cl}(I_{X_0}(x))$ (respectively, $S(y) = \{y\} \subset T(x) \subset \mathrm{cl}(O_{X_0}(x)))$ by (iii) so that $S(y) \cap \mathrm{cl}(I_{X_0}(x)) \neq \emptyset$ (respectively, $S(y) \cap \mathrm{cl}(O_{X_0}(x) \neq \emptyset)$. Hence by Theorem 4.2, T and S have a coincidence and therefore T has a fixed point in C . □

The following minimax inequality which is a generalization of Theorem 3.2 in [16] is an immediate consequence of Corollary 4.3:

Corollary 4.4. Let E be a Hausdorff locally convex topological vector space, $C \subset E$ be non-empty paracompact convex, $X_0 \subset C$ be non-empty compact convex, $K \subset C$ be non-empty compact and $f : C \times C \to \mathbb{R}$. If

(i) $f(x,x) \leq 0$ for all $x \in C$;

(ii) for each $x \in C$, $\{y \in C : f(x,y) > 0\}$ is convex ;

(iii) for each $y \in C$, $x \to f(x,y)$ is lower semi-continuous on C;

(iv) for each $x \in C \backslash K$, $\{y \in C : f(x,y) > 0\} \subset \mathrm{cl}(I_{X_0}(x))$ (respectively, $\{y \in C : f(x,y) > 0\} \subset \mathrm{cl}(O_{X_0}(x))))$;

then there exists an $\hat{x} \in C$ such that $f(\hat{x},y) \leq 0$ for all $y \in C$.

The following result is an improvement of Theroem 3.3 in [16] which in turn generalizes Proposition 2 in [5]:

Corollary 4.5. Let E be a Hausdorff locally convex topological vector space, $C \subset E$ be non-empty paracompact convex, $X_0 \subset C$ be non-empty compact convex, $K \in C$ be non-empty compact, $f : C \to E$ be continuous and $p : C \times E \to \mathbb{R}$, be continuous such that for each $x \in C$ and for each $r \in \mathbb{R}$, the set $\{y \in E : p(x,y) < r\}$ is convex. Suppose (i) for each $x \in C\backslash K$ and for each $y \in C\backslash cl(I_{X_0}(x))$ (respectively, $y \in C\backslash cl(0_{X_0}(x))$), $p(x,y-f(x)) \geq p(x,x-f(x))$ **and** (ii) for each $x \in C$ with $x \neq f(x)$, there exists $y \in C$ such that $p(x,y-f(x)) < p(x,x-f(x))$. Then f has a fixed point in C.

Proof. Suppose f has no fixed point in C; then by (ii), for each $x \in C$, the set $T(x) = \{y \in C : p(x,y-f(x)) < p(x,x-f(x))\}$ is non-empty. Since p and f are continuous, $T^{-1}(y)$ is open in C for each $y \in C$. By assumption, $T(x)$ is convex for each $x \in C$. By (i), $T(x) \subset cl(I_{X_0}(x))$ (respectively, $T(x) \subset cl(0_{X_0}(x))$). Hence $T : C \to 2^C$ satisfies the hypotheses of Corollary 4.3, so that T has a fixed point x_0 in C, i.e. $x_0 \in T(x_0)$ or $p(x_0,x_0 - f(x_0)) < p(x_0,x_0-f(x_0))$ which is impossible. Therefore f must have a fixed point in C. □

Corollary 4.5 together with the proof of Theorem 1 in [5] can be applied to obtain the following slight improvement of Corollary 3.5 in [16]:-

Corollary 4.6. Let E be a Huasdorff locally convex topological vector space, $C \subset E$ be non-empty paracompact convex, $X_0 \subset E$ be non-empty compact convex, $K \subset C$ be non-empty compact, $f : C \to E$ be continuous and $p : C \times E \to [0,\infty)$ be continuous such that for each $x \in C$, the mapping $y \to p(x,y)$ is convex on E. Suppose (i) for each $x \in C\backslash K$ and for each $y \in C\backslash cl(I_{X_0}(x))$ (respectively, $y \in C\backslash cl(0_{X_0}(x))$), $p(x,y-f(x)) \geq p(x,x-f(x))$ and (ii) for each $x \in C$ with $x \neq f(x)$, there exists $y \in I_C(x)$ such that $p(x,y-f(x)) < p(x,x-f(x))$. Then f has a fixed point in C.

Since every metrizable space is paracompact, Corollary 4.6 gives the following new generalizations of the Schauder fixed point theorem, generalizing Corollaries 3.6 and 3.7 in [16]:

Corollary 4.7. Let $(E, ||\cdot||)$ be a normed space, $C \subset E$ be non-empty convex, $X_0 \subset C$ be non-empty compact convex, $K \subset C$ be non-empty compact and $f : C \to E$ be continuous. Suppose (i) for each $x \in C\backslash K$ and for each $y \in C\backslash cl(I_{X_0}(x))$ (respectively, $y \in C\backslash cl(O_{X_0}(x))$), $||y-f(x)|| \geq ||x-f(x)||$ and (ii) for each $x \in C$ with $x \neq f(x)$, there exists $y \in I_C(x)$ such that $||y-f(x)|| < ||x-f(x)||$. Then f has a fixed point in C.

Corollary 4.8. Let $(E, ||\cdot||)$ be a normed space, $C \subset E$ be non-empty convex, $X_0 \subset C$ be non-empty compact convex, $K \subset C$ be non-empty compact and $F : C \to E$ be continuous. Suppose (i) for each $x \in C\backslash K$ and for each $y \in C\backslash cl(I_{X_0}(x))$ (respectively, $y \in C\backslash cl(O_{X_0}(x))$), $||y-f(x)|| \geq ||x-f(x)||$ and (ii) for each $x \in C$, $f(x) \in cl(I_C(x))$. Then f has a fixed point in C.

Question. Can $I_C(x)$ be replaced by $O_C(x)$ in each of the Corollaries 4.6, 4.7 and 4.8?

REFERENCES

1. G. Allen, Variational inequalities, complementarity problems, and duality theorems, J. Math. Anal. Appl. 58(1977), 1-10.
2. J. -P. Aubin, Applied Functional Analysis, John Wiley, 1979.
3. H.F. Bohnenblust and S. Karlin, On a theorem of Ville, Contribution to the Theory of Games, Ann. of Math. Studies No. 24, Princeton University Press, Princeton, N.J., 1950, pp. 155-160.
4. H. Brezis, L. Nirenberg and G. Stampacchia, A remark on Ky Fan's minimax principle, Boll. Un. Mat. Ital. 6(1972), 293-300.
5. F.E. Browder, On a sharpened form of the Schauder fixed-point theorem, Proc. Nat. Acad. Sci. USA 74(1977), 4749-4751.
6. F.E. Browder, Coincidence theorems, minimax theorems, and variational inequalities, Contemporary Mathematics, to appear.
7. K. Fan, Fixed-point and minimax theorems in locally convex topological linear spaces, Proc. Nat. Acad. Sci. USA 38(1952), 121-126.
8. K. Fan, A generalization of Tychonoff's fixed point theorem, Math. Ann. 142(1961), 305-310.
9. K. Fan, Extensions of two fixed point theorems of F.E. Browder, Math. Z. 112(1969), 234-240.
10. K. Fan, A minimax inequality and applications, Inequalities III (ed. Shisha, Academic Press, New York, 1972), pp. 103-113.
11. K. Fan, Fixed point and related theorems for non-compact convex sets, Game Theory and Related Topics (ed. O. Moeschlin and D. Pallaschke, North Holland, Amsterdam, 1979), pp. 151-156.
12. K. Fan, Some properties of convex sets related to fixed point theorems, Math. Ann. 266(1984), 519-537.

13. I.L. Glicksberg, A further generalization of the Kakutani fixed point theorem, with application to Nash equilibrium points, Proc. Amer. Math. Soc. 3(1952), 170-174.
14. B.R. Halpern and G.M. Bergman, A fixed point theorem for inward and outward maps, Trans. Amer. Math. Soc. 130(1968), 353-358.
15. S. Kakutani, A generalization of Brouwer's fixed point thoerem, Duke Math. J. 8(1941), 457-459.
16. H. -M. Ko and K. -K. Tan, A coincidence theorem with applications to minimax inequalities and fixed point theorems, to appear.
17. M. -H. Shih and K. -K. Tan, Covering theorems of convex sets related to fixed-point thoerems, to appear.

PARTICIPANTS

*Alexander, J. University of Maryland, College Park, MD, 20742, U.S.A.

Anichini, G. Ist. Mat., Univ. di Firenze, 50134 Firenze, Italy.

Appell, J. Math. Inst.,Univ. Augsburg, 8900 Augsburg, West Germany.

Bacciotti, A. Politecnino di Torino, 10129 Torino, Italy.

Baillon, J. B. Universite Lyon, Villeurbannes 6900, France.

Balzano, M. S.I. S.S. A. Strada Costiera 11, 34014 Trieste, Italy.

Bartsch, T. Univ. Munchen, D-8000 Munchen 40, West Germany.

Bates, P. Brigham Young Univ. Provo, Utah, 84602, U.S.A.

Battelli, F. Univ. di Urbino, 61029 Urbino, Italy.

Beauzamy, B. Univ. Lyon, Villeurbannes 6900, France.

*Berestycki, H. Universite Paris XIII, 93430 Villetaneuse, France.

Bertotti, M. S.I. S.S. A. Strada Costiera 11, 34014 Trieste, Italy.

Blat, J. Univ. de Plame de Mallorca, Fac. de Cienc., Baleares, Spain.

*Browder, F. E. University of Chicago, Chicago, IL, 60637, U.S.A.

Brown, R. Univ. of California, Los Angeles, CA, 90024, U.S.A.

Canino, A. M. Universita della Calabria, 87036 Arcavacata di Rende (CS) Italy.

Cappozzi, A. Universita di Bari, 70125 Bari, Italy.

Carbone, A. Univ. della Calabria, 87036 Arcavacata di Rende (CS) Italy.

Cerami, G. Univ. di Palermo, 90123 Palermo, Italy.

Chiappinelli, R. University of Sussex, Brighton, BN1 9QH, U.K.

Conti, G. Dip. di Matem., Fac. di Architettura, Univ. di Firenze, Italy.

Coti Zelati, V. S.I. S.S. A. Strada Costiera 11, 34014 Trieste, Italy.

Davies, M. J. Univ. College of Wales, Aberyswyth, Wales, U.K.

De Pascale, E. Univ. della Calabria, 87036 Arcavacata di Rende (CS) Italy.

Downing, D. Oakland University, Rochester, Michigan, 48063, U.S.A.

* Fadell, E. R. University of Wisconsin, Madison, Wisconsin, 53706, U.S.A.

Faure, R. Univ. de Lille, 59655 Villeneuve, France.

Finbow, A. St. Mary's University, Halifax, NS, Canada.

Fisher, B. Univeristy of Leicester, Leicester, LE1 7RH, U.K.

Fitzpatrick, P. M. University of Maryland, College Park, MD, 20742, U.S.A.

Fleckinger, J. 41 rue Boyssonne, 31400 Tolouse, France.

Fortunato, D. Universita di Bari, 70125 Bari, Italy.

Fournier, G. Univ. de Sherbrooke, Sherbrooke, PQ, Canada.

Furi, M. Univ. di Firenze, Viale Morgagni 67/A, 50134 Firenze, Italy.

* Geba, K. Inst. Math., University of Gdansk, 80952 Gdansk, Poland.

Girolo, J. California Polytech. State Univ., San Luis Obispo, CA, 93407, U.S.A.

Gobbo, L. Universita di Torino, 10123 Torino, Italy.

Goerniewicz, L. Inst. Math., Univ. Torun, 87100 Torun, Poland.

Gossez, J. P. Universite Libre de Bruxelles, B-1050 Bruxelles, Belgium.

Grossinho, M. do R. C.M.D.A., Av. Prof. Gama Pinto, 2, 1699 Lisoba, Portugal.

Guzzardi, R. Universita della Calabria, 87036 Arcavacata di Rende (CS) Italy.

Iannacci, R. Univ. della Calabria, 87036 Arcavacata di Rende (CS) Italy.

*Ize, J. I. I. M. A. S., Univ. Nac. Auto. Mexico, Mexico 20 DF, Mexico.

Jayne, J. Univ. College London, WCiE 6BT, London, U.K.

Kelly, M. 32 Phelps St., Binghamton, New York, 13901, U.S.A.

Khodja, M. 25 Fue Villa St. Michel, 75018 Paris, France.

Kielhöfer, H. Univ. Augsburg, 8900 Augsburg, West Germany.

Kirk, W. A. University of Iowa, Iowa City, Iowa 52242, U.S.A.

Kravaristis, D. National Tech. Univeristy of Athens, Athens, Greece.

Kyritsis, C. Nat. Tech. Univ. of Athens, Athens, Greece.

*Lasry, J. M. Univ. de Paris Dauphine, CEREMADE, 75775 Paris, France.

Lau, A. T. University of Alberta, Edmonton, Canada, T6G 2G1.

Lloyd, N. G. Univ. College of Wales, Aberyswyth, Wales, U.K.

Lupo, D. S.I. S.S. A. Strada Costiera 11, 34014 Trieste, Italy.

Madsen, A. Roskilde University, Post Box 260, DK 4000 Roskilde, Denmark.

Mancini, G. Universita di Trieste, 34127 Trieste, Italy.

Marsico, A. Univ. della Calabria, 87036 Arcavacata di Rende (CS) Italy.

Massa, S. Univ. di Milano, Via C. Saldini 50, 21033 Milano, Italy.

Massabo I. Universita della Calabria, 87036 Aracavacata di Rende (CS) Italy.

*Mawhin, J. M. Univ. Catholique de Louvain, B-1348 Louvain-la-Neuve, Belgium.

Merle, F. Ecole Normale Sup. 75005 Paris, France.

Nkashama, M. Univ. Catholique de Louvain, B-1348 Louvain-la-Neuve, Belgium.

Nugari, R. Univ. della Calabria, 87036 Arcavacata di Rende (CS) Italy.

Pacella, F. Dip. di Matematica e Appl., Univ. di Napoli, 80134 Napoli, Italy.

Papini, P. L. Univ. di Bologna, 40127 Bologna, Italy.

Parrott, M. Univ. of South Florida, Tampa, Florida, 33620, U.S.A

Pejsachowicz, J. Univ. della Calabria, 87036 Arcavacata di Rende (CS) Italy.

Perri, U. Univ. della Calabria, 87036 Arcavacata di Rende (CS) Italy.

Rallis, N. E. Boston College, Chestnut Hill, MA, 02167, U.S.A.

Ricceri, B. Univ. di Catania, 95125 Catania, Italy.

Roux, D. Universita di Milano, Via C. Saldini 50, 20133 Milano, Italy.

Ruf. B. ETH-Zentrum, CH-8092 Zurich, Switzerland.

Sanchez, L. C.M.A.D., 1699 Lisoba Cedex, Portugal.

Schirmer, H. Carleton University, Ottawa, ON, Canada, K1A 5B9.

Scholz, K. College of St. Thomas, P.O. Box 4065, St. Paul, MN, 55105, U.S.A.

Sessa, S. Univ. of Napoli, Fac. di Architettura, Napoli, Italy.

Siconolfi, A. Univ. della Calabria, 87036 Arcavacata di Rende (CS) Italy.

Singh, S. P. Memorial University of Newfoundland, St. John's, NF, A1C 5S7, Canada.

Solimini, S. S.I. S.S. A. Strada Costiera 11, 34014 Trieste, Italy.

Steinlein, H. Univ. Munchen, D-8000 Munchen, West Germany.

Stratis, J. Nat. Tech. Univ. of Athens, Athens, Greece.

*Stuart, C. A. Ecole Poly, Fed. Lausanne, CH-400 Lausanne, Switzerland.

Tan, K. K. Dalhousie Univ., Halifax, NS, Canada.

Thomeier, S. Memorial University of Newfoundland, St. John's, NF, A1C 5S7, Canada.

*Toland, J. University of Bath, Claverton Down, Bath, 3A2 7AY, U.K.

Trombetta, G. Dip. di Matematica, Univ. della Basilicata, 85100 Potenza, Italy.

Van Groesen, B. Catholic Univ. Nijmegan, 6533 ZA Nijmegan, The Netherlands.

Vignoli, A. 11(Second) Universita di Roma, 00173 Roma, Italy.

Walsh, J. Memorial Univeristy of Newfoundland, St. John's, NF, A1C 5S7, Canada.

Weiss, M. Univ. Munchen, D-8000 Munchen, West Germany.

Willem, M. Univ. Catholique de Louvain, B-1348 Louvain-la-Neuve, Belgium.

* Main speakers.

SUBJECT INDEX